IMAGE PROCESSING TECHNIQUES IN ASTRONOMY

ASTROPHYSICS AND SPACE SCIENCE LIBRARY

A SERIES OF BOOKS ON THE RECENT DEVELOPMENTS
OF SPACE SCIENCE AND OF GENERAL GEOPHYSICS AND ASTROPHYSICS
PUBLISHED IN CONNECTION WITH THE JOURNAL
SPACE SCIENCE REVIEWS

VOLUME 54
PROCEEDINGS

IMAGE PROCESSING TECHNIQUES IN ASTRONOMY

PROCEEDINGS OF A CONFERENCE HELD IN UTRECHT
ON MARCH 25–27, 1975

Edited by

C. DE JAGER AND H. NIEUWENHUIJZEN
Astronomical Institute, University of Utrecht, Utrecht, The Netherlands

D. REIDEL PUBLISHING COMPANY
DORDRECHT-HOLLAND / BOSTON-U.S.A.

Library of Congress Cataloging in Publication Data

Main entry under title:

Image processing techniques in astronomy.

(Astrophysics and space science library ; 54)
Bibliography: p.
1. Astrophysics—Congresses. 2. Spectrum analysis—Congresses. 3. Astronomical instruments—Congresses.
I. Jager, Cornelis de. II. Nieuwenhuijzen, H. III. Series.
QB461.I48 522 75-23032
ISBN-13: 978-94-010-1883-8 e-ISBN-13: 978-94-010-1881-4
DOI: 10.1007/978-94-010-1881-4

Published by D. Reidel Publishing Company,
P.O. Box 17, Dordrecht, Holland

Sold and distributed in the U.S.A., Canada, and Mexico
by D. Reidel Publishing Company, Inc.
306 Dartmouth Street, Boston,
Mass. 02116, U.S.A.

TABLE OF CONTENTS

INTRODUCTION

The primary inducement for organizing an international Conference on 'Image Processing Techniques in Astronomy' was the fact that the recording microdensitometer VAMP ('Vol Automatische Micro Photometer') of the Utrecht Astronomical Institute was operative for a few years. The necessity of comparing the instrument and its performance with similar instruments nowadays available at many other institutes, was stimulating enough to organize a meeting on the above subject. It took place in Utrecht on March 25, 26 and 27, 1975.

The Scientific Organizing Committee consisted of J. Borgman (Groningen), R.B. Dunn (Sacramento Peak), H. Elsässer (Heidelberg), L.D. de Feiter, T. de Groot, J.R.W. Heintze, C. de Jager, H. Nieuwenhuijzen (Utrecht) and W. Wiskott (Genève).

About 175 scientists from 14 countries participated in the meeting which appeared to be successful and offered a good opportunity of exchanging information and comparing experiences.

The VAMP was bought with financial support of the Utrecht University and the Netherlands Foundation for Scientific Research (Z.W.O.).

The conference was organized with financial support from
The Netherlands Ministry of Science and Education,
The European Southern Observatory,
The Leids Kerkhoven-Bosscha Fonds,
The Astronomical Institute of Utrecht,
to which Institutes and Organisations we express our sincere gratitude.

C. de Jager
H. Nieuwenhuijzen

editors

PART 1

WHAT INFORMATION DO WE NEED, FOR WHICH ASTRONOMICAL PROBLEM?

ASTROMETRY

K. Aa. Strand

U. S. Naval Observatory
Washington, D. C.

INTRODUCTION

Considerable progress has taken place in astrometry over the past two decades.

Along with other fields of astronomy, effective use has been made of new technical developments both in regard to the improvement of performance of existing telescopes as well as in the design of new ones.

New automatic or semi-automatic measuring machines along with high speed computers have increased the accuracy and the ease of obtaining the data from photographic plates.

New areas have been added to astrometry in the form of radio and space astrometry which give promise of high precision. Because of the magnitude of the problems, and the need for global coverage, several international programs have been initiated and others have already been successfully carried out.

Added impetus to astrometry has been received from other fields of astronomy and astrophysics. These fields have generated needs for astrometric data for such programs as stellar evolution and luminosities, stellar kinematics, statistical parallaxes and galactic research. Requirements for increased accuracies of stellar positions and proper motions, especially for fainter stars, have also been generated from satellite geodesy and geophysical research.

In the following, an attempt is made to review the present status of some areas in astrometry especially in need of data and

de Jager/Nieuwenhuijzen (eds.), Image Processing Techniques in Astronomy. 3-10.

to mention the implementation or improvements which might be considered in regard to instrumentation and observations.

REFERENCE SYSTEMS

One of the objectives of astrometry is to make precise observations and to construct catalogs of the positions and motions of stars and solar system objects in order to provide an appropriate reference system of coordinates, including a local standard of rest.

Basic to this problem are the observations of the brighter objects which are carried out at present with transit circles and with astrolabes.

Within recent years there has been a substantial decline in the number of transit circles in operation. It would have been very critical for the continuation of such observations if significant improvements had not been made at the same time in the rate of acquisition and in the accuracy of the observations made with those still active.

In the first half of this century the accuracy of a single observation of good quality was about $\pm 0\overset{s}{.}020 \sec\delta$ and $\pm 0\overset{''}{.}40$ in right ascension and declination respectively (Fricke 1973). In the AGK3R program, where the observations were still carried out with the travelling wire micrometer, the transit circles showing the best performance had internal mean errors of $\pm 0\overset{s}{.}012 \sec\delta$ and $\pm 0\overset{''}{.}21$ (Scott 1967).

Further improvements in the internal accuracy of the conventional transit circle have been obtained by Høg (1972) with the Hamburg instrument located at Bickley, Australia, during the Southern Reference Star (SRS) program using a photoelectric multi-slit micrometer of his own design, and by Requieme (1973) at Bordeaux with his photoelectric tracking photometer.

Considerable improvement in speed and accuracy of the reading of the angular position of the instrument has been attained with the automatic circle scanning system first developed by the Copenhagen astronomers. The miniaturized version of this system now in operation on the 6-inch transit circle at the U. S. Naval Observatory (USNO), scans the circle in six places and produces a digital read out with a repeatability of 10 micro-degrees.

Another angular position system which provides continuous read out of the telescope during the observation is the inductosyn-system (an electronic angular position transducer) which is now being tested on the 6-inch USNO transit circle. It also has a repeatability of 10 micro-degrees. The operation of this telescope in conjunction with an IBM 1800 computer, permits the data to be available the morning following a night's observation.

Besides the present transit circles with their potential improvements in performance, there are three new types in various stages of development. All three are aimed at minimizing the troublesome flexure and thermal problems of the conventional instrument.

The horizontal transit circle at Pulkovo Observatory (Pinigin et al. 1974) is now reported to provide observations in right ascension with mean errors of $\pm 0^{s}.011 \sec\delta$ and to be essentially free from systematic errors.

At the USNO in Washington a 24cm Cassegrain type automatic transit circle which, in the design stage gave promise of accuracies of $\pm 0''.05$ in both coordinates, is approaching completion after long delays by the manufacturer in correcting numerous deficiencies in design and fabrication. The positional accuracy which can be achieved with this instrument is therefore uncertain at this time.

Finally, the most recent design of a new type is the so-called glass horizontal transit circle by Høg (1974) which will be relatively simple in design, have very small flexure, and will require a smaller building than the conventional meridian circle.

Of special importance to future work is the further development of photoelectric micrometers, with promise of the possibility of observing objects as faint as 14th magnitude, thus allowing a substantial number of galaxies to be tied directly to the fundamental system. While the present transit circles of 15 to 20cm aperture have a limiting visual magnitude of 10 using the travelling wire micrometer, and the photoelectric method at present has made possible a gain of approximately one magnitude in sensitivity, the continued improvement in sensitivity of the phototubes seems to make the additional gain of three, perhaps four, magnitudes not unrealistic.

What are the programs in meridian astrometry which need our attention over the next 10-20 years and what means do we have to accomplish them? I believe there is a continuous need for the observation of solar system objects both for the improvement of their ephemerides and the fundamental system. The AGK3R program, which was so successfully carried out in an international cooperation, has a mean epoch of 1959, and should therefore be repeated for a mean epoch near 1980. This would provide an additional interval of some 20 years to the time-base upon which the proper motions are determined. Similarly, a repetition of the SRS program is especially critical because of the paucity of early reliable observations to provide first epoch positions for proper motions.

I hope that the same successful international cooperation can be organized again with the observations carried out either with the new types of telescopes or the conventional ones in their updated, computerized versions.

I am aware of the proposed space projects which promise observations with precision as much as 2 orders of magnitude greater than the ground based equipment just mentioned. There is the project by Lacroute of measuring angular distances of 90° across the sky in either the Spacelab option or in the automatic Tiros Delta satellite option. Another project is associated with the Large Space Telescope (LST) project and involves measuring angles of variable size across the sky with high precision. The Spacelab, scheduled for launch in late 1980 but as yet not funded, is designed for 50 missions of 7 days duration each. It has to date 240 separate proposals for payloads with which the Lacroute project has to compete. The LST project is an approved NASA project but is also unfunded. It will contain an astrometric package for the observation of double stars, proper motions and parallaxes, but the gyro project has been dropped because of costs.

I do not wish to make predictions of the future of these space projects, nor do I wish to dispute their claims for high accuracy, but I am convinced we should continue the ground based observations until such time when they have definitely been superceded by the new methods in accuracy and operational reliability.

ASTROGRAPHIC CATALOGS

I would like to mention next the problem of there being no astrographic catalogs covering the entire sky, which are based on modern observations.

For the northern hemisphere there is the recently completed AGK3 catalog which contains positions near the epoch of 1958 of approximately 180,000 stars in the northern hemisphere to a limiting visual magnitude of 9.5. Proper motions of these stars were derived with the aid of the AGK2 catalog (epoch 1930) which was revised to the FK4 fundamental system. However, since both first and second epoch plates were measured without the benefit of modern automatic measuring machines, the potential accuracy of the plate material was not fully realized, neither in regard to positions nor limiting magnitude.

Plans are already in progress to produce a new, improved photographic catalog for the northern hemisphere with a limiting visual magnitude of at least 11 (de Vegt 1974). The Hamburg Bergedorf Observatory has acquired a new astrograph with an aperture of 23cm for this purpose. It is intended to use the AGK2 plates which will be remeasured as first epoch for the proper motions. The overall positional accuracy from the measurements with a modern automatic measuring machine is estimated at 0".14. With a baseline in excess of 40 years, the centennial proper motions will have an estimated mean error of 0".50 or nearly half the error of the AGK3 proper motions.

Photographic observations of the entire southern sky with the Cape astrographic camera have been completed with a total of approximately 6,000 plates providing a fourfold overlap. These plates have been transferred to Herstmonceux to be measured on the Galaxy machine. As reported by Clube and Nicholson (1974) the anticipated relative positional accuracy is of the order of 0".05 over a fairly wide area of the sky. If this mean error is attainable, the accuracy will be substantially greater than previously obtained in the Cape zone catalogs where the positional mean errors range from ±0".22 in the earliest catalog to ±0".14 in the latest (Dieckvoss 1963).

It is hoped that the necessary support can be obtained for both of the above mentioned programs, thereby providing general catalogs of position and proper motion to a limiting visual magnitude between 11.5 and 12 for both hemispheres, with an average of 8 to 10 stars per square degree.

TRIGONOMETRIC STELLAR PARALLAXES

I would like to call your attention next to the current activities of determining trigonometric stellar parallaxes. We find that most of the observatories which contributed the major share of the earlier parallax series are still active in this area.

With the exception of van Maanen's parallaxes with the Mt. Wilson 60-inch and 100-inch reflectors, the telescopes used were long-focus refractors. These are still providing good results for parallaxes of the brighter stars, but they are not effective for stars fainter than visual magnitude 13, because of excessive long exposure times combined with the substantial number of plates required in modern parallax series.

The 61-inch astrometric reflector of the U. S. Naval Observatory has now been in operation since 1964. Including a third catalog which is now in press they contain parallaxes for approximately 300 stars, of which 75% are fainter than $m_V = 13$.

Since most of the stars for the program were initially selected from the Lowell proper motion survey because of the availability of identification charts, parallaxes have been obtained thus far for only a few stars fainter than $m_V = 16$. Fainter stars selected from the Luyten 48-inch Palomar Schmidt proper motion survey, are now on the program. With exposure times of 50 minutes, a photographic or visual magnitude of 18 is reached.

Including results from the third catalog, 21 stars have been found within 10 parsecs, none of which had previously published parallaxes, and 99 within 20 parsecs. Significantly, four new stars within 5 parsecs have been added to the list of 45 systems of 60 stars previously known (van de Kamp 1971).

The substantial number of parallaxes obtained in the program was predicated on the use of a telescope designed and dedicated primarily to parallax work, an automatic measuring machine, and automation of data handling (Strand 1966, 1971).

At the present time, the only other reflectors being used for parallax work are the one meter aperture Turin astrometric reflector recently installed and the 98-inch Isaac Newton Telescope. With the latter telescope, because of the limited availability of time for such work, a small program restricted to some 20 faint stars between photographic magnitude 16 and 18.5 is being carried out (Murray 1974).

While efforts in the northern hemisphere appear to be satisfactory in regard to parallax determinations, the situation in the southern hemisphere is critical with only the Cape refractor engaged in this area.

There is, therefore, a need to make time available on reflectors of sufficient aperture to carry out the observations of the many low luminosity stars lacking parallaxes. For lack of a reflector specifically dedicated to this work, and the need of a fairly rigid observing schedule, however, it will be difficult to carry out an effective program.

Perhaps the proposed program by Murray (1974) of using the UK Schmidt at Siding Springs might produce results which will promote interest in parallax work in the southern hemisphere.

VISUAL DOUBLE STARS

The observations of visual double stars is another area of astrometry in which progress has to depend upon a very small group of participants in spite of the fact that direct information concerning the masses of individual stars is entirely dependent upon such observations. In addition to the study of the relative motions of the components in binary systems, there is also the need for the determinations of their mass ratios, their parallaxes and proper motions, as well as the search for invisible companions.

Without going into details I would like to make a special plea for the study of the relative motions of binaries with separations less than 2 to 3 seconds of arc, which can not be observed photographically.

The bulk of the observations are still carried out with the conventional micrometer, but there are only 3 institutions in the United States where such observations are made regularly, a few more in Europe, and only one in the southern hemisphere. Clearly this combined effort is insufficient to keep track of some 70,000 known binaries of which approximately 1 percent have determined orbits.

For the very close binaries with separations of a few tenths of a second of arc, there is a problem of systematic errors of their separations as measured with the conventional filar micrometer. The use of interferometry, area scanners, and image tube photography, and other automatic image processing techniques have produced promising results but as yet not in sufficient quantities.

While I realize that observations of double stars are of a long-range classical nature, and often termed pedestrian, they are essential to the progress of astronomy and must be continued.

RADIO ASTROMETRY

A new area introduced into astrometry is radio astrometry. By means of radio interferometric techniques absolute position determinations are accomplished at a level of accuracy which competes with the best optical observations. Accuracies of a few milliseconds in right ascension and a few hundredths of a second of arc in declination have been achieved to date.

The advantage of radio astrometry over optical astrometry is the relative unimportance of atmospheric refraction, the ability to measure declinations with respect to the Earth's instantaneous axis of rotation, and to measure large angles with essentially the same accuracy as small angles.

The disadvantage from the standpoint of fundamental astrometry is the inability of existing instruments to determine the location of the vernal equinox without recourse to optical observations.

Most of the radio sources which are suitable for astrometric measurements with interferometers are extragalactic, with optical counterparts as faint as magnitude 19.

Since these objects are numerous and well distributed over the sky, they are well suited to establish a precise extragalactic reference frame. However, because of their optical faintness, the main problem of tying their positions to the fundamental system is to bridge a magnitude difference of approximately 10 with a minimum loss of positional accuracy. The possibility of doing this has already been demonstrated by Murray, Tucker and Clements (1971) and by others. Clearly the success of obtaining a network of sufficient density of radio sources tied to the fundamental system depends upon the close cooperation between radio and optical astronomers. A program has been initiated at the USNO using the 6-inch Transit Circle and the 61-inch astrometric reflector, while awaiting the operation of an 8-inch double astrograph for the intermediate step.

CONCLUSION

In conclusion, I would like to mention that there are other areas of astrometry which are in need of special attention. Those I have chosen are the ones with which we are concerned at the USNO. Another astrometrist might have chosen other areas to talk about. Generally, we may conclude that existing optical astrometric data in most areas are not adequate to meet modern requirements in spite of progress made in recent years.

REFERENCES

Clube, S.V.M., and Nicholson, W.: 1974, New Problems in Astrometry, IAU Symp. 61, p.97.
de Vegt, Ch.: 1974, New Problems in Astrometry, IAU Symp. 61, p.209.
Dieckvoss, W.: 1963, Basic Astronomical Data, (Ed. K.Aa. Strand), Univ. of Chicago Press, Chicago and London, p.40.
Fricke, W.: 1973, Mitt. Astron. Rechen-Inst. Heidelberg, Ser. A, No. 68.
Høg, E.: 1972, Astron. Astroph., 19, p.27.
Høg, E.: 1974, New Problems in Astrometry, IAU Symp. 61, p.243.
Murray, C.A., Tucker R.H., and Clements, E.D.: 1971, R. Obs. Bull. No. 162.
Murray, C.A.,: 1974, New Problems in Astrometry, IAU Symp. 61, p.151
Pinigin, G.I., Sukharev, L.A., and Timashkova, G.M.: 1975, New Problems in Astrometry, IAU Symp. 61, p.261.
Requieme, Y.: 1973, Astron. Astroph., 23, p.453.
Scott, F.P.: 1967, Astron. J. 72, p.570.
Strand, K.Aa.: 1966, Vistas in Astronomy, 8, p.9.
Strand, K.Aa.: 1971, Publ. USNO, 2nd Ser., XX Part I.
Van de Kamp, P.: 1971, Ann. Rev. Astron. Astroph., 19, p.103.

PHOTOMETRY

J. Borgman

Kapteyn Astronomical Institute
University of Groningen
Groningen, The Netherlands

ABSTRACT. The significance of photometry in astronomy is discussed in relation to three basic parameters: wavelength, spectral resolution and spatial resolution.

1. INTRODUCTION

Photometry is essentially low resolution spectroscopy. However, a separate discussion of photometry appears to be justified in view of the different techniques and the traditional use of the data. With further development of detection techniques the differences between spectroscopy and photometry are expected to become even smaller, particularly if the need for higher spectral resolution photometry can be satisfied at little or no expense to limiting magnitude.

The scope of this conference is quite broad as is illustrated by the wide range of topics suggested by the titles of papers. In an effort to avoid overlap while maintaining a reasonable relation to the general theme of the conference, limits have been adopted for the wavelength range and spectral resolution which are to be considered in this paper; these limits are:

$$1000\ \text{Å} < \lambda < 1\ \text{mm}$$

$$2 < \lambda/\Delta\lambda < 50.$$

This means that I will not limit myself to the domain of

de Jager/Nieuwenhuijzen (eds.), Image Processing Techniques in Astronomy. 11-20.

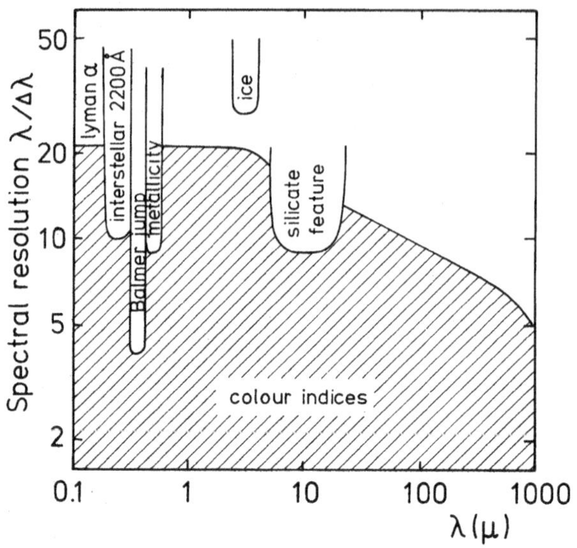

Fig. 1. Observable quantities in astronomical photometry.

classical photometry (3000 Å < λ < 9000 Å, see the comprehensive review by Golay, 1974) but will include the ultraviolet and infrared regions. The spectral resolution limits are chosen to include on the one side wide-band photometry in a full atmospheric window (e.g. 8 - 13 μ) but to stop at the higher resolution where narrow-band photometry yields equivalent widths of individual hydrogen lines. In between lies the realm of photometry with observable quantities as schematically illustrated in Fig. 1.

2. SPECTRAL RESOLUTION, SPATIAL RESOLUTION AND WAVELENGTH

Knowledge of what can be achieved with photometry in the visible part of the spectrum is well advanced. In order of increasing spectral resolution as needed to define observable quantities we mention here:

- the colour index, a temperature indicator but influenced by interstellar extinction.
- linear combinations of colour indices to describe the Balmer discontinuity: Q in the UBV system (Johnson and Morgan, 1953), c_1 in the uvby system (Strömgren, 1966), γ in the Groningen system (Borgman and Blaauw, 1964), etc.
- linear combinations of colour indices, correlated to metallicity: m_1 in the uvby system, α in the Groningen

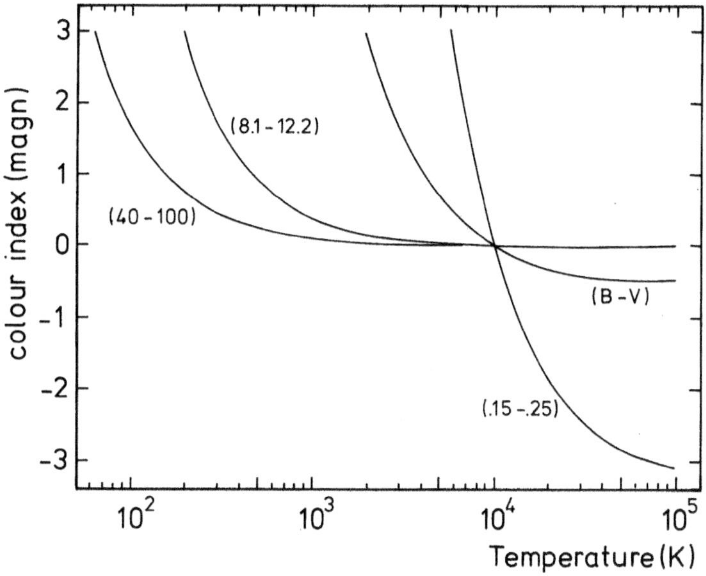

Fig. 2. Temperature dependence of some colour indices.

system, etc.

- linear combinations of colour indices to measure the blanketing by groups of lines due to a single chemical element; e.g. the depression caused by the crowding of hydrogen lines in front of the Balmer discontinuity: β in the Groningen system, [B - L] in Walraven's (1960) system.

With an extension of the accessible wavelength range towards the ultraviolet and towards the infrared and further sophistication of the detection techniques similar developments in terms of observable quantities are taking place in these wavelength regions:

- colour indices are now being measured over the entire spectral interval 1000 Å $< \lambda <$ 1 mm and it is possible to select wavelengths for an optimum combination of temperature sensitivity and available energy (Fig. 2).
- the 2200 Å interstellar extinction feature (Bless and Savage, 1972) can be studied quite well with photometric spectral resolution, e.g. in the ANS system (Van Duinen et al, 1975).
- the 9.7 μ interstellar feature and the 10 μ circumstellar emission can be studied without significant loss of information by photometric techniques with a spectral

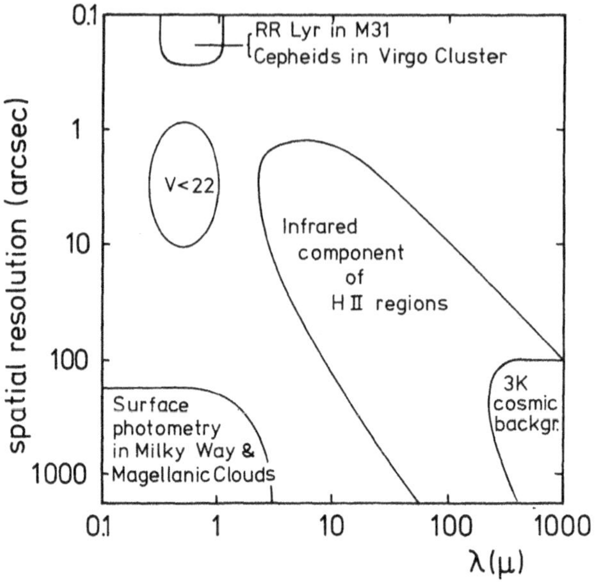

Fig. 3. Typical objects for photometric studies in relation to spatial resolution and wavelength.

resolution of 10 (cf Borgman, 1973)

It is of interest to say something in addition about spatial resolution. The limiting magnitude of photometric work is restricted by the size of the seeing disc. An enormous step forward can be made when large space telescopes with diffraction limited optics become available. This may push the limiting magnitude to V = 27 with adequate detectors. Some typical objects for photometric studies in relation to spatial resolution and wavelength are indicated in Fig. 3.

3. ASTRONOMICAL PROBLEMS AND PHOTOMETRY

The range of problems to which we might address ourselves is very wide. Within the scope and title of this conference we will restrict the present discussion to technical problems which are related to data quality, data abundance, data presentation and data information content.

3.1. Data quality

Photometric data quality is a broad term which may include anything between statistical signal fluctuations and data editing consistency. Here we will only deal with the requirements on photometric reproducability and accuracy.

Requirements on accuracy can vary widely. If a V magnitude is measured with no other purpose than to derive the photometric distance it is evident that in most cases an accuracy of 0.1 mag is ample, considering the uncertainty of the absolute magnitude and reddening correction. On the other hand, if B-V is measured to characterize the colour of a hot star it may be worth while to aim at an accuracy of 0.01 mag, as the temperature sensitivity is only small (Fig. 2). Again, this requirement may be relaxed, e.g. when an uncertain reddening correction has to be taken into account while deriving the intrinsic colour.

In general it is wise to make observations and to handle data while keeping the required accuracy in mind. This works both ways and more often than not there is a valid requirement for more accuracy than can be obtained easily. This is in particular the case for a "clean" photometric parameter with a small range and, more generally, for faint stars. Decisive factors in this respect are the stability and reproducability of the photometric system; these properties can only to a very limited extent be restored or improved during the data handling process, in spite of what is often claimed in the literature.

It is not yet entirely clear whether the high data quality that can be achieved with a one-channel detector will be met by the modern two-dimensional detection systems. However, both the electronographic camera and the television-type devices seem capable of a development which will eventually lead to a combination of the best properties of photoelectric photometry and photographic photometry: high quantum efficiency, excellent linearity and large storage capacity. Several papers at this conference will tell us about the latest developments.

3.2. Data abundance

Schmidt telescopes, large field Ritchey-Chrétien systems, the photographic plate and modern two-dimensional detectors conspire to flood the astronomical community with data. A Schmidt plate can hold photometric information on some 10^9 picture elements. Extraction of such abundant information has always been preceded by a selection process where the judgment of the astronomers came into play. With modern plate measuring machines and automated reduction procedures there is a clear and understandable tendency to shift the step which involves assimilation and judgment by the astronomer to a later phase in the program.

Both the information gathering capability of our instruments and the hesitation or inability to use human judgment to isolate a small number of significant data lead to a mass production of preliminary, redundant or low quality data. The ease with which numbers can be generated in computer printouts is demonstrated by the extensive tables in publications and the growing volume of the A. & A. Supplement Series.

Several aspects of the problems discussed here are reflected by the sophisticated KPNO machine for reduction of electronographic stellar photometry where a number of meaningful shortcuts are introduced by interactive programs, involving graphical and numerical displays (cf Livingston, review paper published in these Proceedings).

In infrared photometry modest efforts are being made to increase data collection efficiency by the introduction of discrete detector arrays. Background problems and the lack of adequate modulation techniques are likely to cause a delay of more efficient data collection until a satellite-borne infrared facility is available. With such a facility single element detectors at 10 μ are capable of a limiting magnitude gain of two orders of magnitude and even more when the spatial and/or spectral information is multiplexed.

3.3. Data presentation

The selection of the most appropriate form of data presentation is becoming increasingly important, both in publications and during analysis. Photometric data are conveniently handled in numerical form but this in itself is not sufficient to justify the publication of extensive tables. Graphical presentation nearly always is to be preferred where it is helpful to demonstrate relations between two or more parameters, e.g. in colour-colour diagrams, lightcurves of variable stars, etc. Only in rare cases large scale numerical presentation is unavoidable, e.g. in catalogues where the discriminate usage of high quality data is left to others.

To illustrate these considerations a variety of photometric

data on one object are shown in Figures 4, 5 and 6. The object is the blue association no. 89 (Lucke, 1974) near 30 Doradus in the Large Magellanic Cloud. The photometry at 1550 Å in Fig. 4 was obtained with the ultraviolet photometer onboard ANS (Netherland's Astronomical Satellite) using a 2.5 arcmin x 2.5 arcmin diaphragm. Each photometric sample contains several stars which contribute in a significant way to the measured flux. The contour presentation has been chosen to combine visual and numerical information in a single diagram. For comparison, a photograph taken in blue light is shown in Fig. 5. The 6 cm contour diagram of Fig. 6 indicates that the region is a weak H II region. The combined information in Figures 4, 5 and 6 indicates the presence of a large young association with relatively low gas content. Observations with ANS at other wavelengths, combined with B,V measurements suggest that the association is reddened with an extinction law which is characterized by the absence of the common 2200 Å bump (Borgman et al, 1975).

This is an example of selective use of data; in such cases it should be good practice that more complete access to the numerical data is not only encouraged but also assured to be useful. It is beyond doubt that many observing programmes are aiming at data which could be retrieved conveniently from existing data files.

3.4. Information content of data

The information content of wide band photometric data is largely restricted to colour indices and brightness. It is evident (cf Fig. 2) that the colour index wavelengths must be chosen to fit the colour temperature of the objects, in order to optimize the information content.

Three-colour photometry allows definition of a parameter which characterizes the Balmer jump. The UBV system and the RGU system are equally useful in this respect. Reddening can be taken into account only under favourable conditions, e.g. in a cluster of young stars with a well defined main sequence.

The use of these three-colour systems to characterize metallicity by an ultraviolet excess (and a much smaller blue-green excess) should in general be avoided. A far better and more unique discrimination can be obtained by measurements in four colours, e.g. the uvby system.

Adequate range of a photometric parameter is quite

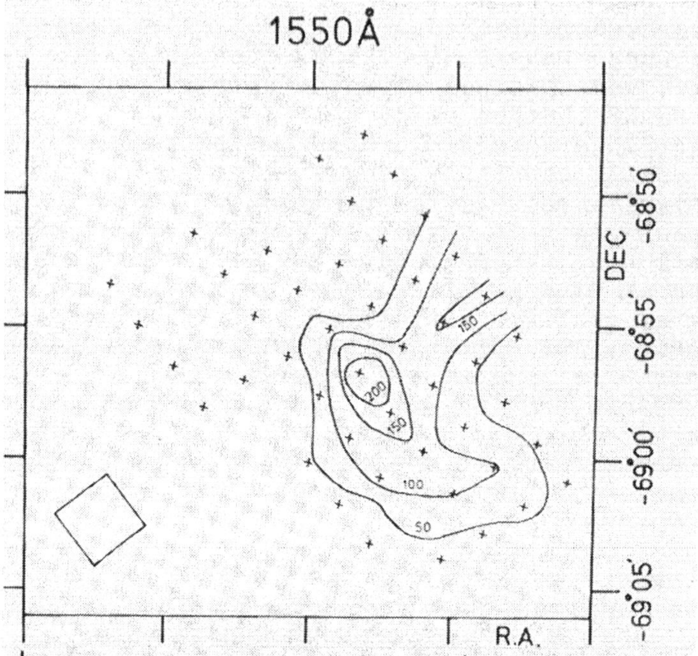

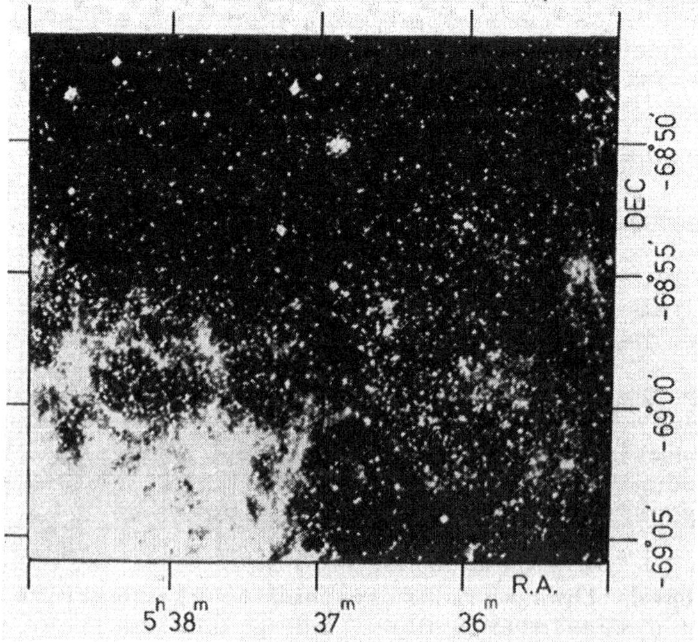

Fig. 4 (top). Contour diagram at 1550 Å of a blue association near 30 Doradus.

Fig. 5 (bottom). Blue photograph of the same area as in Fig. 4; the 30 Doradus Nebula is in the lower left corner.

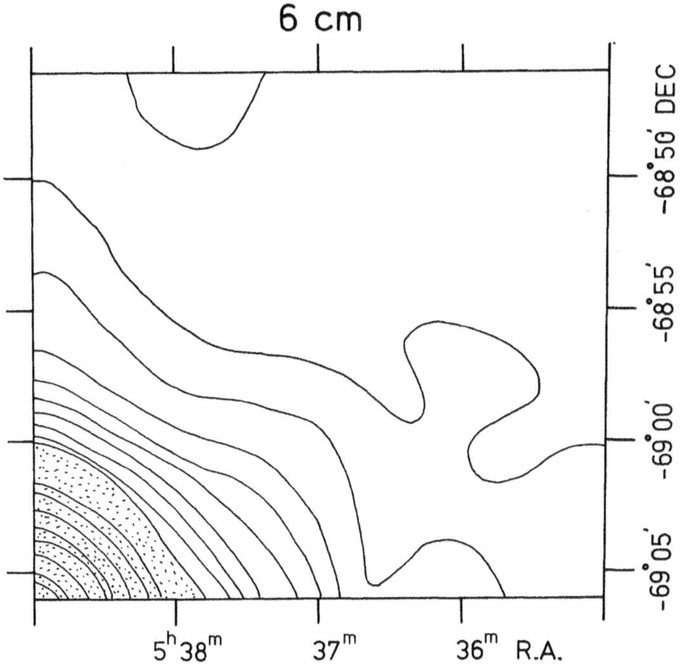

Fig. 6. Contour diagram at 6 cm of the same area as in Figures 4 and 5 (adapted from McGee et al, 1973)

important but it should not be bought at the expense of confusion or insufficient separation of intrinsic properties. A clear example is the use of the equivalent width of H_β and H_γ , together with the Balmer jump, to seperate and to define the colour temperature and luminosity. The equivalent width of H_β or H_γ is conveniently measured with a 30 Å wide filter; a more narrow filter will build up the range of the quantity which defines the equivalent width but errors may arise from line broadening. A wider bandwidth is safer in this respect but the measured effect will lose in range whereas other spectral lines may confuse the information.

It is evident that the effectiveness of any photometric system to be defined in relation to special problems will benefit from a preceding spectroscopic study of a limited number of objects. It is unfortunate that the potential benefits of such studies have not always been fully realized, leading to the existence of some widely used imperfect photometric systems.

CONCLUSION

Photometry will continue to be an important tool to the astronomer. It is likely that direct photography will become less important to quantitative photometry when modern two-dimensional detection systems are sufficiently developed. In the visible and the ultraviolet the application of photometry may yield to low resolution spectroscopy, except for large scale statistical studies and for the faintest stars. In the infrared, apart from high resolution spectroscopy at $\lambda\lambda < 6\mu$, photometry continues to be the unchallenged tool to characterize the spectral energy distribution.

REFERENCES

1. R.C. Bless and B.D. Savage, 1972, NASA SP-310, 175
2. J. Borgman, 1973, Astron.Astroph., **29**, 443
3. J. Borgman and A. Blaauw, 1964, Bull. Astron. Inst. Netherlands, **17**, No. 5
4. J. Borgman, J. Koornneef and R.J. van Duinen, 1975, Astron. Astroph., in print
5. M. Golay, Introduction to Astronomical Photometry, D. Reidel Publ. Comp., Dordrecht, 1974
6. H.L. Johnson and W.W. Morgan, 1951, Ap. J., **114**, 522
7. P.B. Lucke, 1974, Ap. J. Suppl. Ser., **28**, 73
8. R.X. McGee, J.W. Brooks and R.A. Batchelor, 1972, Austr. J. Phys., **25**, 581
9. B. Strömgren, 1966, Ann. Rev. Astron. Astroph., **4**, 433
10. R. J. van Duinen, J.W.G. Aalders, P.R. Wesselius, K.J. Wildeman, C.C. Wu, W. Luinge and D. Snel, 1975, Astron. Astroph., in print
11. Th. Walraven and J.H. Walraven, 1960, Bull. Astron. Inst. Netherlands, **15**, No. 496

THE PHOTOMETRY OF EXTENDED OBJECTS: DIODE-ARRAY DETECTORS AND INTERACTIVE DATA PROCESSING

W. C. Livingston

Kitt Peak National Observatory*
Tucson, Arizona U.S.A.

ABSTRACT. Monolithic silicon diode-array detector technology is reviewed as applied to astronomical applications. The massive data files that can result from diode-array detectors are most conveniently and efficiently digested by the astronomer through computer-coupled interactive picture processing.

The astronomer now possesses two new tools which assist him in the deduction of the physical conditions in extended objects. The first is a piece of *hardware*, the solid-state diode-array detector, which will likely supplant most vacuum devices within a few years. Diode-arrays offer fundamental photometric advantages over television type transducers (for example, SEC vidicons) and even other photoemissive tubes (including electronographic types) because they have a fixed geometry and a determined (and fixed) transfer characteristic. Surface photometry equal in precision to that afforded by the photomultiplier tube becomes feasible. The resulting 2-D arrays of data present some difficulties in interpretation, but their digestion is aided by the second new tool, primarily a *technique*, which we shall call "interactive data processing". Computer generated displays of data fields are presented in pictorial or graphical form. The display is not in hard copy but by a CRT display which is under on-line computer control. The astronomer then must interact with his data within seconds,

---

* Operated by the Association of Universities for Research in Astronomy, Inc. under contract with the National Science Foundation.

de Jager/Nieuwenhuijzen (eds.), Image Processing Techniques in Astronomy. 21-30. *All Rights Reserved.*

or tens of seconds, seeking out areas of interest and proscribing further (on-line) manipulations or computations. Although partly a matter of computer and display hardware, interactive data processing is more importantly a methodology which says the astronomer must repeatedly interact with his observational data in order to derive a maximum of benefit from it.

This paper reviews the characteristics of diode-arrays and briefly outlines some on-going efforts at interactive data processing.

I. DIODE-ARRAYS

Two kinds of solid-state monolithic diode-arrays are at present commercially available: the Integrated Diode-Array (IDA) and the Charge Coupled or Charge Injection Device (CCD or CID). The IDA and CCD devices operate as p-n junction photodetectors in the photon integrating mode (Weckler, 1967). Incident absorbed radiation releases electron hole-pairs (1 photon → 1 electronic charge) which discharge the biased diode during exposure.

How the stored photo-charge is read out distinguishes the different devices. Noise of the readout is a critical characteristic. In the IDA, field-effect transistor switches sequentially connect the diodes to a common video bus and the signal emerges as tiny (~ 20 ns wide) charge packets. The content of these charge packets may be accurately measured by a charge integrating amplifier such as designed by Trumbo (see Livingston et al., 1975). An example of an IDA is the Reticon 512 B, a linear array of 512 elements, which has an elemental diode capacitance $C_D \sim 0.6$ pf and a video line capacitance $C_V \sim 50$ pf. With a 5v bias the saturating charge is $2(10^7)$e. Coupled to the 512 B, Trumbo's amplifier exhibits a noise equivalent charge referred to the input of $1.0 \cdot \sqrt{Hz}$ electrons for Hz > 10^3, where Hz is the bandwidth. For Hz = 10^3 the predicted amplifier noise is ~ 32 e. In practice, at a bandwidth of $2(10^5)$ Hz a noise of 950 e is achieved. This is about twice the predicted noise and implies the existence of excess noise of unaccounted origin.

In the CCD devices the photocharges are transported from element-to-element within localized potential wells by the application of appropriate voltages to overlying electrodes (Barbe, 1975). The charge is finally detected by an on-chip integrating amplifier. An example is the Fairchild 244 (Walsh, 1975), an array of 190 x 244 elements, which has a $C_D \sim .07$ pf, and $C_V \sim .02$ pf. Because the noise of a field-effect transistor amplifier $\propto (C_V + C_D)/gm$, where gm is the transconductance, the low value of C_V for the CCD compared with the IDA, makes the device of special interest for low light applications.

In practice the CCD and CID detectors also suffer from sources of excess noise which, in common with the IDA, seem to place the noise equivalent charge in the range 400 - 1500 e. Reports by members of industrial laboratories at a recent symposium on CCD technology bears this out (Smokler, 1975). Rodgers and Giovachino (1975) of RCA described a 512 × 320 element CCD imager for which the noise equivalent charge was 500 - 1000 e. Michon et al., (1975) of General Electric reported on a 244 × 248 CID imager with noise equivalent charge of ~ 1500 e. In a non-destructive mode of readout they demonstrated that the stored photo-charges could be repeatedly sampled for long periods of time (3 hours!), implying that by averaging, random system noise could be reduced to arbitrarily levels for certain applications. Antcliffe et al., (1975) of Texas Instrument discussed a 160 × 100 CCD with a noise equivalent charge input of 400 e. The bandwidth for the above quoted noise figures was the order of 10^6 Hz.

To summarize the above, diode-arrays have photon charge storage-capacities ranging from 10^5 to $2(10^7)$ e, predicted readout noise ~ 10 - 50 e or less, but achieved noise levels to date ~ 400 - 1500 e. Improvements in detector technology seem likely to close the gap between predicted and achieved noise, but the winning device(s) is unclear in the spring of 1975.

The diode-array which has seen the most astronomical use is an IDA, the Reticon 512. The device is a linear array of 256, 512 or 1024 elements having 25 μm widths. It is available with element lengths of 25 μm or 600 μm, the difference being merely a matter of opaque masking with the electrical properties being the same. The photometric properties of the Reticon 512 are summarized from the work of the following authors: Dravins (1975); Høg and Wiskott (1974); Livingston et al., (1975); Smithson (1975); Title (1974) and Tull and Nather (1973).

spectral response: λ0.39 - 1.2 μm.

dark current: (electrons · sec^{-1} per element)
10^7 @ + 20° C; 10^6 @ 0° C; $6(10^2)$ @ -76° C, dry ice; $2(10^{-2})$ @ -183° C, LO_2. The array functions at LO_2 temperatures.

modulation transfer: For λ < 0.75 μm, diodes act as descrete elements. For λ > 0.75 μm, increasing loss of response, e.g. at λ = 1.0 μm the MTF response is 0.5 the discrete value for a spatial frequency equal to twice the element spacing.

completeness of readout (lag): For a saturated exposure the readout is complete to ~ 1.5%. No bias effect results from repeated dark reading.

uniformity, electronic fixed pattern noise: 5% of saturated exposure. The pattern is very stable and subject to removal by calibration.

uniformity, light dependent: For $\lambda > 0.6$ μm interference (Fabry-Perot pattern) may or may not occur, depending on sample.

uniformity, light response: Diode-diode variation ~ 1% r.m.s.

linearity: Needs to be investigated, the literature contains disagreements.

Examples of performance capability of the Reticon 512 when used with a solar magnetograph are given in Figs. 1 and 2 (from Livingston et al., 1975). Dynamic range and signal/noise are illustrated in Fig. 1. On a single recording, magnetic fields ranging from 2 to 2000 Gauss are measured. The weakest fields represent a brightness detectivity of $\Delta I/I = 3(10^{-4})$. Figure 2 is a full disk brightness picture of the sun at the center of H-alpha ($\lambda = 0.6563$ μm), and indicates the resolution and general quality obtainable with the device.

2. INTERACTIVE DATA PROCESSING

A century ago the astronomer went to the telescope and looked through the eyepiece. He made notes, he made drawings, hypothesis were proposed and accepted, or rejected, based on what he saw "on-line", viewing the universe.

Later photography came into vogue. No longer were decisions made at the eyepiece. Instead he busied himself with the mechanics of guiding, focusing, and developing. The plates were to be measured or photometered in some way and the results thought over, back at the office.

Today we come away from the telescope with large reels of magnetic tape. Owing to the sensitivity and preciseness of our instruments the tapes may contain all kinds of subtle and wonderful information. For example solar magnetographs can detect line-profile changes the order of one part in 10,000. Velocities ~ 5 m/s can be deduced. But to learn about the solar atmosphere the tapes must be reduced to an interpretable form - CRT pictures, contour maps, etc. Furthermore, and most sadly, the computer only answers the questions put to it. Repeated looks at the data to follow up on new ideas is a slow process.

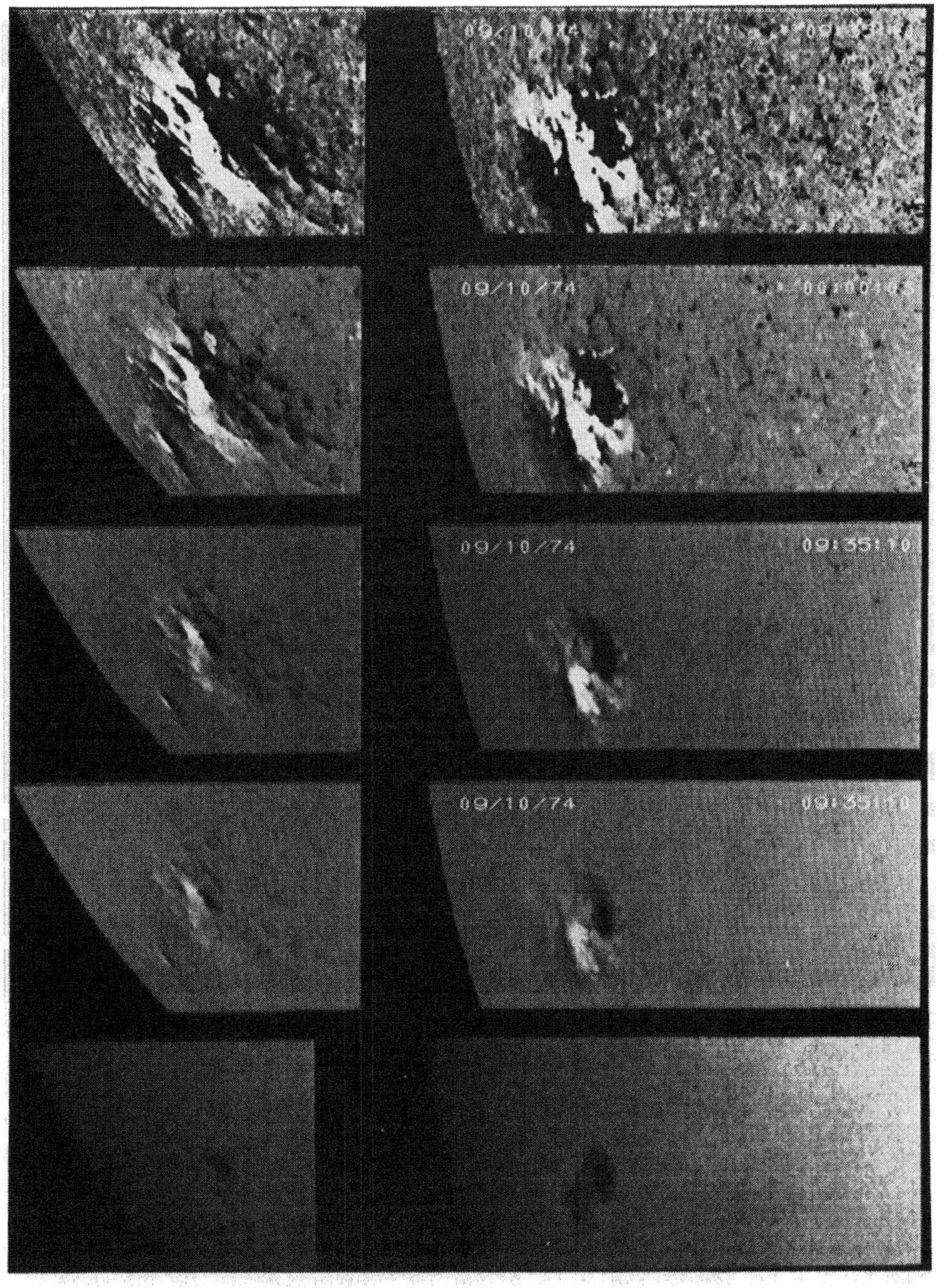

Fig. 1 Multiple CRT displays of an Fe λ 0.8688 μm magnetogram showing two sunspots near the east limb of the sun, 10 September 1974, 16:35 U.T. The display gains are in the ratio 1:2:4:16. Lower frame is the brightness map.

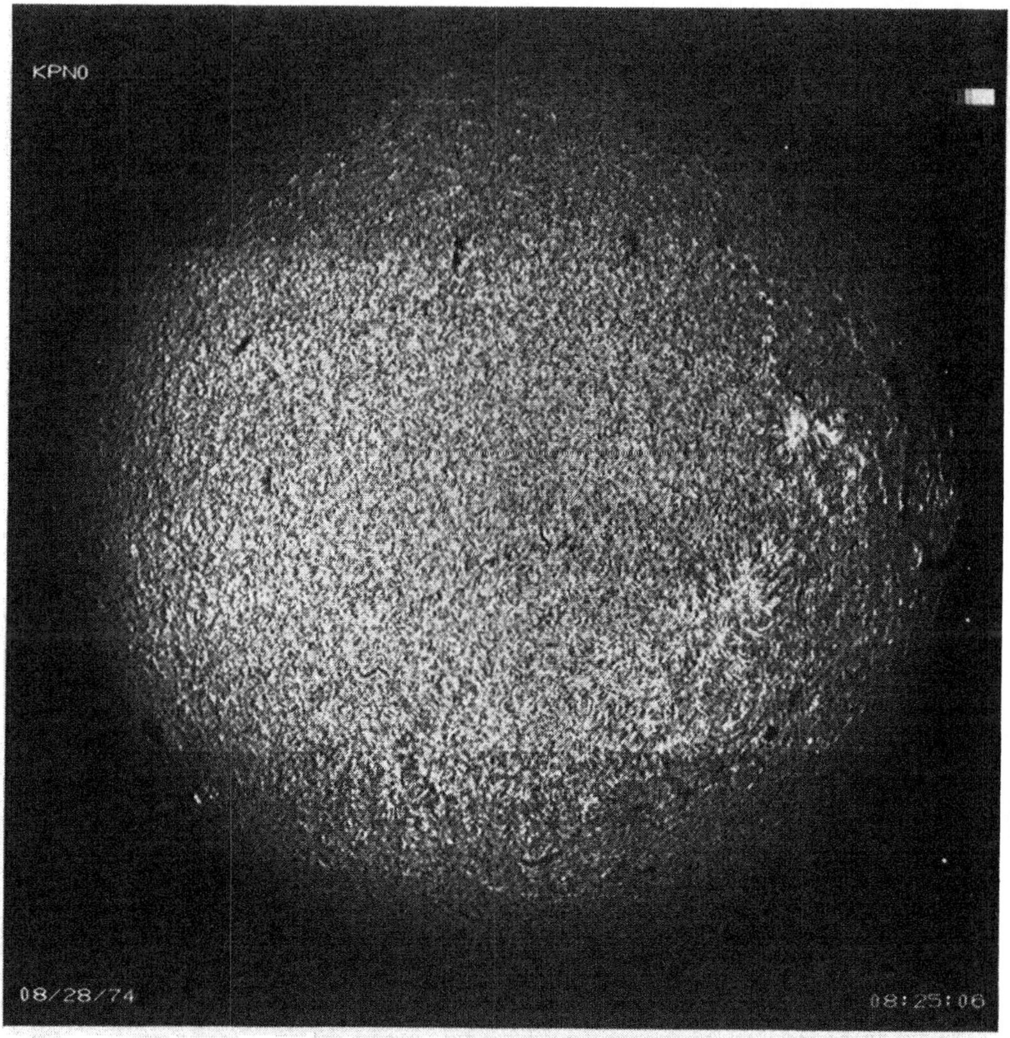

Fig. 2 Diode-array generated CRT pictures of sun in H-alpha, 28 August, 1974. 15:25 U.T.

But now times are changing as something called "Interactive Data Processing" is becoming possible. The astronomer can now sit at a console, load his observational data into the computer, view the data on a screen - graphically or as a continuous tone picture. He may then ask for time or spatial differences or compute some function of the data and view the result. He asks a question - he immediately is presented the answer. He pursues then and there any surprises or hints at the unexpected. Something called the "track-ball" allows him to seek out regions of interest for further processing or questioning. In a very real sense the *astronomer has been put back at the eyepiece*.

J. Harvey, S. Strom and D. Wells have developed an interactive data processing facility at KPNO. A block diagram of the KPNO system is shown in Fig. 3. Data to be analyzed is loaded by magnetic tape into the Varian V74 computer. Usually the data is raw diode-array magnetograms or raster scans made with the PDS microphotometer of photographic material. A 512 x 512 block of data is then transferred to the COMTAL Image Display, the transfer time being ~ 20 sec. The display is in black and white, color, or "pseudo color"; the latter being used to form either isophotes

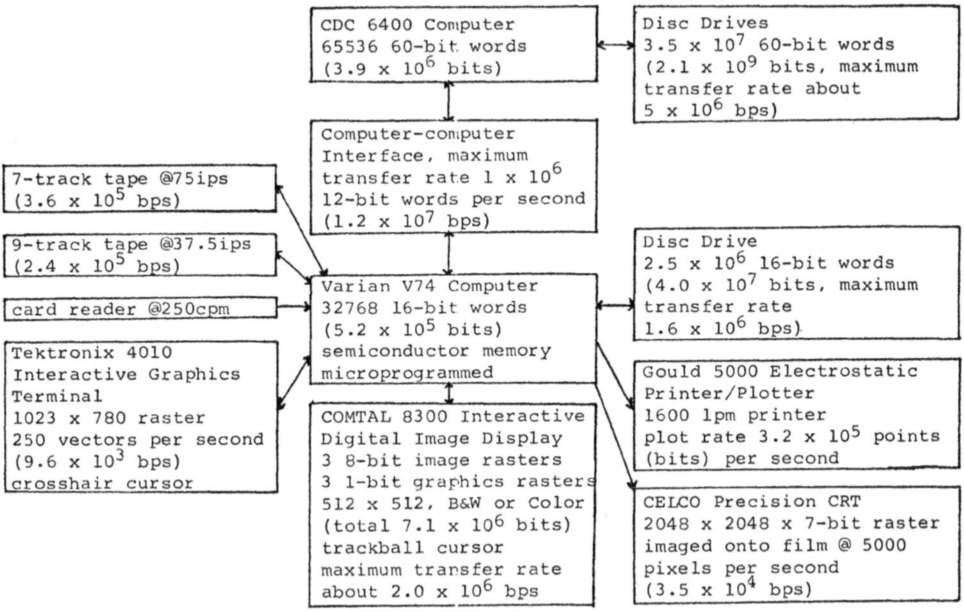

Fig. 3 The hardware configuration of the interactive data processing facility, Kitt Peak National Observatory, March 1975, courtesy D. Wells.

or extend the range of effective display intensity (Fig. 1 having demonstrated the inability of a single CRT picture to accommodate the enormous range of intensity provided by diode-array recordings). Areas of interest are identified by a track-ball cursor, with computations on that area being performed on the CDC 6400 computer, and immediately displayed. Some examples of interactive data processing:

a). <u>Fitting of synthetic to observed spectrum:</u> Multislit image tube spectra of the solar corona taken in Fe XIV (λ=0.5303 μm), with Ne comparison, are scanned with the PDS microphotometer. Synthetic spectra are computed using as free parameters the Doppler width, Voigt damping, and assumed velocity shifts. The synthetic spectrum is fitted to the observed spectrum by a least square process. Synthetic and observed profiles are superimposed on CRT display and parameter adjustments made until the operator is satisfied (Harvey et al., 1974).

b) Comparison of computed magnetic field lines with EUV coronal structures: Given as input a longitudinal photospheric magnetogram. Request CDC 6400 to compute by potential theory a magnetic field line in 3-D and inquire as to coincidence of computed line and EUV loop, or other structure. Use cursor to identify starting point for field line. Importance of the method is that one can concentrate on those field lines that originate at the foot-points of coronal brightness structures (Harvey and Sheeley, 1974).

c) Tracing cloud movements on Venus: Mariner 10 pictures are displayed as Mercator projections. Clouds are identified via the cursor and real motions calculated (Belton and Smith, 1975).

d) Narrow band photometry of planetary nebulae: Alignment of mono-chromatic images of planetary nebulae taken in such lines as Hα, [O III], [S II], for the purpose of mapping ionization and density conditions (Goad, 1975).

e) Spatial structure in Orion nebula: To delineate the reflection component (from continuum pictures), the H II region (Hα), and the ionization boundary ([S II]). Aim is to identify ionization conditions and detection of "bow-wave" phenomena from intensity differences in the three images. Alignment and ratio techniques are involved (Gull, 1974).

f) Time history of solar velocity fields: Registration of individual frames of velocity diode-array data to study variation with time. Difference of two data frames is displayed and operator manipulates fit for minimum error in regions of most interest (Giovanelli and Slaughter, 1975).

These examples fall into different catageories of usage. (a) is a case of interactive programming, i.e., learning how and by what criterion the computed spectrum should be fitted to the observed. (b and c) are examples of the selection of points, or areas, on which further computations are to be made. (d, e, and f) are registration tasks, where, because of variable quality (seeing effects in the velocity data, distortion in the image tube pictures) human judgment is required.

This combination of diode-array image photometry, rapid-scan photographic photometry, and interactive processing of the transduced data can be expected to have a major impact on observational astronomy over the next decade.

REFERENCES

Antcliffe, G. A., Hornbeck, L. J., Younse, J. M. Barton, J. B. and Collins, D. R.: 1975, proceedings of Symposium on Charge-Coupled Device Technology for Scientific Imaging Applications, Jet Propulsion Laboratory Special Publication 43-21, JPL, Pasadena, California 91103.

Barbe, D. F.: 1975, *Proc. I.E.E.E.* **63**, *38*.

Belton, M. and Smith, G.: 1975 (private communication).

Dravins, D.: 1975 (this conference) D. Reidel, Dordrecht, p. 97.

Giovanelli, R. and Slaughter, C.: 1975 (private communication).

Goad, L.: 1975 (private communication).

Gull, T.: 1974, "A Photographic Study of Spatial Structure", proceedings of the 8th ESLAB Symposium, June 4-7, Friscati, Italy, ed. by A. F. M. Moorwood, H II Regions and the Galatic Centre, p. 1.

Harvey, J., Livingston, W. and Doe, L. A.: 1974, *Bull. A. A. S.* **6**, *288*.

Harvey, J. and Sheeley, N. R.: 1974, (private communication).

Høg, E. and Wiskott, D.: 1974, European Southern Obs. Tech. Report No. 5.

Livingston, W. C., Harvey, J., Slaughter, C. and Trumbo, D.: 1975, submitted to *Appl. Optics*.

Michon, G. J., Burke, H. K. and Brown, D. M.: 1975, JPL, Special Publication 43-21.

Rogers III, R. L. and Giovachino, D. L.: 1975, JPL, Special Publication 43-21.

Smithson, R. C.: 1975, *Solar Phys.* **40**, *241*.

Smokler, M. I.: 1975, ed. of proceedings, JPL, Special Publication 43-21.

Title, A. M.: 1974, *Solar Phys.* 35, *233*.

Tull, R. G. and Nather, R. E.: 1973, "Experimental Use of Self-Scanned Photodiode Arrays in Astronomy", eds. J. W. Glaspey and G. A. H. Walker, Univ. of British Columbia, Vancouver, Astronomical Observations with Television-Type Sensors, p. 171.

Walsh, L. R.: 1975, JPL, Special Publication 43-21.

Weckler, G. L.: 1967, *I.E.E.E. Jnl. Solid State Circuits* SC-2, *65*.

TRENDS IN ASTRONOMICAL SPECTROSCOPY

K. O. Wright

Dominion Astrophysical Observatory, Victoria, B.C.

ABSTRACT. A brief survey is made of improvements in spectroscopic data analysis in recent years. The need for accurate observational material is stressed, especially with the highly efficient receivers now available. A few problems that will be studied for many years are considered: stellar evolution and nucleogenesis, stellar atmospheres and chromospheres, double and variable stars, and galactic, interstellar and extra-galactic astronomy. It seems probable that the new major advances will be made by those who can take advantage of the greatly improved modern techniques to observe faint objects that are similar to, but different from, those that are now known.

INTRODUCTION

In a discussion about astronomical spectroscopy of the future, one must obviously consider what has been done in the past. The 1960's have been called the "golden age of astronomy" since it was then that quasars, pulsars, supernova remnants, the high-energy universe of X-rays and γ-rays, the three-degree background radiation and most of the complex molecules of interstellar space were discovered. Most of these discoveries were not made with optical telescopes but from observations made with radio telescopes or from data obtained above the earth's atmosphere. However these discoveries open up new fields for the spectroscopist, who will be able to use the new large telescopes with their modern spectrographs of both high and low dispersion to develop his interests in many fields, including: the analysis of spectra to derive model atmospheres and to analyze the composition of the stars; studies of stellar chromospheres, gas streams and mass exchange in double-star

de Jager/Nieuwenhuijzen (eds.), Image Processing Techniques in Astronomy. 31-48.

systems; spectral classification and statistical analysis to determine the basic structure of our local stellar system and to compare clusters and associations; the discovery and analysis of peculiar and variable stars; and studies of galaxies, their nuclei and star systems, and the red shifts that are so important for cosmology. This, of course, is only a partial list and can be expanded by any astronomer. However it does show the vast range of stellar spectroscopy, yet no mention has been made of the new spectral regions that have been opened up with radio telescopes, observations in the microwave range and the far- and near-infrared regions that have been explored very inadequately. There is the ultraviolet region that can be studied to about 3150 Å at many locations, but rarely is, and also the data obtained from balloons, rockets and satellites from 3000 Å to the Lyman-α region as well as the X- and γ-ray regions. Research in nearly all of these regions has barely begun. Major efforts must be made to improve our present data and expand our knowledge for future generations.

The photometric observers, especially those who observe in several colours, and also the low-dispersion spectroscopists have a very important role to play in the analysis of the galaxy and in discovering objects that may be of interest for detailed studies at high dispersion. Data obtained from the radio, infrared and far ultraviolet regions of the spectrum have given us a completely new picture of the origin and structure of our galaxy. However although many, and possibly most, of the new exciting discoveries may be made in these unexplored regions of the spectrum, observations made with optical telescopes will remain the source of most of our basic information and are required to interpret observations made in other spectral regions. We must continue to develop this field with the instruments, equipment and techniques that will be discussed at this conference.

It is most appropriate that the spectrophotometry of nearly fifty years ago should be mentioned here, since much of the early work was pioneered by Ornstein, Minnaert and their associates at Utrecht. Both laboratory and stellar data were obtained photographically and reduced to intensities with a Moll microphotometer. To show how far we have advanced, one of my most exciting memories is of the day I first traced a three-prism spectrum of α Persei and saw Hγ appear as a broad line with wings extending 30 Å in each direction, an observation that, at that time, was quite unexpected in an F5 star. However the conversion from densities to intensities was a very laborious process as it was necessary to read off each point from the calibration curve. A marked improvement was made in 1939 after Beals had visited the Netherlands and adapted one of the ideas of Minnaert and Pannekoek to make our intensitometer which, using graphical methods, permits the rapid conversion from density to intensity and, if a logarithmic tracing is made, also permits rectification of the continuum. Now, of course, most of the drudgery of intensity conversion is eliminated through the use of computers.

Before considering our main theme I should like to emphasize once more the need to maintain and improve the standards of data collection and analysis. The I.A.U. has set up committees on standards for radial velocities and line intensities, to mention two fields that are related to our spectroscopic studies. The standards proposed by these two committees have been useful for inter-observatory comparisons but the cooperation in achieving the results has never been as great as might be desired. In the photographic process the errors of spectrophotometry are well known but frequently good papers could have been made better with additional or more reliable material; calibrations are not checked frequently enough and the data are often based on one plate instead of several - though the analysis of the data still takes much longer than the reductions. When the more-efficient image intensification techniques are used the sensors are rarely uniform over the field and, although the response may increase almost linearly with the light flux, correction for the non-uniform background must be carefully made. Therefore calibration techniques are just as important as for the photographic process and the reduction procedures are even more difficult. It is fortunate that adequate methods of computer processing have been developed.

To extend this theme a little farther, I have seen papers related to radial velocities where large systematic corrections have been applied. It is well known that the measurement of the broad lines in early-type spectra is difficult when few lines can be seen and, for extended atmospheres, lines of different excitation may give different velocities. However stars that have been accepted as radial-velocity standards are available over most of the sky, and should be observed. Even with the best spectrographs and established wavelength lists, deviations from the accepted velocities do occur but if large differences are observed consistently, the collimation, optics and rigidity of the spectrograph and its mounting should be checked. The same type of check should be made for intensity measurements, though the accuracy of the data for the so-called standard stars is not as high as might be desired.

Recent developments, both in telescope design and in instrumentation have almost revolutionized observational astronomy. The greatly increased efficiency in photon collection through the use of image tubes, electronographic cameras, television techniques etc., as well as improved photographic plates, has been so great that, as Greenstein commented recently, the five-metre Hale telescope is now as effective as a fifteen-metre telescope would have been without the new techniques. The purpose of this talk is not to discuss instrumentation. The question of which system is best remains open since the decision depends on the type of research that is planned. In Victoria we have concentrated principally on improving our spectrographs, using coated optics and image slicers. There has been a gain of two magnitudes in the last ten years.

NEW HORIZONS AND PROBLEMS

When Dr. de Jager asked me to prepare a talk about the information needed for the most urgent problems in astronomical spectroscopy, the challenge intrigued me even though there had been considerable discussion about future research problems at the Sydney I.A.U. General Assembly. In addition the subject was thoroughly discussed at the conference on Research Programmes for the New Large Telescopes at Geneva in May 1974, and the Greenstein report on Astronomy and Astrophysics for the 1970's is a landmark study that surveys the future needs in a remarkably concise presentation. I think that most of the pressing problems have been mentioned in these reports. Therefore I can do no better than to mention a selection of them, with comments and additions in the fields in which I am most interested.

Basically the problems of to-day are extensions of those considered at the beginning of the century when astronomers realized that spectroscopy added another very important dimension to the measures of position and luminosity. These problems can be stated in brief as: to learn about the beginning and end of the universe; to determine whether the same physical laws apply everywhere; to examine and interpret the processes of the evolution of the stars and galaxies; and to study the possibility of life beyond the earth. Spectrographic studies are required for all these topics. In spite of all the efforts that have been made over the past fifty years, more work is required for nearly every problem, though the developments in some fields have been far more extensive than astronomers could have visualized even twenty-five years ago. In some areas, such as energy generation, many of the most exciting questions have been answered but, as in all research, the solution of one problem shows the need for answers to many more. A few years ago it was possible to study only the brightest stars with high dispersion (5 Å/mm or better) and except for the largest telescopes, stars fainter than 15^m could be observed spectroscopically only with objective prisms. With the great improvement in light collecting techniques, the effective apertures of our telescopes have increased several fold, but the most interesting objects always seem to be at the limit of observation and the most exciting explorations will still be made with the largest telescopes and the crucial observations will be made by relatively few astronomers. The extra light-gathering power will permit greater time resolution for rapidly varying objects but the principal value will be in the observation of faint objects and the use of higher resolution in studies of stars in distant galaxies and their red shifts, and of newly-discovered objects such as quasars and pulsars.

STELLAR EVOLUTION

As we all know, the origin and evolution of stars is one of the most important fields in astronomy. In the past fifty years, great advances have been made in the theory of stellar structure and evolution and especially of the sources of energy generation. We think we know a great deal about the processes involved and something about the time scale for stars of different masses. However a great deal more research needs to be done.

Pre main-sequence stars have been studied extensively by Herbig and by M.F. Walker; recently the Stroms have added considerable information especially with their infrared observations. The recently-discovered infrared excess observed for most of these objects is presumably produced by a circumstellar cloud surrounding these evolving stars. Many more traditional spectroscopic observations are also required to determine spectral types and conditions in the outer regions of these ojects, as well as the motions in their atmospheres. Both inflow and outflow of matter in these proto-stars have been observed in different objects, and much more work is required to correlate the spectroscopic, photometric and infrared data.

Studies of stellar evolution depend a great deal on the abundance of helium. The globular clusters are generally considered to have been formed early in the history of the galaxy and their composition may reflect conditions when the galaxy was forming. It is still uncertain whether the helium content of stars of Populations I and II is different, and spectroscopically the solution is very difficult since helium lines appear only in the spectra of the relatively few hot stars. An average value of about one-quarter the amount of hydrogen is rather generally accepted, but it is important for theories of stellar evolution that the ratio be determined accurately.

The theory of the post main-sequence evolution of the stars has been developed in considerable detail in recent years. The chemical composition of the stars, that can be inferred from the composition of their atmospheres, is fundamental for the theory, but most of the changes occur deep in the interiors. However some effects can be detected when the hydrogen core has been depleted and helium produces most of the energy. It is important to determine conditions at the "instability strips" where the Cepheids and RR Lyrae pulsating variable stars are found in order to compare theory with observation and to determine the ages of these stars.

Iben (1974) has given as excellent summary of this evolutionary phase and concludes that we still may not know enough about the mixing processes in the interior of the sun since fewer neutrinos are produced than are required by theory. Once a star has left the main sequence, the trend of its evolution and the time scale depends on the mass, which is difficult to obtain from observations, though it can be derived from theory and the mass-luminosity law. The basic determination of the composition of main sequence stars remains a most important field since real differences require explanations in terms of origin and evolution, which may vary from one galaxy to another. The RR Lyrae and Cepheid variable stars are important bench marks in evolutionary development and data obtained from their spectra are essential.

When the evolution has progressed to the red giant stage, further development depends on the mass. It seems probable that considerable mass is lost at this stage and some confirmation is obtained from atmospheric motions derived from the radial velocities of these stars. However the observed magnitude of this effect seems smaller than is required by theory.

The late stages of evolution are of great interest but the stars are intrinsically very faint and the observations are difficult to obtain. Greenstein has made a great effort to observe as many white dwarfs as possible and his work will remain a milestone for many years to come - but we do not yet know much about the magnetic fields and polarization effects that seem to be present in these dense stars. At least some information concerning these stars can be obtained from studies of double stars, one of which is an evolved star which may be too faint to be observed. Scorpio X-1, HZ Herculis and Cygnus X-1, observed as X-ray sources, are probably neutron stars or black holes.

NUCLEOGENESIS

The origin of elements and their development throughout the evolution of the galaxy is a major field of research, which will continue for a long time to come. The basis for these researches was provided by Burbidge et al. (1957, 1958) when they explained most of the processes involved in stellar evolution in terms of nuclear reactions in the core during and after the hydrogen- and helium-burning stages. They gave examples of representative stars in various stages of evolution. Recently Truran and Cameron (1971) have traced the evolutionary process for stars $4 \leq M \leq 8M_\odot$ that become supernovae and leave neutron stars as remnants. They predict that 20% of unevolved stars should have $Z \simeq 0.1\ Z_\odot$, which is inconsistent with present observations. The answer could lie in an earlier pre-galactic violent generation of more massive stars that evolved to form 25% of the heavy elements in the solar system and many black holes. For the light elements, Reeves (1974) has compared observation with theory and suggests that the "big bang" hypothesis can explain the observed distribution of the light elements, hydrogen, deuterium, helium and lithium, while spallation induced by galactic cosmic rays resulting from supernova explosions produces the heavier elements, beryllium and boron. Deuterium has not yet been observed in the optical region but molecular radio-frequency lines have been observed and an interstellar ratio of $D/H = 1.4 \times 10^{-5}$ has been derived from observations of Lyman α. It is generally accepted that $^{4}He/H = 0.1$, or $Y = 0.29$ for the primeval abundance. Additional checks on these observations are needed.

Arnett (1974) recently reviewed the arguments for explosive nucleosynthesis as the source of the heavy elements in stars and concluded that the abundances from neon to nickel support this hypothesis in spite of the fact that the metals seem to be relatively abundant even in the youngest stars. He feels that many questions remain to be answered - the s-process, cosmic ray spallation, the r- and p- processes etc.

STELLAR ATMOSPHERES

The study of the physics of stellar atmospheres has been a major effort in astronomical research for more than fifty years. The composition of the stars is important since the theories of nucleogenesis depend on it and also since, if we know the mass and the radius, we can estimate the rate of energy production and thus establish the star's position on the evolutionary sequence. The variation of the temperature, pressure and velocity field throughout the atmosphere is required not only for theoretical investigations of stellar structure, but also to provide the basis for the theory of energy transfer in the stars.

The analysis of stellar atmospheres developed rather slowly at first during these years partly because the physics is very complicated but also because it was laborious as well as difficult to obtain accurate line intensities. Here again I should like to mention the pioneer work of the Utrecht physicists and astronomers and their efforts to develop both the theory and measurement of line intensities. The curve of growth is still a useful relation to convert observed stellar absorptions to the number of atoms producing the line of a given atom in a given stage of ionization in the atmosphere. Since for many types of stars the theory of the structure of the atmosphere is now well enough known to permit the calculation of line profiles and total intensities for a grid of models with varying parameters, it has become even more important that the data be determined accurately. The photographic plate remains the best information collector when a large range of wavelengths is being studied, but photoelectric scans of shorter regions should be used to check the accuracy of the measurements. The status of transition probabilities and f-values is gradually improving; the early data for the higher excitation lines of neutral iron have been shown to be in error and the discrepant results for temperature and electron pressure for different stages of ionization and excitation have now been reconciled for the different regions of the solar and other stellar atmospheres. For many years most physics laboratories considered that the theory of spectra had been solved about 1928. However astronomers and a few physicists persisted in their interest in intensity problems and I was very pleased to find that Kurucz' (1974) recent analysis of the Fe I spectrum showed very good agreement between observation and theory.

The theory of stellar atmospheres has also progressed very rapidly during the past few years as a result of work by Mihalas, Strom, Avrett, Pecker and many others. For main-sequence stellar models it is now possible to trace the flow of radiation through the outer layers of a star and to obtain good agreement for the atmospheric parameters for the different stages of ionization for each element - which is the usually accepted criterion that the computed data represent a real stellar atmosphere.

There have now been several hundred analyses of stellar atmospheres, most of which are based on the curve-of-growth technique. The data obtained more than ten years ago must be given low weight since inaccurate f-values were used, too simple models were employed in the interpretation of the observations and/or the observational material was too uncertain. A number of computer programs for model atmospheres are now available. There are also programs listing all possible lines and their intensities, molecular as well as atomic, in the photographic region of the spectrum for a range of typical temperatures, e.g. Bell and Rodgers (1969). A very useful project might be to re-analyze the published intensity data for many of these stars in order to obtain at least statistical results for most spectral classes before deciding that new and better material is required for every star.

The sun may, or may not, be a typical, normal G2 V star. Books have been written about the sun and several satellites have been launched with experiments designed to study most of the observable phenomena, while at least one journal is devoted to the publication solely of solar research. No wonder that stellar astrophysicists sometimes wonder whether they are out of their depth when they try to explain the observed stellar spectra in terms of theories that leave out explanations for granulations, sun-spots, prominences, flares, limb-darkening, the corona, stellar winds, magnetic fields, etc. No doubt some or all of these phenomena are present in solar-type stars, and many are probably enhanced in stars with more extended atmospheres, yet it is almost impossible to detect any of these phenomena from integrated spectra of the solar disc. Certainly the simple LTE theory of radiative transfer is only a first approximation; this is presumably the reason why Pecker began to look for non-LTE effects in the solar spectrum ten or more years ago. However the differences are very small in the sun and can only be observed for a few selected lines on high-dispersion spectra taken at different points on the solar disc. Therefore it is not surprising that details of stellar atmospheres are still so poorly known.

As an example of line-profile studies, consider the paper by Altrock and Canfield (1974) on the Mg lines, λλ 2852, 5172 and 4571 in the solar spectrum; they determine an electron density distribution using the Harvard-Smithsonian Reference Atmosphere, and find that the microturbulent velocities vary from 2 km/sec. in the photosphere to 0 km/sec at 400 km and then increase to 4 km/sec at 900 km. By combining data for both line cores and wings, they consider that solar microturbulence is real and isotropic and different from the macroturbulence produced by the solar granulation and the five-minute oscillations. Although detailed studies such as this cannot be made for the stars, since this paper is based on very accurate profiles determined at the centre of the disc and near the limb, such solar data should be useful in preparing programs for stellar observations.

The question of turbulence in stellar atmospheres continues to be an active one. Ever since Struve and Elvey (1934) found that stellar line intensity observations could not be fitted to theoretical curves of growth until an empirical velocity parameter was added, thus raising the horizontal portion of the curve, both theoretical and observational astronomers have been trying to explain this phenomenon. Kuhi (1974) notes that Smith's (1973) observations of sharp-lined A-type stars can be explained by lowering the turbulent velocity from a previously suggested 7 km/sec to 4 km/sec without any need to consider non-LTE effects for these dwarf stars. However, for supergiant stars, Kuhi comments that there does seem to be a general velocity field, with velocities increasing with height in the atmosphere, with luminosity and with temperature. This problem is related both to theory and observation and should continue to receive high priority since it is important not only for the understanding of the basic physics of stellar atmospheres, but also for the determination of the stellar abundances that are fundamental for our understanding of stellar evolution.

Although Kuhi (1974) and others consider that non-LTE effects and uncertainties in the effects of microturbulence may vitiate some of the abundance determinations made in the past, there is no doubt that stars do show large differences in composition, e.g. the Population I stars contain a larger percentage of metals in their atmospheres than do those of Population II. Mihalas and Athay (1973) have surveyed the field relative to departures from LTE. For the sun they note that the profiles of the strong resonance lines are quite different from those calculated for LTE, and also agree better with observations. In the spectra of the hot stars, the differences are even greater for lines of H, He I and He II, but can be detected for lines of other elements; they suggest that in some cases the LTE abundance calculations may be in error by a factor of ten or more. Not only will the transfer problems for spherically symmetrical extended envelopes have to be solved, but hydrodynamical phenomena must also be considered. We have just reached the stage where accurate profiles can be determined to check the theories and to derive meaningful abundances from these profiles.

In the peculiar and metallic-line A- and F-type stars, the abundance anomalies from Al to Hg vary markedly even for stars within the same group, as do some of the isotope ratios, especially for Hg. Major differences in the abundance ratios have been found by different investigators; some of these differences may be related to the lines that have been studied - e.g. O I lines in different regions of the spectrum seem to give different abundances. There is also the possibility that some stellar atmospheres change with time, for it is known that some of these stars have long periods. The atmospheres of these stars require extensive investigation since the peculiarities are so extreme. There is not yet complete agreement about proposed atmospheric models; the oblique rotator model is accepted by many, but the diffusion theory also has proponents. Better observational material, both equivalent-width measurements and accurate line profiles is required.

STELLAR CHROMOSPHERES

Among the developments of the past few years has been the acceptance of the fact that many stars have extensive chromospheres. By analogy with the sun and observations made at solar eclipses, this phenomenon was to be expected but the observational evidence was difficult to obtain. The ζ Aurigae eclipsing systems, with a giant primary star that eclipses a small hot companion, have been shown to have chromospheres since lines arising from low-excitation levels are greatly enhanced close to eclipse and decrease in strength rapidly as the line of sight of the hot star passes beyond the outer atmosphere of the giant primary star. Some of the best-analyzed data are those for ζ Aurigae made by Wilson and Abt (1954), though similar observations for 31 and 32 Cygni have been obtained at Victoria for a number of years. We have finally completed a

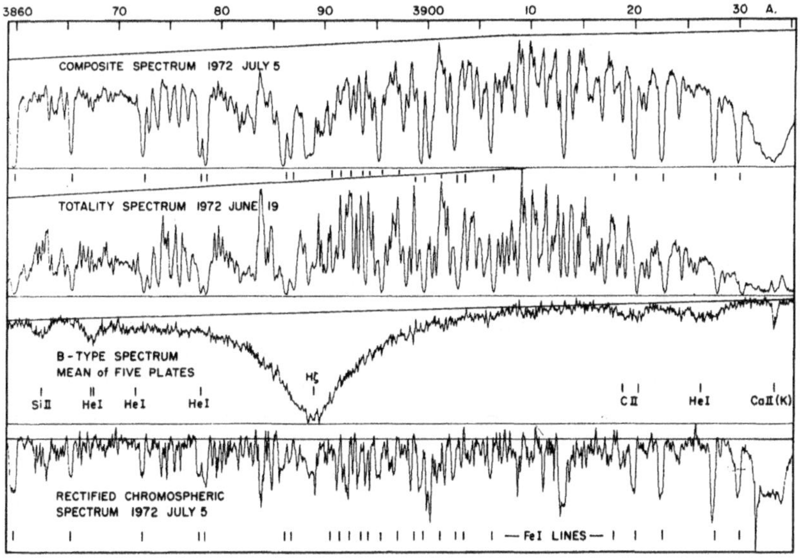

Fig. 1. Intensity Tracings of 31 Cygni at the 1972 Total Eclipse

computer program that separates the spectra of the two stars and, after further analysis, the final tracing shown in Figure 1 is, we believe, the pure chromospheric spectrum of the K star.

A summary of other chromospheric problems under investigation was given at the NASA colloquium on Stellar Chromospheres (1972), and they were also discussed by Kuhi at the 1972 Meudon conference on "L' Age des Etoiles". Wilson has been studying the chromospheric cores of the Ca II emission lines observed in spectra later than F5 for many years, and now finds that there is a dependence between age and luminosity. These results are being extended to fainter stars and to members of young clusters through the use of image-tube spectra. Observations of the He lines, λλ 5876 and 10830, should show chromospheric effects in late-type stars but the results obtained by Vaughan and Zirin (1968) were not as promising as had been hoped.

In many stars, chromospheres and extended atmospheres may be almost identical phenomena. The strong helium line cores in spectra of supergiant B-type stars have been shown by Underhill to be produced by non-LTE effects in the outer regions of these stars. In late-type stars, violet-displaced absorption lines have been observed by Weymann (1962) and others. Since mass-loss phenomena are important for evolutionary theories, such observations should be extended. Emission lines are also indicators of stellar envelopes and, of course, have been observed in many variable stars such as T Tauri and the RV Tauri stars. The source of the envelope is of considerable interest. Is it contracting or expanding? When polarization effects are observed, the explanation is usually given in terms of stellar envelopes since it is thought that, in this case, they arise from electron scattering of dust particles. More observations are needed.

DOUBLE STARS

Double stars have been a fruitful field for spectroscopic investigation for many years, and their importance seems to be increasing. Over the years much large-telescope time has been spent in observing double stars because the orbital data provide the best estimates of stellar masses. For important eclipsing systems observations are continuing and efforts are now being made to combine visual and spectroscopic data to obtain results for long-period pairs. However the domain of double stars has become even more fruitful in the past thirty years since Struve discovered that mass transfer and gas streams are common phenomena, and are important in the evolution of close binary systems. Theoreticians such as Paczynski, Plavec and Kippenhahn and their groups have shown that the theory of stellar structure can be applied to double stars; in particular the Algol paradox seems to have been explained satisfactorily as the result of mass transfer through the Lagrange point from primary to secondary so that the roles are reversed and an underluminous, massive apparently unevolved star is the primary companion of a luminous low-mass evolved star. According to Paczynski (1971), R Canis Majoris type stars, and detached systems with undersize subgiant stars are two classes of low-mass binary systems that have not yet been explained satisfactorily. A beginning has been made in studies of the cataclysmic variables that are double stars, such as novae and U Geminorum type variables; Kraft (1965) has suggested that they originate as W Ursae Majoris type stars.

Recently binary star evolution has become a topic of great current interest with the evidence that for many of the low-mass evolved systems involving mass exchange, the secondary components are neutron stars (HZ Herculis, Crampton and Hutchings, 1974) or even black holes (Cygnus X-1, Hutchings et al., 1974). This is a very rapidly expanding field and shows that optical spectroscopy combined with X-ray and other data may provide the answers to some of the major problems of modern astrophysics.

The question of how widely separated two stars can be before being considered close enough to interact with each other has not yet been solved completely, but Paczynski has computed models for stars with periods of more than 20 years and I.A.U. Commission 42 has changed its name from Photometric Double Stars to Close Double

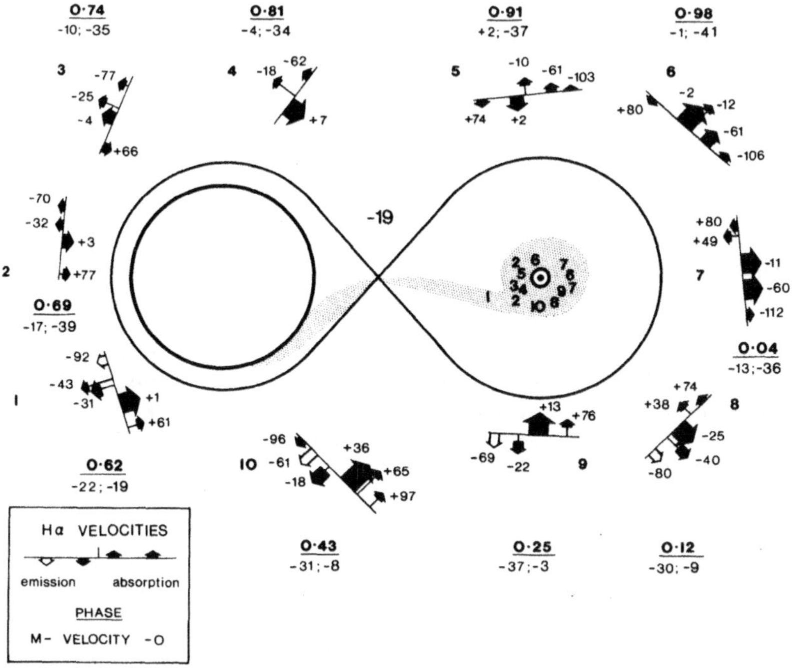

Fig. 2. Model of VV Cephei showing observed velocity distribution with phases.

Stars. One example of this type, where high-dispersion spectroscopy has been used is the VV Cephei system with a 20-year period and a separation of 10 A.U. The Victoria data give a mass of 20 $M_\odot$ for each component and there seems to be good evidence for a mass flow from the primary M-type supergiant through the Lagrangian point and around the secondary early-type star which is surrounded by a hydrogen-emission envelope. A model that agrees with most of the observations is shown in Figure 2.

VARIABLE STARS

Variable stars have always been a favourite field for spectroscopists since there are so many examples and so many explanations can be given for their variation. In the past, the problems of variable stars seemed almost insoluble because, for long-period variables, the stars were faint and high-dispersion studies could be made for only a few stars at maximum light, and for short-period variables, the exposure times were often an appreciable fraction of the period. With large telescopes and new techniques, the necessary observations can be obtained much more quickly and therefore, both types of problem can be studied under much more favourable conditions.

As Feast (1973) comments in his I.A.U. report on Variable Stars, these objects offer an unrivalled opportunity to study physical conditions in stars. In the case of stellar pulsation, he feels that both observations and theory are becoming sufficiently refined for definite explanations for the observed phenomena to be attainable. However there are still difficulties for both Cepheids and β Canis Majoris stars, as well as for stars with multiple periodicities. In red variables, RV Tauri and R Coronae Borealis stars, the ejection of clouds does coincide with the observed pulsations. Other observations can be related to circumstellar matter and mass-loss phenomena, especially for variables that are contracting towards the main sequence. Many more observations are required to determine rates of mass loss, expansion etc. - and these remarks apply also to novae, eruptive binaries, flare stars, magnetic variables, symbiotic stars and pulsars, to name only a representative selection.

One star that is not typical, but that has attracted much attention is FG Sagittae (Kraft, 1974) where the spectrum has changed markedly in the past fifteen years and the luminosity has increased by four magnitudes during the last fifty years. It may be the central star in a planetary nebula; it is a cool star and seems to be changing from a normal F-type giant to a barium- or even an S-type star, cooling by 250° per year and increasing the abundance of the heavy elements in its atmosphere. Studies of stars such as FG Sagittae must begin as soon as the peculiarities are discovered since from such studies, our information about the processes of evolution and nucleosynthesis will be greatly increased.

GALACTIC AND INTERSTELLAR ASTRONOMY

The study of the Galaxy was greatly advanced by radio astronomy and observations of the H I 21 cm line and, more recently, molecular lines. Star clusters and hot early-type stars have been used as indicators of the extent and structure of the Milky Way. To these have been added Cepheid variables, supergiant M-type stars and H II regions. The random motions of these objects have made the interpretation difficult, but the new techniques should extend our knowledge to 8 or more kiloparsecs from the sun.

The study of interstellar lines at both radio and optical frequencies is a most absorbing subject and many detailed studies are required. In the optical region, Hobbs' (1969) high-resolution investigations with a Fabry-Perot interferometer have greatly extended our data on interstellar clouds derived from the multiple components of interstellar lines; these observations provide basic information about their motions, densities, compositions and distances. The diffuse optical absorption lines, of which the strongest is λ4430, have not yet been identified, though Herbig has found some correlations that suggest that many of them are related, and should eventually be identified with a moderately simple molecule. Some of the molecules that have been identified in the radio- and millimetre- regions of the spectrum are quite complex and if others, similar to those observed on the earth and planets, are found, there could be interesting implications concerning life in the universe. Some of the lines have been observed in dark nebulae and the derived temperatures give important information concerning conditions in these clouds. These vast clouds that seem to be in the centres of star formation usually show molecular lines produced by OH, CO and H_2O, and from studies of these and the atomic lines, definite conclusions concerning the state of these nebulae, H II regions and other features can be obtained.

The galactic centre can, of course, best be studied with radio observations, especially with the lines of H I, H_2CO and CO, and the observations indicate that this centre contains the largest collection of molecules in the Galaxy; the density of the core has been quoted as 10^6 cm^{-3} and the mass, 3 x $10^6 M_\odot$. Zuckerman and Palmer (1974) suggest that OH and CO lines may be the best indicators for further studies of galactic structure. They also review isotopic abundances of C, O and S and find that the interstellar abundance ratios are remarkably uniform and in general agreement with terrestrial ratios. However there are some differences, in Orion A for example, where DCN/ $H^{12}C^{15}N$ is greater than the terrestrial value. Further work in this area is clearly indicated; optical data might be obtained eventually.

EXTRA-GALACTIC ASTRONOMY

Many large-scale features in our own Galaxy have not yet been related to structures in other galaxies, but in our own neighbourhood we sometimes cannot see the forest for the trees. In extragalactic systems, their faintness and small size, which are directly related to their distances, make observations difficult. However Sandage is finally obtaining a consistent value for the Hubble constant, which is fundamental in deriving distances of galaxies from their red shifts. Much more work remains to be done on the quasi-stellar objects, though these programs have been given high priority at the large telescopes in order to complete surveys to 18^m. The different velocity systems found between emission and absorption lines should not be considered unreasonable since the lines may not be formed at the same place, and especially since similar differences have been observed even in stellar atmospheres; however the explanation for these differences must be found. The problem of the apparent bridges between galaxies and QSO's of different velocity shifts may be fundamental, but will undoubtedly have been discussed in earlier reviews.

As E.M. Burbidge remarked in her talk at the ESO Instrumentation Conference (1972), it is important to know whether the stellar populations and their distributions are the same in other galaxies as in our own, since these comparisons are closely related to our theories of evolutionary processes. At the moment few galaxies can be studied in detail with conventional spectrographs, but the possibilities for obtaining the necessary data are almost limitless. Similarly the rotation of external galaxies and studies of differental rotation in their spiral arms, which was begun by Lindblad forty years ago, is a project of more than casual interest.

In the nuclei of galaxies, large-scale motions have been detected (Burbidge, 1970) that suggested violent explosions, possibly at several epochs. In addition, continuum studies and radio observations indicate the presence of non-thermal radiation. Galactic nuclei seem to be of several types, including the Seyfert and Markarian galaxies, but there remain others that must be labelled peculiar. Thus the need for spectroscopic observations of high quality is obvious.

SUMMARY

In the above brief survey of spectroscopic developments in astronomy I have tried to mention many of the more important fields where advances will be made. The most exciting discoveries of the past fifteen years have come as a result of the increased efficiency of our receivers and the opportunity to observe fainter objects. In addition, new regions of the spectrum - radio and microwave frequencies, the infrared and also the whole new field of ultraviolet, X- and γ-ray astronomy have opened up and expanded our horizons almost beyond our dreams. Although I have mentioned these areas, I have intentionally limited the discussion to background material since our discussions here will probably concentrate on the processing of data from telescopes available to the average astronomer. Fortunately a great many astronomers are now able to use the large telescopes that are coming into operation as a result of major efforts that have been made in several countries.

No doubt some of you will wish to add to my list of important areas of future research and others will probably say that I have been too conservative in my projections. This is to be expected since astronomy is now very much a science developed by young researchers who expect to follow their own selected fields rather than participate in the large-scale observatory programs that were a feature at many observatories near the beginning of the century. While I have tried to mention a number of research areas that are obviously ripe for exploitation with the new receivers, I have also emphasized some of the conventional fields where more work should be done, even though it may not be so exciting, before major advances can be made. My aim has been to stress the need for careful, accurate measurements based on established standards and calibration procedures in order to ensure that future developments are based on firm foundations. One can never predict the future, especially in astronomy where so many unexpected discoveries have been made in recent years. However I feel that at least some of the next advances will be made by those who can take advantage of the greatly improved modern techniques to observe faint objects that are similar to, but different from those that are now known.

The value of a conference such as this is to permit workers in different fields to discuss their methods and to learn how they can improve their techniques. I hope that this review has provided a broad enough base to be useful for discussions of the techniques that are the principal object of our meeting.

REFERENCES

1974 Proc. ESO/SRC/CERN Conference on Research Programmes for the New Large Telescopes, A. Reiz, Ed., Geneva

1973 Stellar Chromospheres Coll., S.D. Jordan and E.H. Evrett, Eds., N.A.S.A., Washington

1972 Proc. ESO/CERN Conference on Auxiliary Instrumentation for Large Telescopes, L. Laustsen and A. Reiz, Eds., Geneva

1972 L' Age des Etoiles, G. Cayrel de Strobel and A.M. Delplace, Eds., Meudon

1972 Astronomy and Astrophysics for the 1970's, J.L. Greenstein, Nat. Acad. Sci., Washington

Altrock, R.C. and Canfield, R.C. 1974, Astrophys. J. 194, 733

Arnett, W.D. 1973, Ann. Rev. Astron. Astrophys. 11, 73

Bell, R.A. and Rodgers, A.W. 1969, Mon. Not. R. Astr. Soc. 142, 161, et. seq.

Burbidge, E.M., Burbidge, G.R., Fowler, W.A. and Hoyle, F. 1957, Rev. Mod. Phys. 29, 547

Burbidge, G.R. and Burbidge, E.M. 1958, Handb. Phys. 51, 134

Burbidge, G.R. 1972, Ann. Rev. Astron. Astrophys. 8, 369

Crampton, D. and Hutchings, J.B. 1974, Astrophys. J. 191, 483

Hobbs, L.M. 1969, Astrophys J. 157, 165

Hutchings, J.B., Crampton, D., Cowley, A.P., Fehlmann, G., Glaspey, J.W. and Walker, G.A.H. 1974, Astrophys. J. 191, 743

Iben, I. 1974, Ann. Rev. Astron. Astrophys. 12, 215

Kraft, R.P. 1965, Astrophys. J. 142, 1588

Kraft, R.P. 1974, Sky and Telescope 48, 18
Kuhi, L. 1974, Highlights of Astronomy 3, 121, G. Contopoulos, Ed., Reidel, Dordrecht
Kurucz, R.L. 1974, Smithsonian Astrophys. Obs. Sp. Rep. N. 359
Mihalas, D. and Athay, R.G. 1973, Ann. Rev. Astron. Astrophys. 11, 187
Paczynski, B. 1971, Ann. Rev. Astron. Astrophys. 9, 183
Reeves, H. 1974, Ann. Rev. Astron. Astrophys. 12, 437
Smith, M.A. 1973, Astrophys. J. 182, 159
Struve, O. and Elvey, C.T. 1934, Astrophys. J. 79, 409
Truran, J.W. and Cameron A.G.W. 1971, Astrophys. Sp. Sci. 14, 179
Underhill, A.B. 1972, Vistas in Astron. 13, 169, A. Beer, Ed., Pergamon
Vaughan, A.H. and Zirin, H. 1968, Astrophys. J. 152, 123
Walker, M.F. 1972, Astrophys. J. 175, 89
Weymann, R. 1962, Astrophys. J. 136, 844
Wilson, O.C. and Abt, H.A. 1954, Astrophys. J. Sup. 1, 1
Zuckerman, B. and Palmer P. 1974, Ann. Rev. Astron. Astrophys. 12, 279

P A R T 2

ACQUISITION AND STORAGE

INFORMATION RETRIEVAL BY MICRODENSITOMETRY

T. de Groot

Astronomical Institute, University of Utrecht, Utrecht, the Netherlands

In the ideal case the output $I_n(x,y)$ of a microdensitometer can be considered as a true image, i.e. the final product of a linear optical system "pointed" at an incoherent object brightness distribution $I_o(x,y)$ in the sky or elsewhere. Since a photographic emulsion is a non-linear storage element for intensities (apart from noise and granularity effects), and since the linearity of the scanning system is not a matter of course, this final link in such an optical chain is the weakest.

Linearity is preserved only if an overall point spread function (instrument profile) $P(x,y)$ exists such that

$$I_n = I_o \otimes P \qquad \text{(convolution)}$$

or alternatively, in fourier- (aperture-) space, an overall optical transfer function (OTF) $T(\xi,\eta)$ exists such that

$$J_n = J_o \cdot T$$

Here $T(\xi,\eta) = f^{-} P(x,y)$ and $J_i\ (\xi,\eta) = f^{-}\ I_i(x,y)$
(f^{-} fourier transform operator)[1]

Only in this case the original intensity distribution can be restored using some deconvolution method, provided the instrument profile is known with sufficient accuracy.

With proper scaling factors the object and image spaces can be mapped on to a flat surface the points in which have coordinates (x,y) c.q. $\left((x-x_o),(y-y_o)\right)$. This presupposes distortionless imaging, and generally defines the boundaries of a

de Jager/Nieuwenhuijzen (eds.), Image Processing Techniques in Astronomy. 51-55. *All Rights Reserved.*

restricted region of validity, within which moreover the shape of the diffraction pattern around any point should be virtually independent of its position (isoplanatism patch cf. e.g. Linfoot, 1964). This guarantees the linearity of the system as a mapping device. The two questions now are how microdensitometric methods perform in this respect and within what limits and under what circumstances the transfer function is sufficiently independent of the intensity I , more specifically, the density D . We first discuss the second question.

Low contrast case

If the contrast function, defined by

$$K(x,y) = \left(I(x,y) - \langle I(x,y) \rangle\right) / \langle I(x,y) \rangle$$

is sufficiently less than unity in an image section of such an extent as to be still of interest, the density $D(x,y)$, (likewise a measure for the number density N of the developed grains), can practically be considered to depend linearly on the photon flux density to which the plate has been exposed (small signal linearity), according to

$$\Delta D = D - \langle D \rangle = 0.4343 \gamma \; \Delta I / \langle I \rangle$$

which implies an analogous linear transformation of the small signal OTF already present. In practice ΔD should be less than 0.5 (Dainty, 1971). A more important consequence of the small signal condition is that it can be made to cover the case in which the photographic noise is statistically independent of the signal such that a unique photographic point spread function $P_{ph}(x,y,\langle D \rangle)$ and its corresponding transfer function $T_{ph}(\xi,\eta,\langle D \rangle)$ can be defined in a limited density region ΔD around a mean density $\langle D \rangle$. It is found that T_{ph} , apart from accounting for scattering (more general: for "turbidity") and granularity, may also include the moderate adjacency effects, that might still be present in regions of low contrast. The photographic spread function can thus be related to a linear, but noisy filter, introducing frequencies not present in the signal besides acting as a low pass filter for signal frequencies.

The low contrast T_{ph}'s and the corresponding extent of the allowable contrast for a certain emulsion-development combination at different levels of background illumination can be determined by imaging sinusoidal gratings or interference patterns, and scanning the result in an "ideal" microdensitometer. Conversely one dimensional transfer functions can be obtained as fourier transforms of differentiated scans of edge images. This method has error amplifying phases. The threshold resolution and the fine detail performance of photographic emulsions as quasi linear detectors, in the presence of noise, can be studied, using Fourier methods on low contrast images. How to tackle in general the problem of restoring data from information that passed a noisy filter has been extensively enunciated by Brault and White (1971) in a now already classical paper.

Any contrast case

In most astronomical and spectrographic applications it is necessary to use the full density range of the photographic emulsion along with the D - log E , or characteristic curve with its limited reliability. This curve is the only means to linearize the output in the full range of intensities which the emulsion is capable to transform into significantly distinguishable density levels.

This "calibration" curve can formally be used to save the transfer function concept but it is of limited value since the result is a rather arbitrary average of all small signal transfer functions. The corresponding point spread function is a mixture of the various density dependent broadening and granularity effects, not to speak of the even more untraceable influence of the adjacency effects. These have been found in cases to be responsible for contrast reversal at spatial frequencies of about 15 ℓℓ/mm causing $T_{ph} > 1$ in this region, and a locally negative P_{ph} . These conclusions remain valid when the γ of the film happens to be minus one so that the relationship between the transmittance τ and the exposure $E = It$ is linear.

In the practice of reducing astronomical plates, and more specifically stellar spectra, slit widths are used down to the micron region i.e. of the same order as the photographic line spread function. In order to retrieve true spectral line profiles it is certainly necessary to take account of this broadening effect. For this purpose factory specified data are only of indicative value, the more so as they are given for a specific developer and way of processing.

Microdensitometer non-linearity

The only quantity that is allowed to influence the transmittance during a scan is the grain density within the boundaries of the scanning spot. The least trivial causes of non-linearity, i.e. of a variable system transfer function, are defocussing during the scan and coherence effects. Others, such as "flare" and stray light are controllable and can be kept within acceptable limits.

The danger of defocussing is always present with films and large size plates. It is only recently that autofocus systems have been incorporated, to continually control the position of the exact focus with respect to the emulsion layer (cf. the contribution of J.R.W. Heintze).

The coherence function in the image of the first slit on the plate is the fourier transform of the effective exit aperture of the condensing lens. The coherence interval is thus proportional to the effective wavelength, divided by the numerical aperture of that lens. In practical cases its length is somewhat between one and ten microns.

Partial coherence between points in the scanning spot introduces the possibility that the measured transmittance becomes appreciably sensitive to phase lag variations suffered by the light wave in gelatin and film- and plate material. In other words in this case the transfer function T_2 of the optics that images the spot on the second slit is variable. In a traditional two-slit instrument the influence of this can be minimized by taking care that the effective scanning spot is limited, and determined, by the first slit, contrary to common practice, where thus the system transfer function is virtually determined by the variable T_2.

Strictly speaking this condition means that the ratio of the numerical apertures of first imaging (condenser) lens and second imaging lens is infinite. Taking into account the maximum frequency in the object f_{obj} and the cut-off frequency f_2 of the imaging lens, this condition becomes less strict (Swing, 1970):

$$(NA)_1 \ / \ (NA)_2 \ > \ f_{obj} \ / \ f_2.$$

The linear microdensitometer

Since the above constraint blocks the way to high resolution and high precision, in modern microdensitometers the second imaging system has been abandoned, and replaced by a large aperture light collecting system. In this case the system transfer function depends only on the optics preceding the plate and it is insensitive to coherence in the scanning spot, which may even be formed by a laser beam. Such a system has been proposed, and realized for the first time by Weingärtner et al. (1971) and further e.g. by Reynolds and Smith (1973) and of course in the microdensitometer of the Utrecht Observatory.

REFERENCES

1. J. Becherer and G.B. Parrent Jr., J. Opt. Soc. Am. **57**, 1479, 1967.
2. M.J. Beran and G.B. Parrent Jr., Theory of Partial Coherence, 1964.
3. P. Bouchareine, Optica Acta **17**, 585, 1970.
4. J.W. Brault and O.R. White, Astron. and Astrophys., **13**, 169, 1971.

5. D. Grimes, J. Opt. Soc. Am., **61**, 1263, 1971.
6. E.H. Linfoot, Fourier Methods in Optical Image Evaluation, 1964.
7. G.O. Reynolds and A.E. Smith, Appl. Optics, **12**, 1259, 1973.
8. R.E. Swing, J. Opt. Soc. Am., **60**, 1551A, 1970.
9. R.E. Swing, J. Opt. Soc. Am., **62**, 199, 1972.
10. I. Weingärtner, Optik, **32**, 5, 508, 1971.
11. I. Weingärtner, W. Mirandé, E. Menzel, Optik, **34**, 1, 53, 1971.
12. J.C. Dainty, Optica Acta, **18**, 795, 1971.

FOURIER TECHNIQUES

P. Connes

Circe, Faculté des Sciences Orsay, France°

Fourier Techniques in astronomy have been used so far mostly for spectroscopy; in this case one is not processing an image but rather eliminating it completely: The spectrum is produced without any use of spatial dispersion and an intermediate stage. The basic advantages will be reviewed and some of the most striking recent results discovered.

The main field of Fourier processing in imagery is of course aperture synthesis at radio frequencies but this is outside the field of this review since it is covered by Dr. W.N. Brouw. At optical or infrared frequencies several different lines of approach exist; each one tries to capitalize on a different possible advantage.

Firstly one can try to do the exact equivalent of Fourier spectroscopy for infrared images, i.e. improve point by point image scanning techniques analysing the entire image. In this case the enemy is detector noise. However, a 'multiplex advantage' will be realized only if detector size is not increased. This severely limits the field of application and we know of no successful attempts so far.

Secondly one can try to improve upon the spatial resolving points of any technique (scanning or photographic) in any spectral range insofar as it is limited by the atmosphere. Here one is not trying to gain in speed and the enemy is not photon or detector noise but 'atmospheric noise', i.e. distorsion of wavelengths.

In this line one has to mention first techniques which are

° Present address: University of California, Berkeley, U.S.A.

de Jager/Nieuwenhuijzen (eds.), Image Processing Techniques in Astronomy. 57-58.

simple modern implementations of Michelson's well known fringe visibility and lead to stellar diameter, limb darkening, or separation of close binary measurements. In the simplest case two small apertures in front of the telescope mirror are used. At the same time long path interferometers are being built or planned by several groups for working in the 10 μm window.

TELEVISION DETECTOR TECHNIQUES

A. Boksenberg

Department of Physics and Astronomy,
University College London, London, England

ABSTRACT. A review of image detection by television techniques is given in the context of optical and ultraviolet astronomy. The method of image photon counting, now used routinely in optical astronomy, and the detector system of the International Ultraviolet Explorer observatory satellite, are discussed as representative examples in each field.

1. GENERAL METHODS

A simplified diagrammatic organisation broadly indicating the methods available for intensifying and recording photon image information is given in Fig. 1. Many of these will be discussed in this conference so I will not dwell on them now. Instead, I will discuss representative examples in two main categories: analogue television and digital television.

2. ANALOGUE TELEVISION DETECTORS

In this section I touch on the use of television detectors as inherent photometric devices. In general, with the use of commercially available television camera tubes current at any epoch, compromises and adaptations must be made if they are to be applied successfully in photometrically demanding roles such as in astronomy. There are several cases where this has indeed been done, both in optical and ultraviolet astronomy, using SEC and SIT tubes and solid state arrays. However, it is only at the level of detail that one works when developing such a system that all the problems and limitations of a device come to light: these

de Jager/Nieuwenhuijzen (eds.), Image Processing Techniques in Astronomy. 59-78.

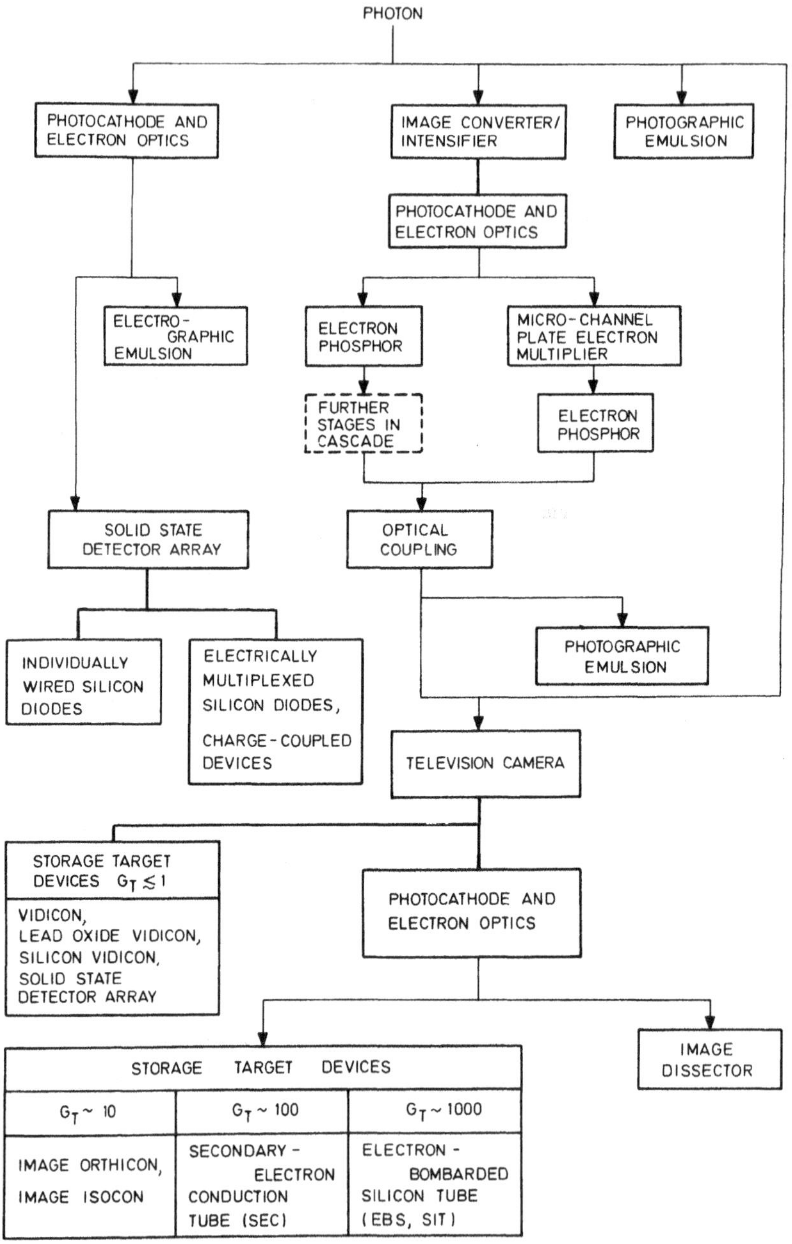

Fig. 1. Image intensification and storage methods.

include noise (random, and 'fixed-pattern'), signal-induced background, non-uniformity, geometrical distortion, small dynamic range, non-linearity, ageing effects, and so on. I give an example of a detector system for which such limitations have been carefully studied and satisfactorily accounted for in the overall design of the detector and preceding optical system, in the following section.

2.1. The detector system of the International Ultraviolet Explorer satellite

The development of the International Ultraviolet Explorer (IUE) satellite is being undertaken jointly by the U.S.A., the U.K. and ESRO. The satellite is intended as a general observing facility for ultraviolet astronomy. Spectral measurements both at high and low resolution will be made in the range 1150-3200Å using two spectrographs each covering approximately half the spectrum. Guest astronomers will be able to carry out their observing programmes from the two observatory control centres established in the U.S.A. and in Europe.

An illustration of the IUE satellite is shown in Fig. 2. The configuration of the spacecraft is designed to meet the requirements of the single large instrument it carries: basically a telescope with two spectrographs. Mainly for reasons of data handling and spacecraft control, IUE will be put into a geosynchronous orbit; at all times it will be in continuous contact with one or both of the control centres and all observing will be done effectively in real time.

In the high dispersion mode of operation, the spectrographs each produce a two-dimensional trapezoidal spectral array, as indicated in Fig. 3, by means of an echelle grating crossed with a conventional concave grating used as an order sorter and camera mirror. When the spectrographs are used at low dispersion for faint objects, a plane mirror is placed before each echelle and one-dimensional spectra due only to the concave gratings are displayed on the camera tubes.

The various camera tubes and means of wavelength conversion that have been studied for use in the IUE spectrographs are illustrated chronologically from top to bottom in Fig. 4. All the camera tubes are of the SEC type since long exposures required for IUE can only be achieved at a convenient spacecraft temperature with a detector having the high resistivity SEC target. The original plan was to use a standard type of visible sensitive SEC tube with electrostatic image stage having a luminescent phosphor coated on its fibre optic faceplate to convert the incident ultraviolet radiation into visible radiation for detection by the tube.

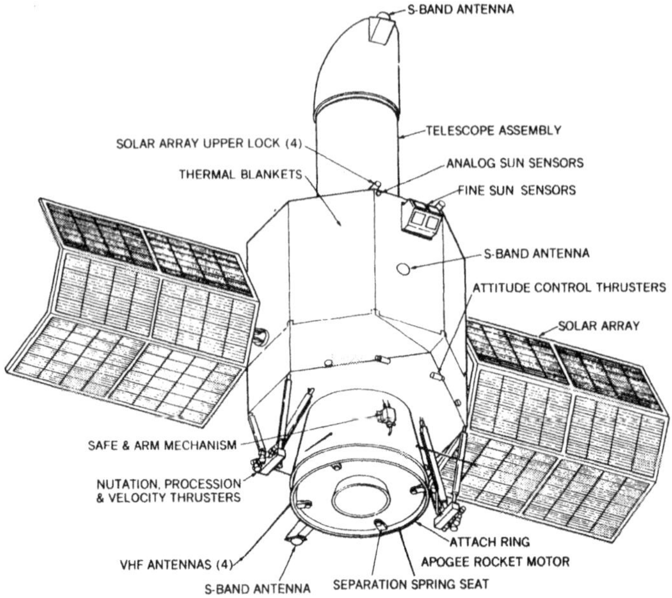

Fig. 2. The IUE spacecraft and scientific instrument.

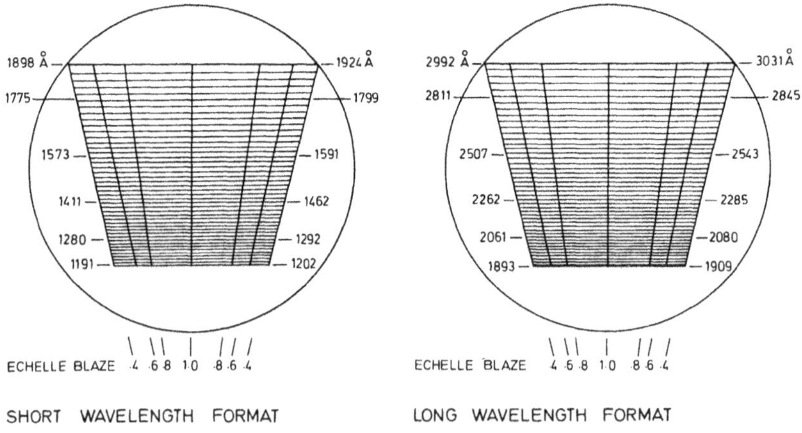

Fig. 3. The two echelle spectrum formats showing the regions of free spectral range: these regions have been geometrically optimised in scale and position on the detector faceplate to give least inter-order overlap and highest overall spectral resolution in the presence of the radially decreasing spatial resolution exhibited by the detector.

The actual tube considered was a space qualified version of the WL30893, made by Westinghouse Electric Corporation, treated to give negligible background in an integration time of at least one hour and having no suppressor mesh. This tube was built to a special spec ication for the IUE project and is designated WX32224. It has a faceplate diameter of 25 mm, suitable for the proposed spectrum format, a bialkali photocathode, and is compatible with the size and weight restrictions obtaining. Suitable luminescent phosphors are sodium salicylate, often used as a reference standard, and tretaphenyl butadiene. However, the use of such a phosphor results in a substantial sensitivity also in the visible region, which is undesirable when observing certain astronomical objects in the ultraviolet in the presence of a strong visible component.

This problem is largely overcome if a 'solar blind' photocathode is used in the primary detecting stage. Three approaches to this, which are geometrically compatible with the plane spectrum images, are depicted in Fig. 4. The first one investigated was the use of an SEC camera tube with proximity focussed image stage having a flat magnesium fluoride window and caesium-tellurium photocathode. This tube, called the UV Proxicon and designated WX31999, also was specially made for IUE by Westinghouse and was based on a visible sensitive version intended for broadcase use. Subsequently, although the tube operated satisfactorily in most respects, its successful production could not be guaranteed by Westinghouse. It was therefore dropped from the programme in favour of the third option shown in the figure: the use of the WX32224 with a proximity focussed photoelectronic image converter to take the place of the original phosphor layer. Suitable converters are made by Galileo and ITT; these have the same input window and photocathode as the UV Proxicon, and a P-11 phosphor on a fibre optic output window. The latter allows direct optical coupling with the fibre optic input window of the WX32224. This combination of image converter and SEC camera tube is the form of spectrograph detector now intended for IUE. However, a fourth configuration, proposed by J. Lowrance of Princeton University, also was considered, and for completeness is worth mentioning here. Again this has a magnesium fluoride window and caesium-tellurium photocathode, but magnetic not proximity focussing was used in the image stage. This tube was expected to have significantly higher spatial resolution than the current spectrograph detector, which, however, still is satisfactory in this respect. Unfortunately, for reasons of finance and programme scheduling, the development of the magnetic tube could not be pursued beyond a few pilot models. The current detector components are shown separately in Fig. 5.

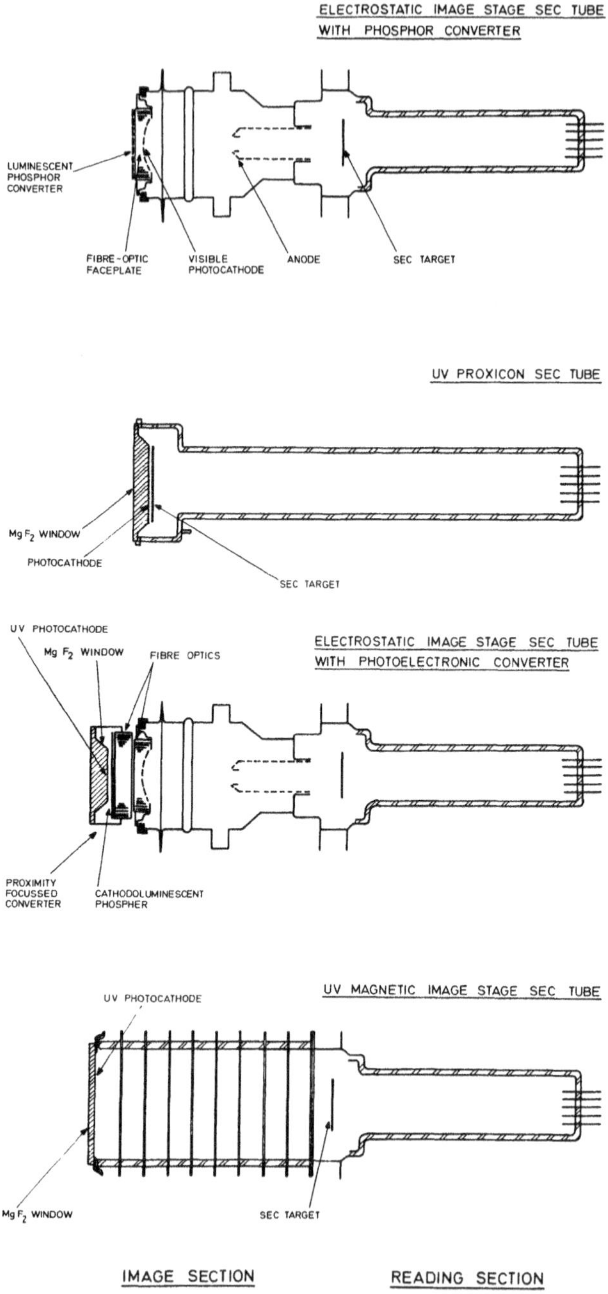

Fig. 4. Illustration of the various spectrograph camera tube options studied for IUE.

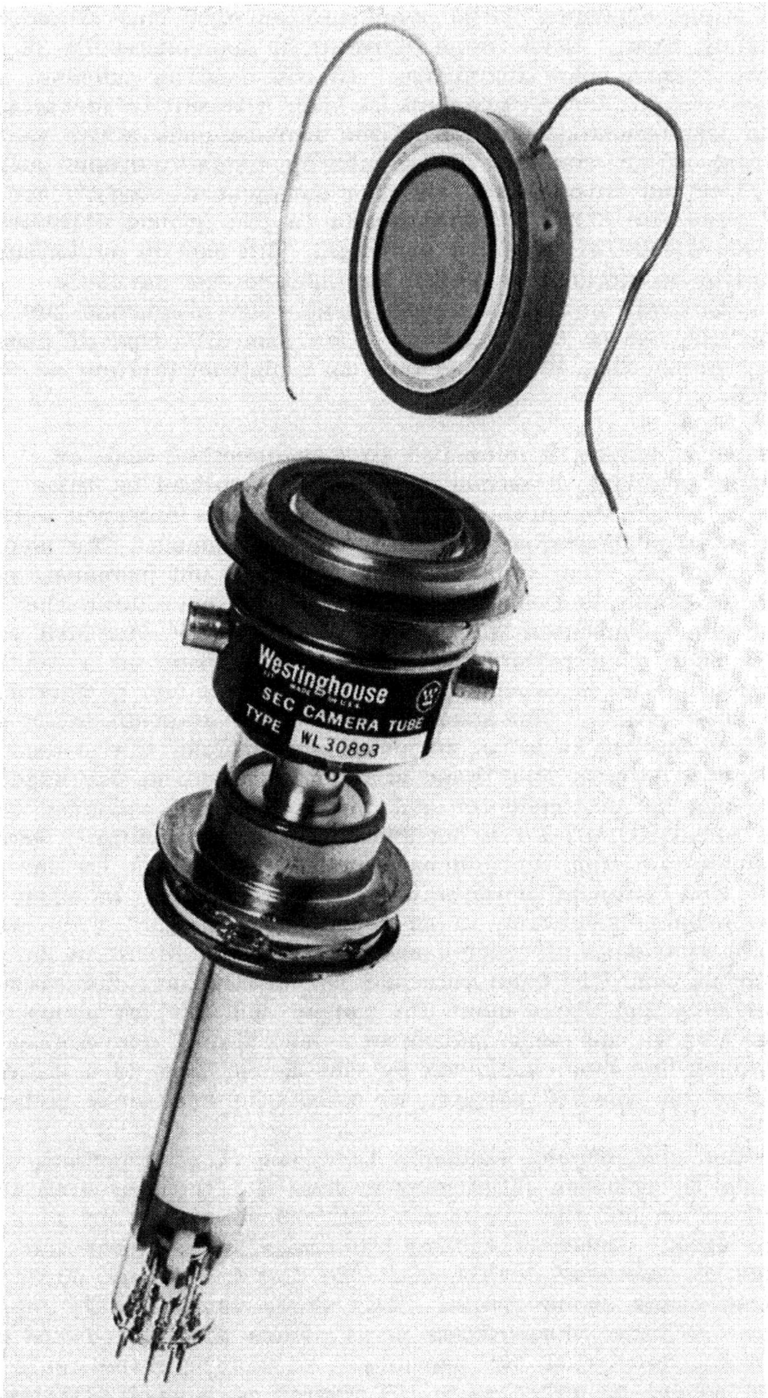

Fig. 5. The IUE spectrograph detector components finally chosen.

The spectra are scanned digitally in a rectangular array of 768 x 768 image elements to be synchronized with the telemetry data handling rate. Each image element is approximately 33 x 33μm on the camera tube faceplate. In the reading process, the accumulated signal charge present in each element is detected by pulsing on the reading beam for a few microseconds after each incrementing of the scans. The resulting signal current pulses are amplified and integrated, and the integrated signals are then digitised for direct transmission to the ground without any intermediate stage of on-board storage. The use of optimised pulse shaping as applied in pulse amplifiers for particle detection achieves an r.m.s. noise level <300 electrons per image element at the camera tube target. For the SEC type of camera tubes used here, this is equivalent to 5 photoelectrons at full target gain.

The camera system is operated in a sequential mode of exposing and reading, a method particularly suited to this application, which features long exposure times compared with normal television practice, possibly several hours. The sequential operation has four separate modes: erase and prepare, expose and store, readout, and standby. Such operation allows the removal of the suppressor mesh, which is normally included in SEC camera tubes to protect the target from damage as a result of exceeding the first secondary emission crossover potential at high exposure levels. The sequential mode is operationally safe because the scanning beam is switched off and the field mesh held at low potential while the image stage is operating for exposing, then the image is switched off and the field mesh restored to its operating potential of a few hundred volts for reading. Removal of the suppressor mesh introduces several advantages in the context of the intended application: (i) a decrease in target shunt capacitance resulting in an increase in signal-to-noise ratio, (ii) avoidance of mesh-target impact on vibration during the launching, and (iii) an increase in resolution, due to the higher deceleration field near the target and the avoidance of lenslet action in the mesh apertures. Additional protection from crossover has been achieved by the deposition of a thin gold surface layer on the SEC target, to raise the crossover potential.

The erase and prepare sequence involves first exposing the photocathode to diffuse illumination from a floodlamp with the image section on and the field mesh at +10 volts acting as a suppressor mesh. After switching the image section off the reading gun is operated, still with the low field mesh potential, over a three-frame erase cycle. This whole procedure is repeated twice more. A final three-frame scan occurs with the field mesh at the correct potential for reading. An active erase cycle such as this is needed to obliterate all memory of a previous image residing in the volume of the target layer. During the expose phase the reading gun is off, the image section is on and the field mesh is again at +10 volts. During readout, the image section is off and the reading gun is on with the field mesh at the correct potential. In addition, the target bias is raised approximately 1 volt above that used for the final erase and expose phases, this being necessary to improve beam acceptance on little exposed areas of the target. Poor beam acceptance is caused by the small fraction of beam electrons having high thermionic emission energies which drive the target surface below gun cathode potential during erasing. If the positive charge that develops in the target during exposing is small, it may not be sufficient to raise the surface potential sufficiently positive relative to the reading gun cathode to permit readout by the majority of the electrons in the beam. Biassing the target further during readout ensures that threshold signals are not compressed or lost. During standby, both the reading gun and the image section are off, and to extend the heater life the heater supply is reduced but not off; the latter also is done during exposing to avoid spurious background illumination.

An extensive testing, evaluation, optimisation and, ultimately, calibration programme has been in progress for several years at University College London for the various camera tubes and converters related to IUE. One of the most interesting results we have obtained concerns the fundamental resolution performance of proximity focussed image stages such as in the Proxicon and the wafer image converters.

Initial measurements on the UV Proxicon indicated a marked decrease in resolution near 1600Å relative to 1216Å and 2537Å. This was soon traced to a fundamental property of the caesium-tellurium photocathode and a programme of measurement and computation was set up to confirm this. The results of this work are summarised in Fig. 6. For the moment ignore the continuous lines drawn in the figure. It may be seen from the experimental points that the spatial response is markedly dependent on wavelength and photocathode potential. The wavelength dependence is shown graphically in Fig. 7 for two spatial frequencies:5.5 lp/mm and 7.0 lp/mm. The observed variation in spatial response with wavelength therefore indicates a related variation in spectral resolution when the detector is used in the spectrograph. In order to understand these results obtained, computations were carried out after the manner of Needham and Thumwood (1969) and Grant (1966) to determine the theoretical performance of the proximity focussed image stage. In these computations the photoelectrons leaving the photocathode are assumed to begin with a Maxwellian distribution in energies and a Lambertian distribution in the direction of the velocity vectors, and the appropriate sine-wave responses are obtained for the various parametric conditions. Strictly, these should be converted into square-wave responses

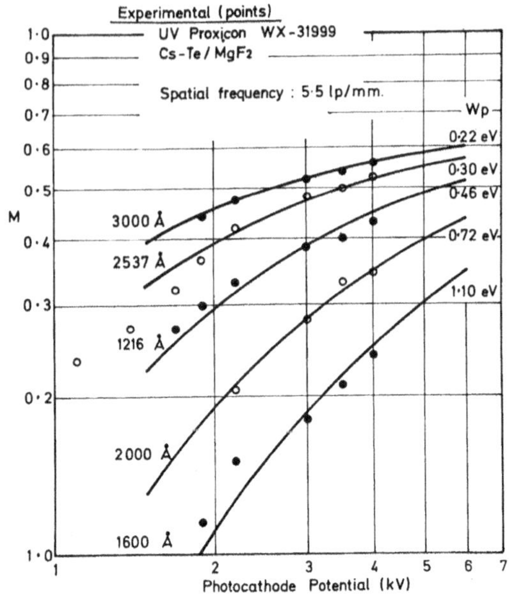

Fig. 6. Comparison of measured and calculated spatial response as a function of photocathode potential and wavelength for a camera tube with proximity-focussed image stage. The continuous lines are computed data as discussed in the text.

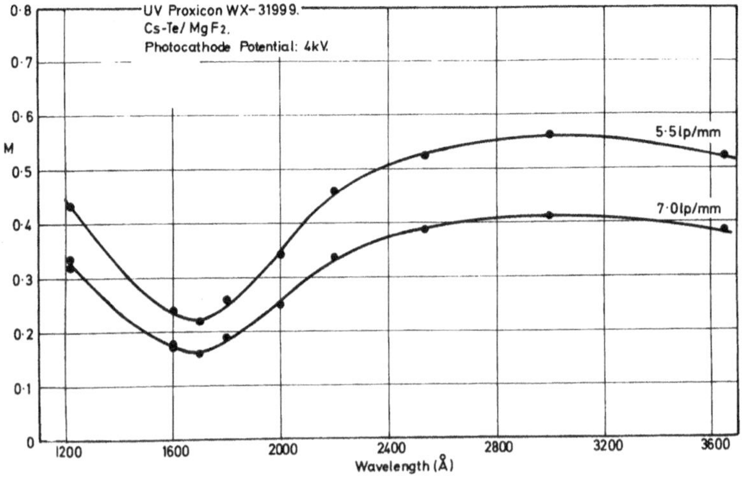

Fig. 7. Square-wave modulation as a function of wavelength for two spatial frequencies.

before comparing with the observed data, but the introduction of the simplifying assumptions make this unwarranted. Results obtained for 5.5 lp/mm are plotted as the continuous lines in Fig. 6. The values of most probable initial photoelectron energy in the several computed curves were selected to give good correspondence with the measured points, and the curves include a factor 0.7 to account for the square-wave response due to the reading process (Beyer and Goetze, 1966). Allowing for the adopted simplifications, the resulting values of most probable energy are in good agreement with published data for caesium-tellurium above 1700Å (Taft and Apker, 1953) and the trend below 1700Å conforms with the production of hole-electron pairs in the semiconductor photocathode (Spicer, 1961).

Explaining the above further, if the photon energy is less than or approximately equal to twice the band gap of the photocathode material (for wide gap materials) an excited electron will lose little energy before escaping, so its energy in vacuum can be relatively large. However, when the photon energy is greater than about twice the band gap, the primary electron can excite a secondary electron, and in the process, lose most of its kinetic energy. Since the spatial resolution decreases as the kinetic energy of the photoelectron increases, the resolution should decrease steadily as the photon energy increases until the threshold for pair production is reached, then the resolution should improve as the photon energy increases further. The band gap for caesium-tellurium is about 3.5 eV, so the worst resolution should occur at about 1700Å. This is precisely as is observed.

In contrast, the square-wave modulation transfer function for the SEC tube coated with 1 $mg.cm^{-2}$ tetraphenyl butadiene showed little difference between results for different photocathode potentials and was sensibly independent of wavelength over the whole range of interest.

3. DIGITAL TELEVISION DETECTORS

The fundamental limitation in detecting radiation at very low levels is, of course, due to the quantum nature of the radiation itself. Ultimately, the information contained in an optical image can be expressed as the spatial and temporal variation in number of photons. The problem of detecting and recording such an image then is essentially one of counting the number of photons in each image element.

It is easily possible to register single photoelectron events with currently available cascade or microchannel plate image intensifiers, but in the normal methods of recording, by means of photographic emulsion or analogue television, the usual shortcomings of these devices are inherent and are somewhat aggravated by the effective reduction in storage capacity experienced when operating with sufficient gain to register individual photoelectron events efficiently. Also, temporal variations are obscured in analogue recording.

Such shortcomings are avoided if the process of recording is separated from that of detection by the use of an external digital memory. There are several approaches to this now in routine astronomical use: the Lick Observatory Image Tube Scanner(Robinson and Wampler, 1972), the Digicon (Beaver and McIlwain, 1971), and the University College London Image Photon Counting System (Boksenberg, 1972). I will now describe the latter as an example of the technique.

3.1. The University College London Image Photon Counting System.

The principle of the image photon counting technique is to detect individual photon events in a two-dimensional image by means of a high-gain image intensifier optically coupled to a continuously scanning television camera (acting as a spatially sensitive one-frame buffer store), and to record their central positions in a digital electronic memory associated with a small on-line computer. An image is then integrated as a digital accumulation of the photon events registered in every television frame, with each image element and its corresponding location in the computer memory acting as an elemental digital photometer chain. A two-dimensional array of about 10^6 simultaneously counting elements can be provided by the standard image intensifier and television camera components used in the system. The amount of digital storage actually employed then defines how many of these elements can be used at once.

An important feature of the system is a special pattern processor which is interposed between the television camera and the computer. This identifies the central position of each photon event despite the degradations imposed by the photoelectronic components, and enables the events to be recorded: (i) with equal weight, (ii) without system noise, and (iii) with a large increase in spatial resolution over that obtained by conventional analogue recording. True photoelectron noise limited performance is achieved and the responsive quantum efficiency of the first photocathode is the basic figure of merit of the system.

Other advantages of the system arise from the use of the online digital memory and computer: (i) storage capacity, in practical terms, is unlimited, (ii) rapid time varying effects can be accommodated, and (iii) for high efficiency of telescope usage the data can be displayed while being accumulated to give immediate feedback to the observer.

The approach we have adopted at UCL is to use well-developed commercially available, image intensifier and television camera tube components and to overcome their practical limitations and deficiencies through the treatment of the image information at the quantum level by means of sophisticated electronic processing. The underlying requirement on the combination of image intensifier and television camera tube is to have sufficient overall gain to enable photon events to be recorded easily and unequivocably. Many suitable combinations can be made up of readily available components, ranging from those in which the required gain is provided more in the camera tube than the coupled intensifier, as in the case of the SIT tube preceded by a low gain intensifier, to those having all their gain in the intensifier and none in the camera tube, as in the system we are using at UCL. The choice of the actual configuration is somewhat governed by the application and may involve considerations of photometric performance, size, weight and cost. Purely in the interest of photometric performance we have chosen to avoid the use of devices with structured components such as fibre optic coupling plates and faceplates, channel plate intensifiers and silicon diode targets. Our system has an EMI magnetically focussed integral four-stage cascade intensifier lens-coupled to a Philips Plumbicon television camera tube, as shown in Fig. 8.

The EMI intensifier (designated 9912) has an S-20 photocathode on a faceplate of diameter 48 mm, employs P-11 phosphor screens, and achieves a blue light gain greater than 10^7 with 40kV applied over the four stages. In contrast with electrostatically focussed tubes having a fibre optic faceplate, the faceplate of the EMI tube is highly transparent in the ultraviolet down to the atmospheric cut-off near 3000Å. When operated in conventional fashion the limiting resolution of this tube is about 40 lp/mm.

The Philips Plumbicon (a lead-oxide vidicon), a standard camera tube used widely for colour television, is basically a high quality vidicon in which the photo-sensitive medium is a thin, polycrystalline lead oxide layer having the characteristics of a reverse-biassed wide-gap nearly intrinsic photodiode with blocking contacts. However, unlike the array of discrete silicon diodes in the silicon vidicon, the Plumbicon layer is formed structureless. Compared with a conventional vidicon the Plumbicon has high (nearly unit) peak quantum efficiency, very low dark current, linear transfer characteristic and very low lag. The actual Plumbicon used is a 30 mm tube designated XQ1020b. Its sensitivity is peaked in the blue to match the output of the P-11 phosphor of the intensifier.

The lens for coupling the intensifier and camera tube has been specially designed by Professor Wynne of Imperial College, London. It is optimised in the blue, has an aperture ratio f/2 and, at a working magnification of 0.59, gives respective fields at the intensifier and camera tube of 35 mm and 21 mm. Taking into account the quantum efficiency of the first photocathode of the intensifier, a blue light gain of 10^7 means that on average about 10^8 photons are generated at the final phosphor screen for each primary photoelectron. About 1% of these are transferred by the coupling lens to the Plumbicon tube where about 70% of those arriving are effective in generating a signal electron in the target. Thus, on average, an individual primary photoelectron results in the generation of about 7×10^5 electrons in the Plumbicon target. This gives an abundant video signal for the purposes of the processing functions outlined above. Compression of weak signals is overcome by means of bias illumination, which improves beam acceptance by building up the target potential to about 1 volt between readouts.

An isometric display of raw photon event signals delivered by the television head amplifier from an area less than 1% of the whole format is shown in Fig. 9. The brightness distribution for single photon events recorded after full signal processing and location of event centres is shown in Fig. 10; note the favourably peaked nature of the differential distribution. A portion of an arc spectrum taken with the event centering logic both active and passive is shown in Fig. 11, indicating the large increase in spatial resolution that is obtained by this pre-recording processing technique. Two examples of data obtained with the system are given in Figs. 12 and 13. The first is a low dispersion spectrum of a faint, normal galaxy taken on the Palomar 200-inch telescope with Bev Oke. The wavelength coverage is 3000 - 83000Å at 250Å mm^{-1}. The spectrum indicates certain features of the system: its reproducibility and linearity. The top trace shows galaxy + sky, the middle trace sky alone, and the bottom trace is the sky-subtracted spectrum of the galaxy. This ability of accurate sky-subtraction is very important for faint-object spectroscopy. The second example is a spectrum of the nucleus of NGC 4151 obtained at 30Å mm^{-1} on the Isaac Newton Telescope with Mike Penston and others. The upper trace is the sky-subtracted spectrum of the region containing Hβ and the forbidden lines [OIII] 4959 and 5007 (these are on the right in the figure). The lower trace is a vertical expansion of the

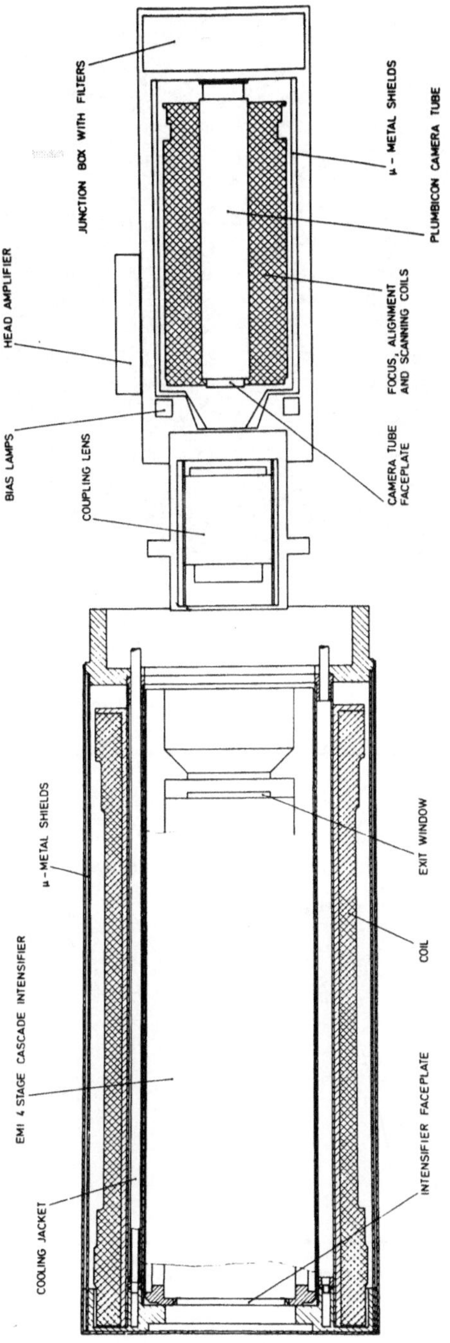

Fig. 8. Sectional view of the image intensifier, coupling lens and television camera head for the UCL Image Photon Counting System.

Fig. 9. Raw video output from <1% of total frame showing single photon events in an isometric display (IPCS).

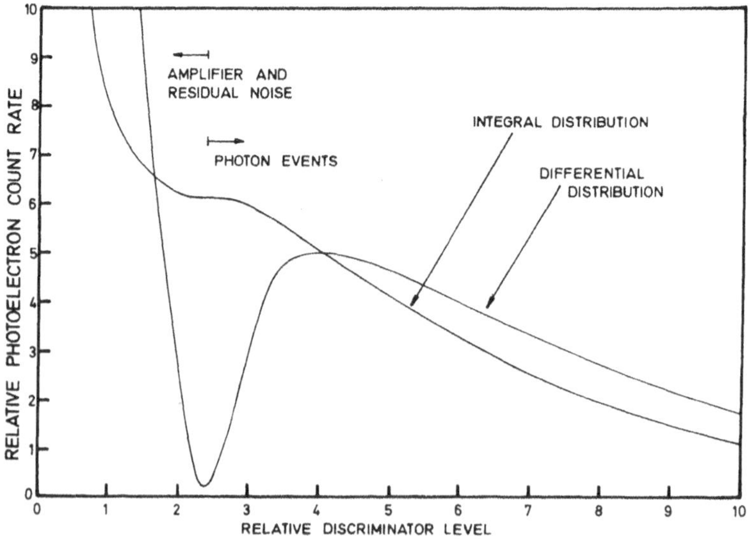

Fig.10. Brightness distribution for single photon events recorded after full signal processing and location of event centres (IPCS).

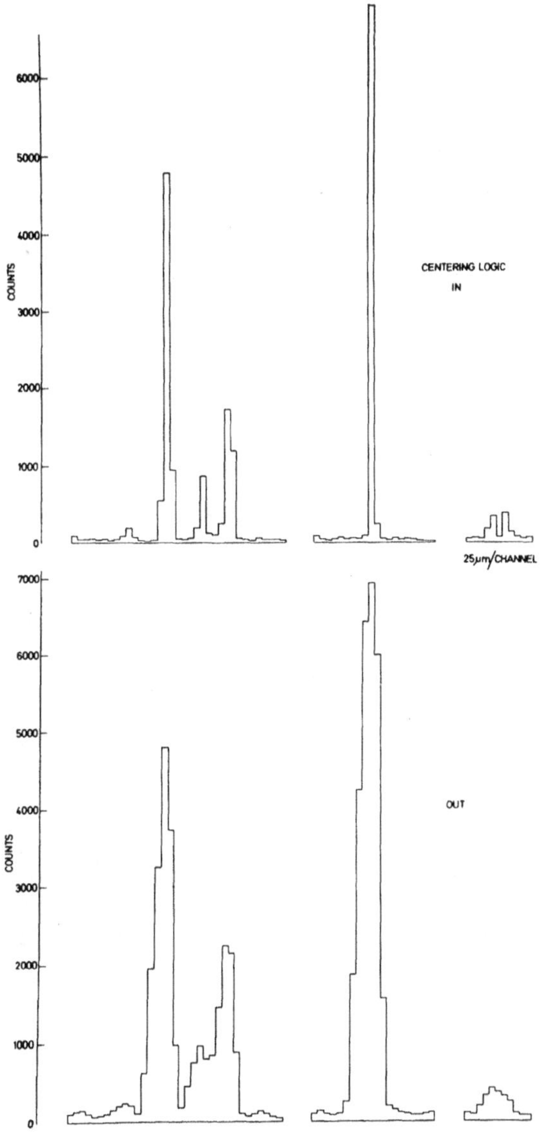

Fig. 11. Portion of an arc spectrum with event centering logic active (upper) and passive (lower) obtained with the IPCS.

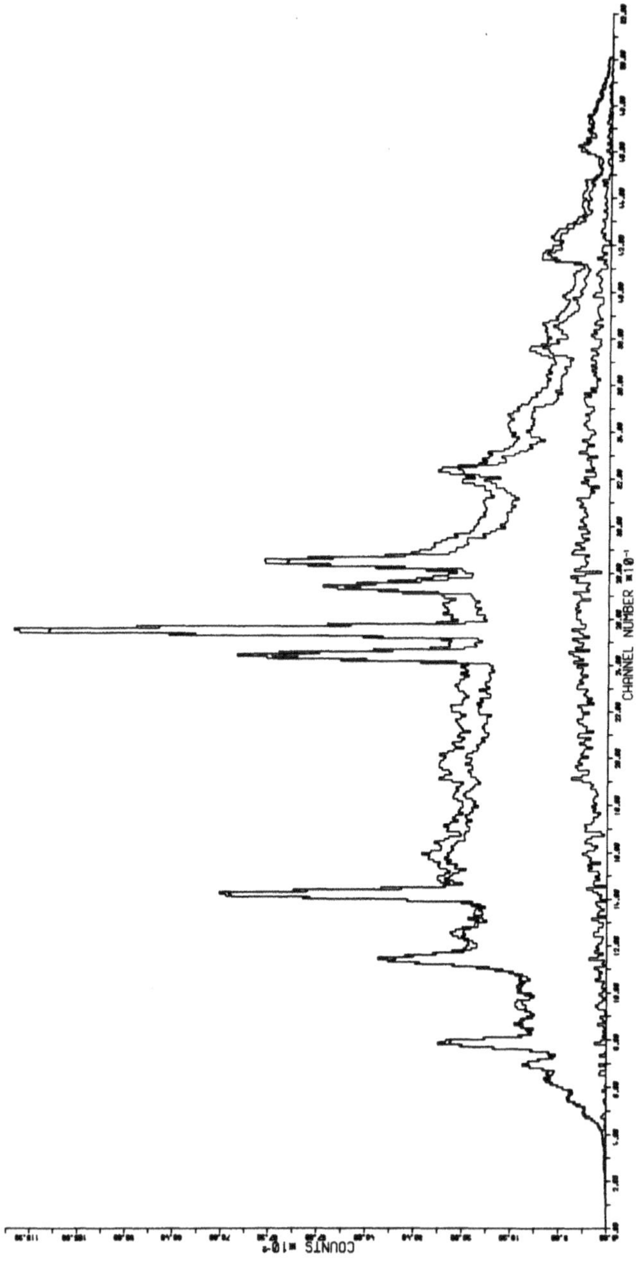

Fig. 12. Spectrum of a faint, normal galaxy obtained with the IPCS on the Palomar 200-inch telescope at 250Å mm^{-1}. The top trace is object + sky, the middle is sky alone and the bottom is the sky-subtracted spectrum of the galaxy.

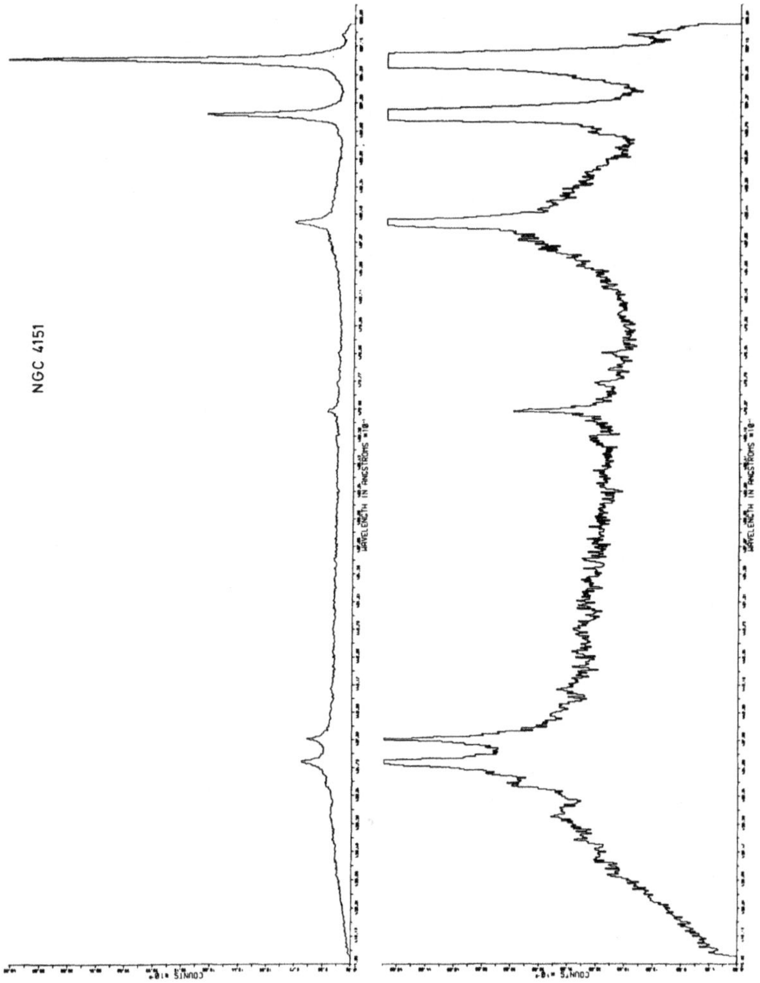

Fig. 13. Spectrum of NGC 4151 obtained with the IPCS on the INT at 30Å mm^{-1}. The upper trace is the sky-subtracted spectrum of the object. The lower is a vertical expansion of the upper to bring up the fine detail.

upper to bring up the fine detail. Note the two-component profile of Hβ, whose core has the same appearance as the [OIII] lines.

REFERENCES

Beaver, E.A. and McIlwain, C.E: 1971, Rev. Sci. Instr., 42, 1321.
Beyer, R.R. and Goetze, G.W: 1966, Adv. in Electronics and Electron Physics, 22A, 241.
Boksenberg, A: 1972, Proc. of ESO/CERN Conference:'Auxiliary Instrumentation for Large Optical Telescopes', ESO, Geneva, 295.
Grant, J.M: 1966, Proc. I.E.E.E., 54, 801.
Needham, M.J. and Thumwood, R.F: 1969, Adv. in Electronics and Electron Physics, 28A, 129.
Robinson, L.B. and Wampler, E.J: 1972, Publ. Astr. Soc. Pac. 84, 161.
Spicer, W.E: private communication.
Taft, E. and Apker, L: 1953, J. O. S. A. 43, 81.

ON LINE DIGITAL CORRELATION OF PHOTON COUNTING TV IMAGES FOR STELLAR INTERFEROMETRY

A. Blazit, L. Koechlin, J.L. Oneto

Observatoire de Meudon, France

ABSTRACT. Stellar interferometry images using one large or several smaller telescopes require on line statistical analysis of photon events. We built a system giving real time two dimensionnal autocorrelations of photon counting TV images, countaining each up to two hundred photon events. The present system is only partly hard wired, a software algorythm being used to derive the autocorrelation. Plans for a fully hard wired digital processor are also given. Such systems are expected to extend interferometric observations to faint objects: m_v=20 for Palomar, m_v=I5 for multi telescopes interferometers.

I. INTRODUCTION

The first technique (I) used in speckle interferometry was to take films from video sequences and transform the images later in a laser beam (two dimensionnal Fourier transform, FT). Now, as new receptors become available (photon counting TV camera) it becomes possible to compute two dimensionnal autocorrelations (AC) on line and integrate them in a digital memory. This technique is best adapted for faint objects yelding few photons per frame and also has real time advantages if one has a computer near the telescope.

2. DATA ACQUISITION FROM TV IMAGES

The photon counting TVcamera used consists of a microchannel intensifier followed by a SIT tube. The images

de Jager/Nieuwenhuijzen (eds.), Image Processing Techniques in Astronomy. 79-84.

represent a few white dots (the photons) on a black background.These images are digitized and sent to the computer by the mean of an interface.

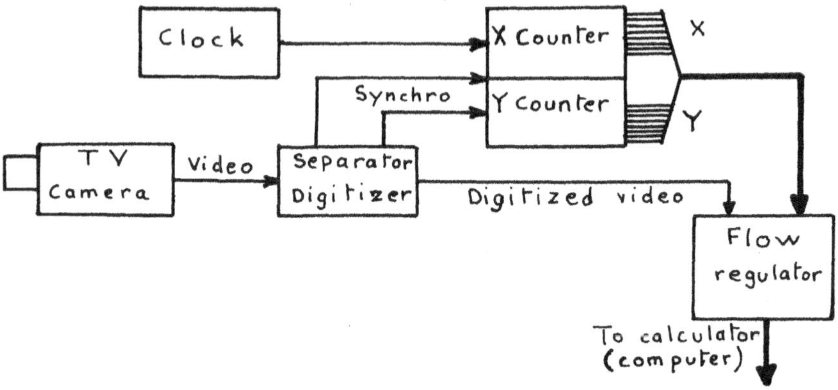

Fig. I. Logical diagram of the interface.

The coordinates of the photons are digitized on 2x8 bits (256x256 pixels).We use a clock to get the x coordinate: at the beginning of each line a counter starts counting clock pulses.When a photon pixel shows up on the video signal,its x coordinate is the content of the counter.The y coordinate is the number of the line.

the successive pixels are spaced by I50ns on the video signal,but the computer is not fast enough and we could loose the photons close to each other;then we must regulate the coordinates flow at the output of the interf. The photons coordinates are stored in a latch memory wher up to 4 points can be stored.This memory is read by the computer at its own speed.

3. PROGRAMMED PROCESSING

The first data processing by this method has been made at the Nice observatory,running AC programs on a PDP 8/E computer.The TV images,recorded on video tapes coming from observations at Mnt Palomar are digitized and fed into the computer by the interface.

3.I Transfer of coordinates

As we do not have a DMA in the computer,each transfer of 2 coordinates take 22μs.It limits us to an absolute maximum of 800 photons per frame.Further limitations due to computation speed are explained later.

3.2 Autocorrelation algorythm

The processing is done in real time;this means the computation time for one frame is less than 20ms.

One can demonstrate the AC can be computed by substracting from the coordinates sf one photon the coordinates of each other and incrementing the memory location whose address is given by the result of the substraction:

$$AC(h,l) = \sum_{i=1}^{N} \sum_{j=1}^{N} (X_i - X_{j-l} , Y_i - Y_{j-h})$$

N is the number of photons in the frame,
X_i , Y_i the coordinates of the ith photon.

The computing time is then:

$$t = \alpha N(N-I)/2 \qquad \alpha \text{ depending on the computer.}$$

The y coordinates are given in increasing order,due to the TV scanning.As we consider only a window in the AC, we can delete further computation as soon as difference in y falls out of the window.We now get, assuming photons are evenly spread:

$$t = \alpha N(N-I) L_w / L_i$$

L_w is the window lengh,L_i is the image lengh.

For our computer =60 s,L_i=256 and L_w=64 .
Then t=I.5 N^2 s,thus every image can be processed as long as N 30 photons.

3.3 Signal over noise ratio

Actually,for more than 30 photons per frame,we still do real time,but taking one image each time the treatment of the previous is completed.Let R be the S/N ratio in a given point of the AC of one frame,assuming the noise in two different frames is statistically independent,for N_f ACs integrated:

$$S/N = R\sqrt{N_f}$$

For low levels of light,R is proportionnal to the number of photons per frame (2,3):

$$S/N = \rho N_{ph} \sqrt{50t} \quad \text{(assuming 50 frames per second)}$$

If the computer is not saturated, for a given observing time:

$$S/N=\rho N_{ph}\sqrt{N_f}$$

If the computer is saturated:

$$S/N=\rho\sqrt{t/\alpha}$$

In that case S/N does not depend on the number of photons per frame. For bright images, the photons overlap and yield grey levels. We then adjust the threshold in the interface to see just the peaks of the speckles. The AC is then distorted, but it is weak in our case.

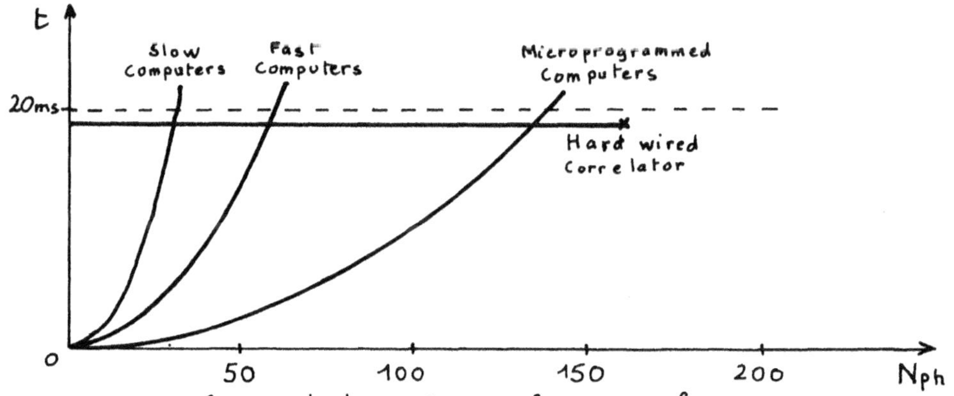

Computation time for one frame

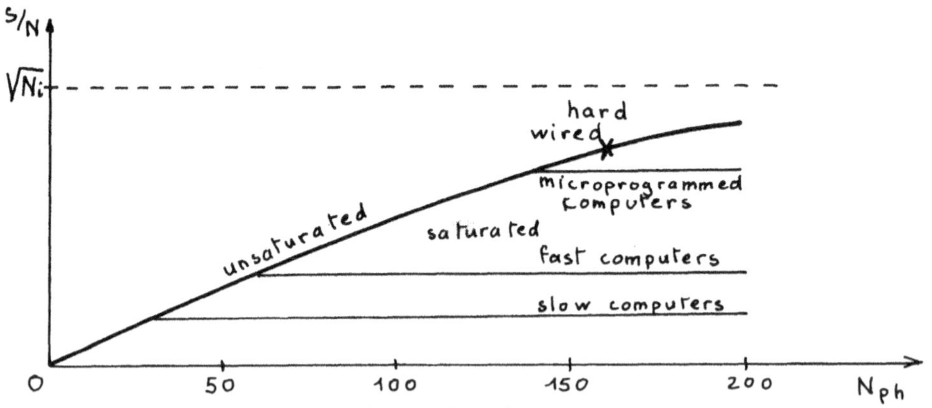

S/N for a given observing time

4. HARD-WIRED AUTOCORRELATOR

The autocorrelator uses the same algorithm as the compute

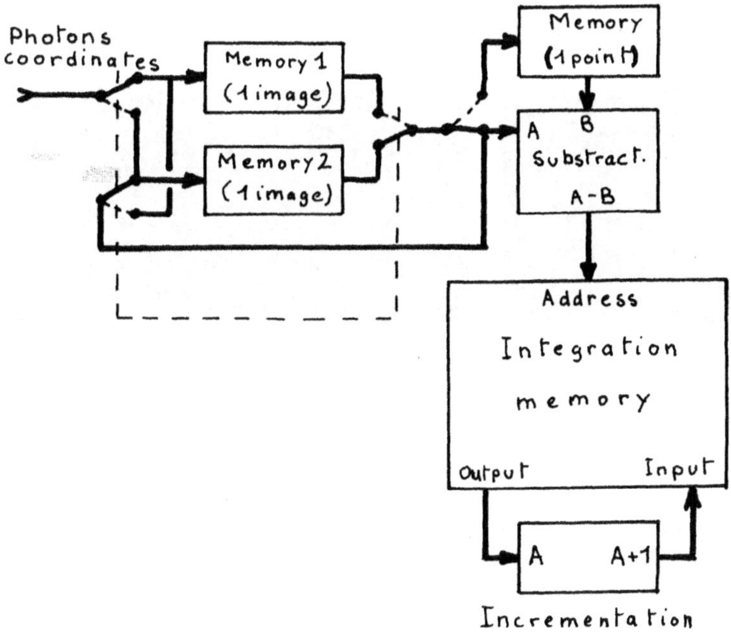

Fig. 2. Logical diagram of the autocorrelator.

The photons coordinates of an image are recorded into a first in-first out memory(FIFO).At the same time,the last image,recorded in the other FIFO memory,is processed. The coordinates of the first point are stored in a latch memory,and are substracted to the ones of the others points by shifting the content of the memory.All the points, except the first one,are rewrited into the fifo memory. Then the second point is stored in the latch memory, and so on.

The contents of the integration memory whose addresses are the differences,are incremented;so we got the AC function.The integration memory has 8KxI6 bits.

5. ACTUAL RESULTS

AUR resolved as double star,separation =5.7 IO^{-2} "

This star has been first resolved optically by Anderson with his interferometer in I922;it has been previously known as a spectral binary with a period of I04 days. The ACs done corroborate the results obtained by the FT method.

ORI possibly resolved;one can see a very weakly contrasted peak just below the central peak.The separation would be 3.9 IO^{-2} ".This is confirmed by the optical FTs

made from the same video sequence.This feature and sever
others need further observations before a conclusion can be made.

α AUR λ=.42 μm β ORI λ=.48 μm

Fig. 3. AC functions of two double stars.

REFERENCES

I. Labeyrie,Bonneau,Stachnic,Gesari,ApJ,<u>I94</u>,I47,I974
2. Dainty,Mon.Not.R.astr.Soc.,<u>I69</u>,63I,I974
3. Roddier,preprint.
4. J.Max,Methodes et techniques de traitement du signal.
5. D. Bonneau,These.

PHOTON-COUNTING DETECTOR ARRAYS BASED ON MICROCHANNEL ARRAY PLATES

J.G. Timothy

Center for Astrophysics, Harvard College Observatory and Smithsonian Astrophysical Observatory, 60 Garden Street, Cambridge, MA. 02138

ABSTRACT. The recent development of the channel electron multiplier (CEM) and its miniaturization into the microchannel array plate (MCP) offers for the first time the possibility of fully combining the advantages of the photographic and photoelectric detection systems. The MCP has an image-intensifying capability and has the potential of being developed to yield signal outputs superior to those of conventional photomultipliers. In particular, the MCP has a photon-counting capability with a negligible dark count rate. Furthermore, the MCP can operate stably and efficiently at extreme ultraviolet and soft x-ray wavelengths in a windowless configuration or can be integrated with a photocathode in a sealed tube for use at ultraviolet and visible wavelengths. The operation of one-and two-dimensional photon-counting detector arrays based on the MCP at extreme ultraviolet wavelengths will be described and the design of sealed arrays for use at ultraviolet and visible wavelengths briefly discussed.

1. INTRODUCTION

Instruments for photometric studies at ultraviolet and x-ray wavelengths have traditionally been divided into two distinct classes: photographic and photoelectric. Photographic instruments, employing film as the detection system, have the great advantage of an image-storing capability. It is therefore possible to use this type of instrument to record a very large amount of data with a single exposure. However, photographic film has a number of major disadvantages. First, the sensitivity is considerably lower than that of a photoelectric detector; at ultraviolet wavelengths, where special emulsions are required, the quantum efficiency is typically about 1% that of a photoelectric

de Jager/Nieuwenhuijzen (eds.), Image Processing Techniques in Astronomy. 85-90.

detector. Second, the response is non-linear as a function of the incident energy. Hence, photometric calibration is a difficult and time-consuming procedure, and the resulting accuracy is poor. Furthermore, the emulsion is sensitive to a very wide range of wavelengths; accordingly, the elimination of background fog levels is extremely difficult. Finally, the output signal is not electrical in character. The use of photographic film in space experiments on orbiting satellites thus implies either recovery (as in Skylab) or a complex onboard processing and video transmission system (as in the Lunar Orbiter Spacecraft).

Photoelectric instruments, on the other hand, are more sensitive, have a greater stability of response, and provide a linear output as a function of the incident energy. The output data format is also fully compatible with spacecraft data-handling and telemetry systems. However, since most photoelectric detectors do not have image-recording capabilities, the data must be read out sequentially, point by point. Consequently, the over-all speed of the system is considerably slower so that a photographic instrument, with all its inherent disadvantages, may be more efficient.

The recent development of the channel electron multiplier (CEM) and its miniaturization into the microchannel array plate (MCP) offers for the first time the possibility of fully combining the advantages of both the photographic and the photoelectric detection systems. The MCP can be operated as an image intensifier and has the potential of being developed to yield signal outputs superior to those of conventional photomultipliers. In particular, the MCP has a photon-counting capability and a negligible dark count rate.

These devices can operate stably and efficiently at extreme ultraviolet (EUV) and soft x-ray wavelengths in a windowless configuration or can be installed with a photocathode in a sealed tube for use at ultraviolet and visible wavelengths.

2. MICROCHANNEL ARRAY PLATES

The concept of combining the secondary-emissive dynode with the resistive potential-divider chain in an electron multiplier was first proposed by Farnsworth[1] in 1930. Since that time there have been many attempts to produce continuous-dynode multipliers, but it was not until 1956 that the first successful results were obtained by Goodrich and Wiley[2] with the magnetic electron multiplier (MEM). However, the MEM has a number of disadvantages, particularly the large magnet structure and the limitation on gain imposed by ion-feedback problems arising from the straight geometry of the dynode strip. The continuous channel electron multiplier (CEM), which does not require a magnetic field to

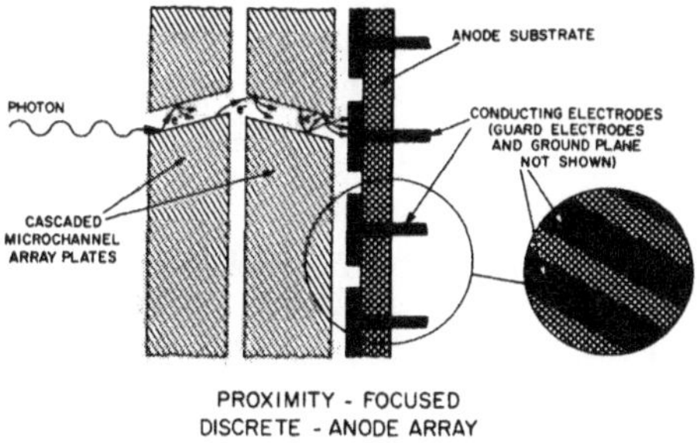

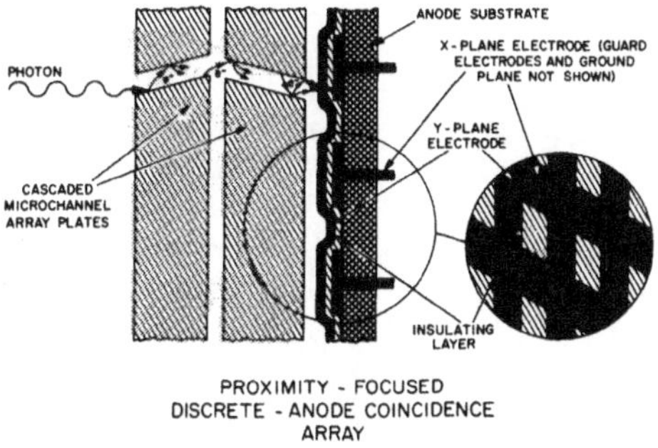

Figure 1. Proximity-focused discrete and coincidence MCP detector arrays.

focus the secondary electrons, was first developed by Goodrich and Wiley[3] in 1958. Parallel development of the CEM was undertaken by Bendix in the US and by Mullard in the UK from 1960 onwards.

The basic CEM consists of a semiconducting glass channel having an internal diameter of a few mm and a length-to-diameter ratio of the order of 50:1. The channel has a resistance of about 10^9 Ω, and the semiconducting inner surface can be used as the photocathode at EUV and soft x-ray wavelengths. A photon or charged particle striking the wall of the channel releases an electron that is accelerated along the channel axis and drifts across to strike the wall with sufficient energy to release secondary electrons. This process is repeated throughout the length of the channel, the output pulses with an amplitude of the order of 10^7 to 10^8 electrons being collected at the anode.

Since the performance of a CEM depends solely on the length-to-diameter ratio of the channel and not on the absolute physical dimensions, the CEM can be reduced in size to a limit set by the available glass technology. Large numbers of these microchannels can then be assembled in a billet to form a microchannel array plate (MCP). Since each channel is an independent photomultiplier, the MCP is a detector with an imaging capability similar to that of a photographic plate, but with sufficient sensitivity to permit photon counting.

However, a basic problem with the operation of current MCP's is that the microchannels have a straight geometry and are hence prone to ion-feedback instabilities at high operating potentials and high ambient pressures. For this reason the gain of a single plate is generally limited to about 10^4, and the device can be operated only at pressures below 10^{-6} Torr. However, it is possible to couple two MCP's in cascade, as shown in Fig. 1. MCP's can be obtained with the microchannels set at any bias angle in the 0-25° range; hence a multiplier can be constructed with suitable bias angles and plate orientations so that positive ions are trapped at the interface between the two plates. The total voltage across the plates can then be increased to a level at which a saturated output pulse-height distribution is obtained.

We have evaluated the performance of cascaded MCP's at EUV wavelengths[4] and have investigated techniques to eliminate the problems of ion-feedback. The angled-field MCP produced by Galileo, Inc., shows promise but suffers from problems of gain reduction because of the accumulation of space-charge on the insulating strips in the channel wall used to establish the angled electrostatic field.[5] We are currently evaluating the performance of MCP's supplied by Mullard, Ltd., which employ curved microchannels to inhibit ion-feedback. The initial results of these tests have proved highly successful.

3. PROXIMITY-FOCUSED MCP ARRAYS

In order to exploit the full sensitivity, dynamic range and photometric stability of the MCP, it is necessary to employ pulse-counting readout systems working directly at the anode. Accordingly we have set out to develop photon-counting detector arrays of high dynamic range in which each image element is connected to an individual amplifier and counting circuit. A breadboard linear array consisting of 64 electrodes, 1.3 mm in length and 25 microns in width, set on 50-micron centers, has been constructed and the performance evaluated in the laboratory over the wavelength range from 461 Å to 1216 Å.[6] Although the dynamic range of the detector array was severely limited by ion-feedback effects in currently available MCP's, the theoretical spatial resolution of 50 microns was achieved, as shown in Fig. 2. A two-dimensional array, in which the output charge from the MCP is detected simultaneously on two linear arrays of electrodes set at right angles (see Fig. 1), has also been constructed and evaluated at extreme ultraviolet wavelengths. This system, employing coincidence circuits to detect the output charge, produces a two-dimensional array which

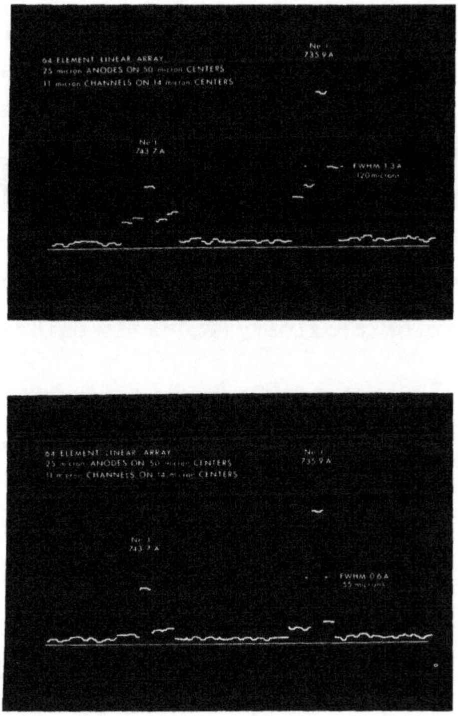

Figure 2. Spectra of Neon recorded with breadboard 64-element linear MCP array.

requires only (n + n) amplifiers and counting circuits to read out information from (n x n) image elements. The breadboard 32 x 32 element system has again demonstrated the theoretical resolution of 50 x 50 (micron)^2 in the laboratory tests.[7]

4. SEALED MCP ARRAYS

We are now constructing second-generation arrays which can be used in both open and sealed configurations. In particular, the construction of proximity-focused "wafer" tubes (Fig. 3) is now in progress. These tubes will have the capability of operating with a variety of anode formats from 1 x 160 discrete elements to 80 x 80 coincidence elements. The production of an extended linear array of 2 x 1024 coincidence elements with an S-20 photocathode for operation at wavelengths between 3000 Å and 8000 Å is the first priority. This tube will be used for photon-counting spectrophotometry at ground-based telescopes. Additional tubes will be produced with CsI photocathodes and MgF_2 windows for use over the wavelength range from 1200 Å to 2500 Å.

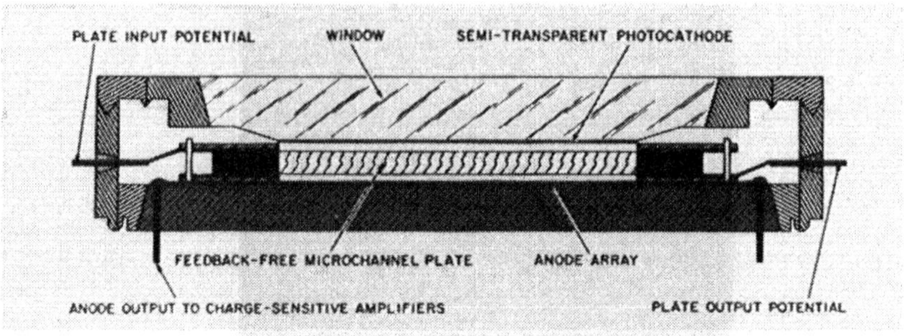

Figure 3. Schematic of proximity-focused sealed MCP detector array.

This work was supported in part by NASA under contract NAS5-3949.

REFERENCES

1. P.T. Farnsworth, US Patent 1 969 399 (1930).
2. G.W. Goodrich and W.C. Wiley, Rev. Sci. Instrum. 32, 846 (1961).
3. G.W. Goodrich and W.C. Wiley, Rev. Sci. Instrum. 33, 761 (1962).
4. J.G. Timothy and L.B. Lapson, Applied Optics 13, 1417 (1974).
5. J.G. Timothy, Rev. Sci. Instrum. 45, 834 (1974).
6. J.G. Timothy and R.L. Bybee, Applied Optics in Press (1975).
7. J.G. Timothy and R.L. Bybee, to be submitted to Rev. Sci. Instrum. (1975).

PERFORMANCE OF THE MIT SILICON VIDICON IMAGING SYSTEM AT THE TELESCOPE*

T.B. McCord, J.P. Bosel, and M.J. Frankston

Center for Remote Sensing, Department of Earth and Planetary Sciences, Massachusetts Institute of Technology, Cambridge, Massachusetts 02139

ABSTRACT. A two-dimensional integrating silicon vidicon imagery system has been developed at MIT and used to record images and spectra of astronomical objects. The system is photon noise limited above the ∿500 charge carrier preamplifier noise. The response is linear with $\gamma=1.00$ over a ranging 10^5.

After the initial effort to show that silicon diode array target vidicons could be used in an integrating mode for groundbased telescope observations of celestial objects (McCord and Westphal, 1972), we at MIT have continued to develop techniques for using the devices and have used them to obtain useful data. We have carried out both direct imaging through interference filters and two-dimensional spectroscopy using a prism spectrometer. This article describes the performance of silicon vidicons (RCA-4532) we have experienced while working with an operating system in the field at the telescope. Several groups have recently reported their experiences (Westphal, 1973; McCord and Bosel, 1973; Honeycutt and Burkhead, 1973; Hunten, 1973; Crane and Davis, 1974; McCord and Frankston, 1975).

The imaging system (Bosel and McCord, 1975) (figure 1) used to obtain the test results was designed and built in our laboratory. The system is composed of a control and data handling system, a camera head, and a photometer or spectrometer.

The silicon diode-array tube is operated like stan-

* This research was supported by NSF Grant GP31516.

de Jager/Nieuwenhuijzen (eds.), Image Processing Techniques in Astronomy. 91-96. *All Rights Reserved.*

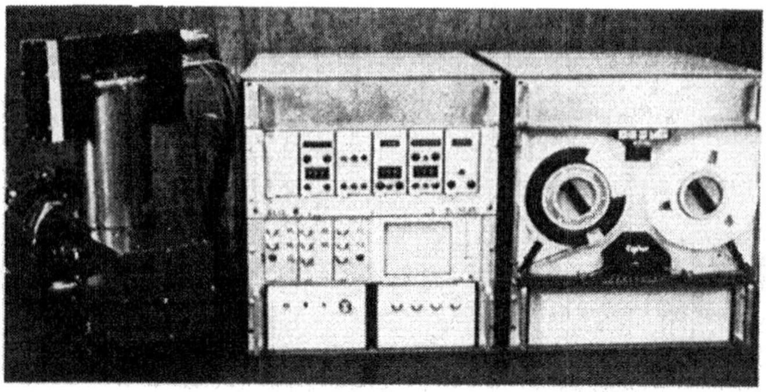

Figure 1

dard vidicons, using magnetic focus and deflection, and is cooled with dry ice.

The digital scan configuration is 256 horizontal elements by 250 lines. A frame of 64,000 pixels is read in about 4.5 seconds in 70μm steps. Output from the system is displayed on a CRT monitor, which is refreshed by a scan converter memory (Brooke, 1975) and also is recorded on standard 9-track magnetic tape.

The vidio signal is encoded to 12 bits precision yielding data numbers (DN) ranging from 0 to 4095. Usually system gain is set so that target saturation ($\sim 10^6$ charge carrier/pixel) is about 10% above DN=4095.

Target preparation is required in order to obtain reproducible photometric images, and the preparation must be carried out in exactly the same way for each image. An image recording routine goes as follows: (1) target is saturated by exposing it to bright light, (2) target is erased 10 times, (3) cathode heater is turned off to prevent target backlighting, (4) shutter is opened to begin exposure, (5) shutter is closed, (6) cathode heater is turned on (a pause is allowed while heater warms up), (7) target is read out once.

The target-to-cathode-potential is adjusted between erase and read to place the image signal on a pedestal signal equal in amplitude to about 10% target saturation.

Backlighting of the infrared responsive silicon target by the cathode heater contributes a non-spatially-uniform signal on the order of 1% of saturation in a minute of warm operation at 6.5 volts. The exact amount depends on the tube used.

The amount of signal remaining on the target with each erase cycle decreases rapidly and is reproducible from erase sequence to sequence to at least within 0.1% of saturation signal level.

Two types of calibration images are required to produce an image which is photometrically precise. A dark field (DF) is made by carrying out the procedures for making an exposure except for opening the shutter. A flat field (FF) is made by imaging a uniformly illuminated field. The flat field must be made using any filters in the optical train and the same f/ratio beam as used for the raw data image. The DF frame is subtracted from both the object and the FF frame. The object is divided by the FF frame.

A contour plot and an intensity traverse for a typical DF and a FF when the full vidicon target is scanned is shown in figure 2. The frequency distribu-

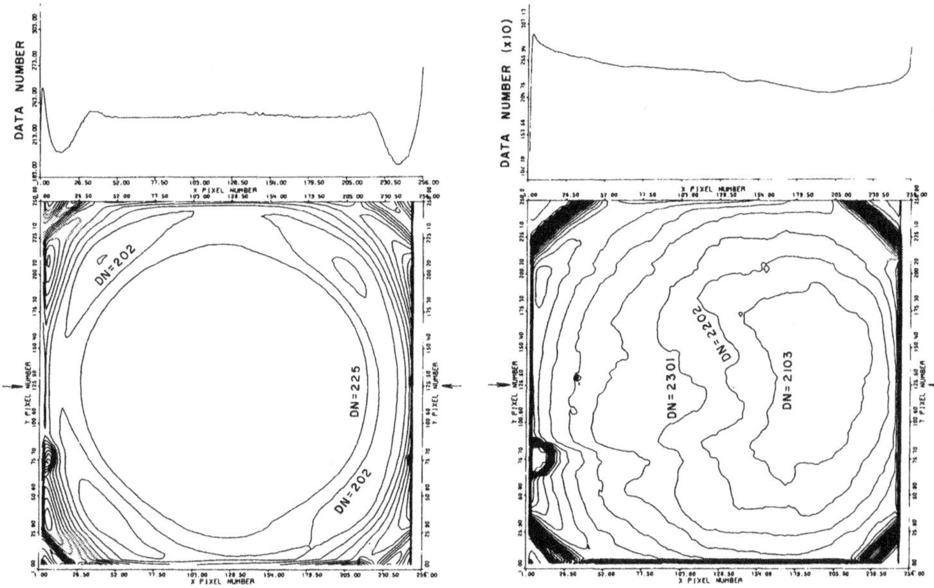

DARK FIELD Figure 2 FLAT FIELD

tions of data numbers for a 40x40 pixel portion of the center for six typical DFs (figure 3) indicates a pixel to pixel r.m.s. deviation of about 0.2% of full scale (4095DN) or about 400 charge carriers per pixel. This noise appears to be due entirely to the preamplifier.

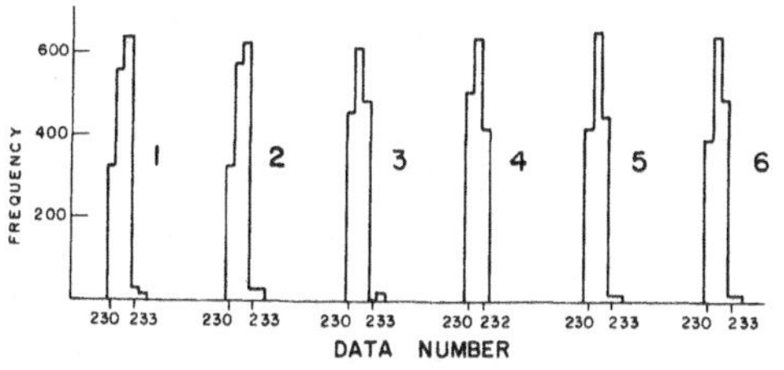

FRAME NO.	MEAN DATA NO.	PIXEL TO PIXEL R.M.S. DEVIATION	% FULL SCALE
1	232.277	.863	.021%
2	232.289	.883	.022%
3	232.064	.847	.021%
4	231.967	.812	.020%
5	232.067	.859	.021%
6	232.116	.849	.021%

Figure 3

Frequency distribution for the full 64,000 pixels for three typical FFs (top of figure 4) shows the effect of shading across the frame. Removal of this shading by division of three FFs by the fourth FF yields the three lower histograms. Statistics for these three calibrated FFs (bottom figure 4) show pixel r.m.s. deviation of about 0.04% of full scale (4095DN) or about 1000 charge carriers. The expected r.m.s. deviation for this sample $\sim 10^6$

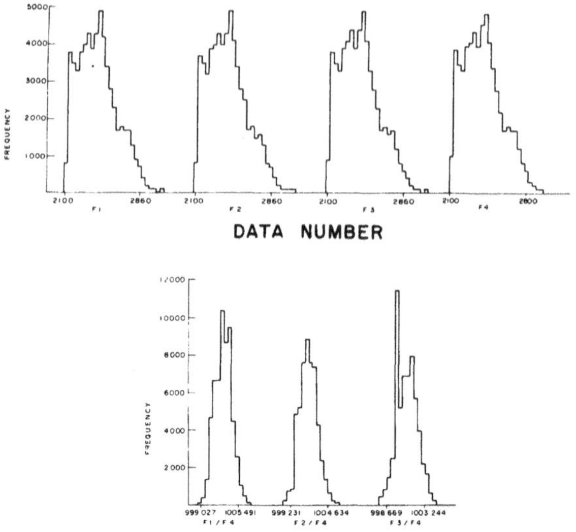

FRAME NO.	MEAN DATA NO.	PIXEL TO PIXEL R.M.S. DEVIATION	% FULL SCALE
1	2366.628	1.8	.044%
2	2367.127	1.7	.042%
3	2365.800	1.6	.039%
4	2363.617	1.8	.044%

Figure 4

events (charge carriers creation in the target) plus 400 charge carriers per amplifier noise is about 1500 charge carriers including the effects of the division. The difference could well be within the uncertainty of our calculation. Thus the imaging system is apparently photon noise limited above the effects of a ∿500 charge carrier preamplifier noise

The response of the system is linear (figure 5), i.e. γ=1.00. The r.m.s. deviation of the measured points from the γ=1 line is 0.44%. When the pedestal is not used the target response becomes non-linear as shown by the bottom curve in figure 5.

A check of the overall sensitivity of the silicon vidicon imaging system was run using a standard star for which spectral flux outside the earth's atmosphere have been derived. A basic system peak to peak noise of about 400 charge carriers was derived.

The accumulation of dark current is about 4% of full scale saturation in one hour when the tube is at dry ice temperature. The dark current across the field reproduces from frame to frame to at least 10% of the dark current level, and perhaps much better.

This imaging system has been used to obtain measurements of objects for scientific purposes. For example a program is under way to image areas of the lunar surface at several wavelengths between 0.35 and 1.10μm (Pieters, et al., 1974). These images are calibrated using dark fields and flat fields. Then ratio images are made by dividing one calibrated image at, say, 0.35μm by another at, say, 0.56μm. These ratio (or color difference) images are contrast enhanced to bring out color

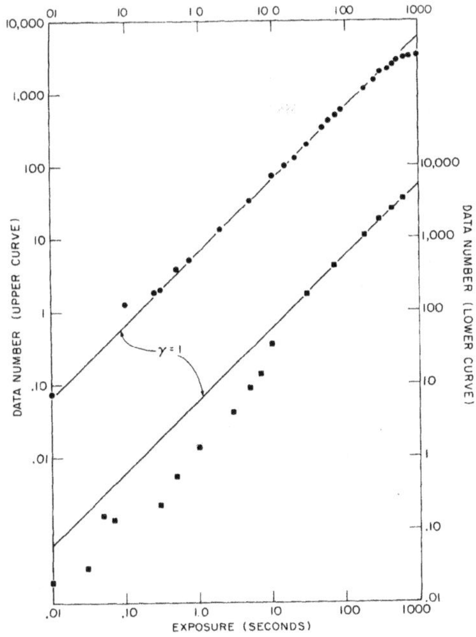

Figure 5

features of 1% contrast or lower. A number of spots within several images have been observed using the vidicon system match those obtained using the photoelectric system to within 1% consistency. Subtle color features of 1% contrast reproduce with repeated measurements except at the very edge of the tube target.

Repeated measurements of stars intensities have been made with reproducibility greater than 0.5%. Also infrared images near one micron have been made of galactic nuclear regions to the sky limit to define structure (Schild et al., 1975).

REFERENCES

1. J.P. Bosel and T.B. McCord, The MIT MKIII Silicon Vidicon Imaging System: A Description, in preparation.
2. M. Brookes, A Digital Solid State Scan Converter with Unusual Display Capabilities, Applied Optics, submitted, 1975.
3. P. Crane and M. Davis, Characteristics of the Silicon Diode Vidicon, pub. Astron. Soc. Poc., in press, 1974.
4. R.K. Honeycutt and M.S. Burkhead, Adaption of a Commercial Silicon Vidicon Detector System for Astronomical Spectroscopy, proceedings of the symposium "Astronomical observations with television-type sensors", U. of B.C. Vancouver, 15-17 May 1973.
5. D.M. Hunten, A Silicon Vidicon Camera System, Quarterly Bulletin, Kitt Peak National Observatory and Cerro Tololo Interamerican Observatory, Oct., 1973.
6. T.B. McCord and J.P. Bosel, Silicon Vidicon Astronomical observations with television-type sensors", U. of B.C. Vancouver, 15-17 May 1973.
7. J.B. Oke and R.E. Schild, The Absolute Spectral Energy Distribution of Alphs Lyrae, Astrophysical Journal 161, 1015, 1970.
8. C. Pieters, T.B. McCord, M.P. Charette, and J.B. Adams "Dark Mantling Material in the Apollo 17 Soil Samples",

Science, 183, 1191-1194, 1974.

9. R. Schild, D.M. Peterson, and J.B. Oke, Effective Temperatures of B- and A- type Stars, Astrophysical Journal, 166, 95, 1971.

10. R. Schild, M.J. Frankston and T.B. McCord, "Observation of a Red Nuclear Halo in NGU 4565", in press, Ap. J. Letters, 1975.

11. J.A. Westphal, Application of the SIT Vidicon to Astronomical Measurements, proceedings of the symposium "Astronomical observations with television-type sensors", U. of B.C. Vancouver, 15-17 May 1973.

A SELF-SCANNED SILICON DIODE ARRAY FOR ASTRONOMICAL PHOTOMETRY

Dainis Dravins

Lund Observatory, S-222 24 Lund, Sweden

ABSTRACT. A digital light detector system using a self-scanned silicon photodiode array with 128 elements has been developed for use in a solar spectrograph and a photographic-plate photometer. Noise patterns are reduced in software, making the signal-to-noise level essentially limited by the 10-bit hardware resolution.

1. THE DIODE ARRAY DETECTOR SYSTEM

Solid-state image sensors hold great promise as detectors for astronomical photometry. Advantages include: very high quantum efficiency, geometric stability, smallness of sensor elements and that the sensor can safely be handled in full daylight. The compactness possible led us to design a detector system that can easily be moved from one instrument to another or between different geographical locations. The key hardware units are shown in fig.1.

The light sensor is a monolithic MOS (Metal Oxide Semiconductor), self-scanned silicon diode array with 128 elements in a 6.5 mm long row with each element 50 μm wide and 25 μm high. Although such arrays are available from several manufacturers, we chose one from Reticon Corp. (Sunnyvale, Calif.) because we had access to previous tests of their arrays. An array of "only" 128 elements was chosen because the tests by Høg and Wiskott (1974) showed that the uniformity decreased beyond the first 128-diode block.

The diodes are beneath a quartz window on the silicon chip which also includes a parallel shift register. At the beginning of an exposure, the diodes are charged to a preset level. Incoming photons generate electron-hole pairs and discharge the diodes (up

de Jager/Nieuwenhuijzen (eds.), Image Processing Techniques in Astronomy. 97-102.

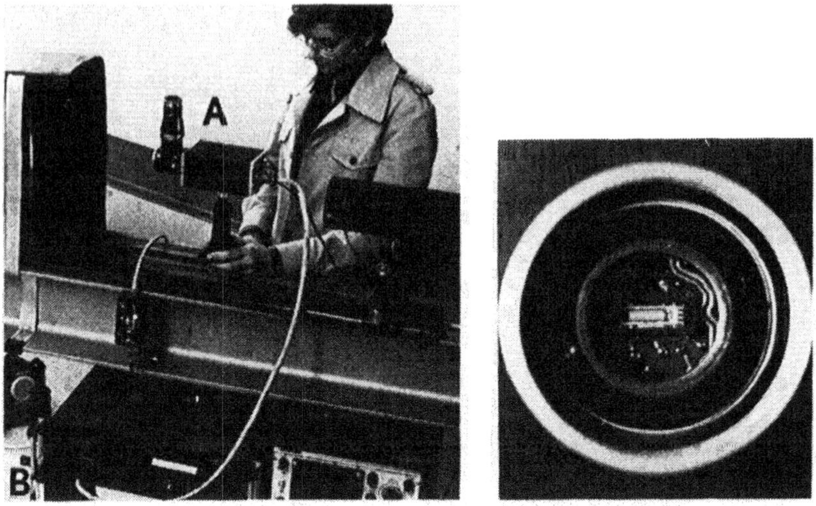

Fig.1. (a)The diode camera (A) is here equipped with a reflex finder to easily locate features. At B is a rack unit with external controls. The signal can be inspected on the adjacent oscilloscope. (b)Front end of the camera; the rectangular chip with the diode array is seen in center. Diameter of inner circle = 25 mm.

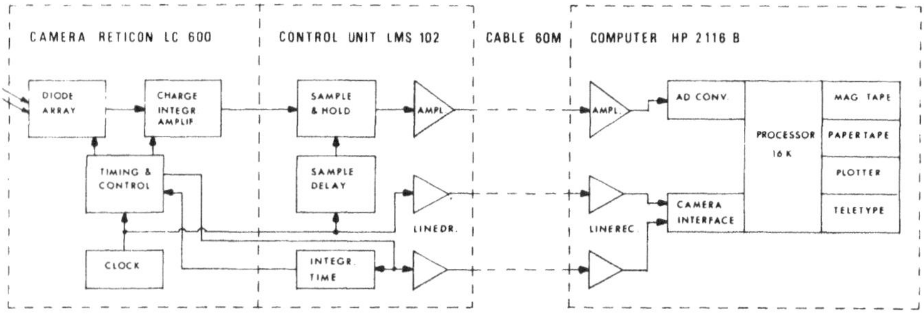

Fig.2. Block diagram of the diode array detector system.

to a saturation level corresponding to complete discharge of the diodes). At the end of the exposure, the readout sequence is initiated by a start pulse. Now each of the diodes is accessed in sequence by the shift register, driven by a square-wave clock, which in turn opens and closes the gate of an MOS switch that connects each photodiode to the common output video line. Each diode is recharged to the video line potential and the magnitude of the charge thus replaced is the measure of the exposure received. Upon the next clock pulse the next diode is read out while the former is left open-circuited and immediately begins light integration for the next exposure. The output thus consists of a string of 128 short (100 ns, say) pulses, separated in time by the inverse clock frequency (500 μs,say). Such a readout then lasts 128x500 μs=64 ms and is repeated after each integration (of 1 s, say).

In order to save development time, we chose to use the commercially available Reticon LC600 line scan camera with video pulse train output as a basic unit (instead of completely designing our own circuitry) but we made our own external controls for exposure time and signal amplification. A sample-and-hold analog storage circuit (Datel SHM-2) samples the voltage level near the top of each video pulse and holds this level so that the signal can be sent over line drivers in an up to 100 m long cable to the 10-bit (1024-level) analog-to-digital converter and the on-line computer. The camera is continuously scanning (it was found that frequent restarting caused drifts) and the scans to be read are selected by an assembler-language routine, callable from BASIC-language.

2. PHOTOMETRIC PROPERTIES

The high quantum efficiency is obtained thanks to the ease with which photons generate electron-hole pairs in silicon. Unfortunately such pairs are also easily generated thermally and this dark signal limits the time that an exposure can last before the diodes become saturated (fig.3a). Although one can drastically reduce the dark signal by cooling, this is not needed for sources that in a few seconds generate a signal that is a reasonable fraction of the saturation level (saturation exposure is ≃ 0.6 lux-seconds). Such is the case for any laboratory photometer or for our solar spectrograph. However the dark charge must be corrected for. It is characterized by a slowly varying "fixed pattern" on which is superposed a level difference between odd- and even-numbered diodes (fig.3b). This pattern is very reproducible. The fixed pattern is apparently due to an inter-diode variation in dark charge but the odd-even pattern is caused by asymmetries in the clock drive signals during the readout sequence. It can usually be adjusted to be quite small. The array permits a choice of readout frequencies up to 10 MHz but we are using the much lower frequency of 1 kHz in order to get some time for on-line reduction.

The dark signal (averaged for 10 successive readouts) has been studied; it was found that:

(a) Short-term variability (minutes) is limited by a noise that is uncorrelated over adjacent diodes and is, in absolute units, independent of exposure time. For individual diodes the rms noise is $7 \cdot 10^{-4}$ of the saturation level. This is identical to the noise in any actual exposure and apparently is a readout noise.

(b) Long-term reproducability (hours) is in addition influ-

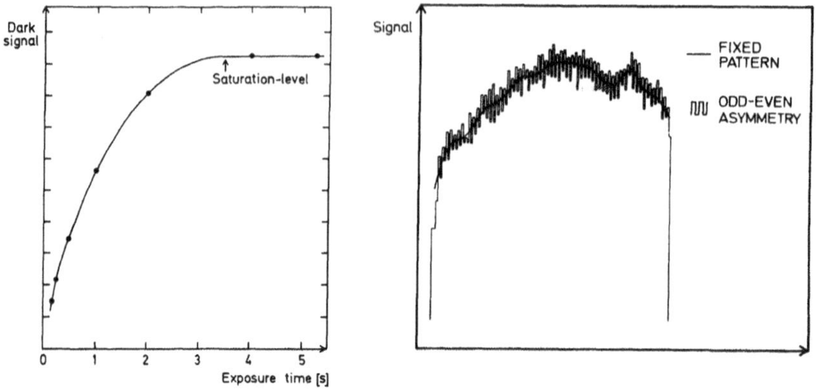

Fig.3. The dark signal; (a)magnitude as function of exposure time in room temperature +21°C, (b)morphology, here greatly enlarged.

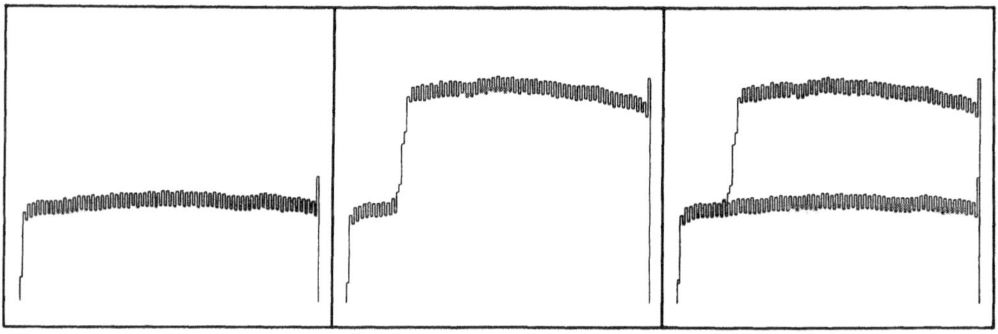

Fig.4. (a)dark signal, (b)array illuminated with a mask shadowing the leftmost part, (c)same with normalized dark signal superposed. The difference between these curves gives the light signal.

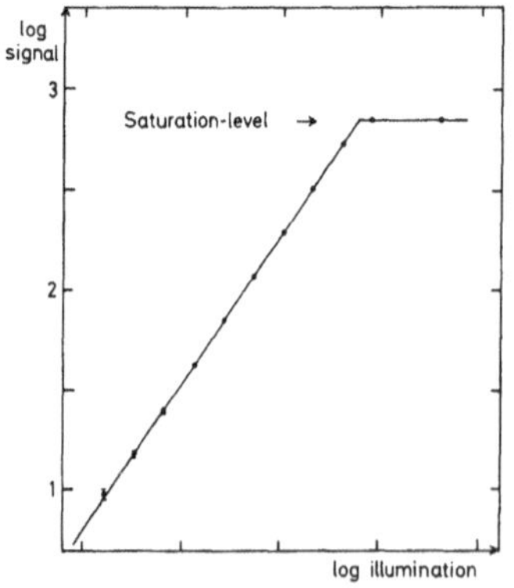

Fig.5. Linearity of response, signal averaged over 100 diodes.

enced by gradual (probably thermal) drifts in the whole dark signal level. However, if the dark signal is normalized to a standard value, the noise is reduced to the above $7\cdot10^{-4}$ rms. This normalization can be made using only the first 20, say, diodes in the array since these drifts are coherent over the whole array. An opaque mask was therefore placed in front of the array, shadowing some 20 diodes, who are thus always giving the dark charge level and leaving 100 diodes usable for photometry. Thus the need for temperature control or for frequent dark-signal sampling is eliminated.

The linearity was tested by imaging a stable light source on the array through neutral density filters with known (diffuse) densities (fig.5). As is seen, the response is completely linear up to the saturation level, i.e. $\Delta\log(\text{Signal}) = K\cdot\Delta\log(\text{Illumination})$. To determine the exact value of the constant K is more difficult since the laboratory setup illuminated the filters in a (semi-) specular manner, giving densities proportional to, but slightly different from the diffuse ones. The filters were remeasured in another photometer and thus we find K equal to unity within experimental limits. The difference in the linearity coefficient K across the array is extremely small and at our measuring limit. All these measurements are with a 1 s exposure time where the dark signal is considerable - the lowest signal point in fig.5 is only 2 % of this dark signal.

The inter-diode variation in sensitivity was measured by moving the array on a micrometer stage and exposing each diode in turn to the same illumination level. However the results are not conclusive because: (a)Here the array must be operated without a lens and is exposed to air. Due to the smallness of the diodes even microscopic dust particles may shadow a measurable fraction of it. (b)There is reflected or scattered light on a 1 %-level, possibly in the quartz window covering the diodes. In any case,the sensitivity changes are slow along the array and are probably 1-2%.

3. ASTRONOMICAL APPLICATIONS

The solar telescope at Lund Observatory has a 30 cm polar coelostat that is feeding a 25 cm refractor. After two plane mirrors and an adjustable secondary lens, a solar image of between 8 and 18 cm diameter is projected onto the entrance slit of the 4 m horizontal Littrow spectrograph which gives a dispersion of ≃ 1 Å/mm. If the array is placed parallel to the direction of dispersion in the focal plane of the spectrograph, about 6 Å is covered with 50 mÅ resolution per diode. For a 12 cm solar image, the diode height of 25 μm corresponds to 0.4 arcsec on the Sun. Exposure times of the order 1 s are required (fig.6).

A new two-coordinate, digitally controlled plate measuring

engine has been built at Lund Observatory which can use this diode array as a detector. The array system can be moved from the solar telescope to the laboratory photometer in a matter of minutes.

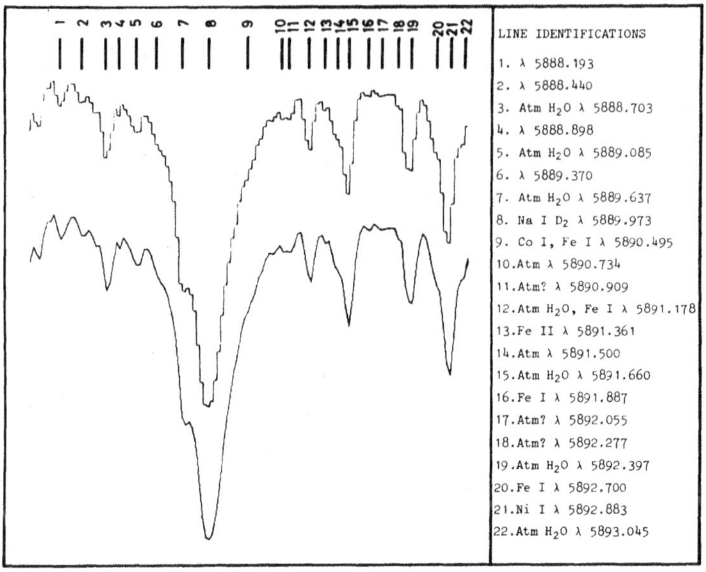

Fig.6. Region around the Na I D_2 line in the solar spectrum. Top: histogram-curve plot, bottom: same data in continuous-curve plot.

ACKNOWLEDGEMENTS

Several persons at Lund Observatory have contributed to the diode array detector system development but especially G. Arinder, L. Lindegren, H. Lindgren, P. Marsäter and T. Wiesel.

BIBLIOGRAPHY

Early, J.M. et.al.:1973, in Advanced Scanners and Imaging Systems for Earth Observations, NASA SP-335, Washington, p.305
Høg, E., Wiskott, D.:1974, Automatic Measurement of Photographic Plates with a Photo-Diode Array, ESO Tech.Rep.No.5
Horlick, G., Codding, D.:1973, Anal.Chem. 45, 1490
Melen, R.:1973, Electronics 46, 106
Smithson, R.C.:1975, Solar Phys. 40, 241
Tull, R.G., Nather, R.E.:1973, in J.W.Glaspey and G.A.H.Walker (eds.):Astronomical Observations with Television-Type Sensors, Vancouver, p.171

IMAGE DETECTORS FOR RECORDING ASTRONOMICAL ULTRAVIOLET SPECTRA

W.M. Burton

Appleton Laboratory (Astrophysics Research Division),
Culham Laboratory, Abingdon, Oxfordshire, England.

ABSTRACT. Various image tube devices have been evaluated as possible detectors for astronomical spectroscopic studies. An electrostatic image intensifier coupled by fibre-optics to photographic film has been used to record echelle format spectra. Ultraviolet sensitivity is obtained by coating the detector faceplate with a fluorescent layer. An improved image intensifier incorporating a microchannel plate to give increased gain has enabled the detection of individual photon events. Magnetically focused electronographic systems have also been used to record ultraviolet spectra. The detector systems are being developed for use in instrumentation carried on rocket vehicles or on the Shuttle-borne 'Spacelab' manned orbital platform for astronomical ultraviolet spectroscopic studies.

1. INTRODUCTION

When compared with TV and similar electronic image recording systems, the more conventional methods of photography and electronography have some disadvantages but they do provide relatively simple methods for integrating photon events in large area formats with many image elements. The dynamic range and linearity of photographic emulsions are rather limited but electronographic recording offers significant gains in both respects, also providing better spatial resolution together with a high quantum efficiency similar to that obtained in photoelectric devices. This paper describes several image detector systems which use photographic or electronographic film as the final image storage medium for recording ultraviolet spectra in the wavelength range 900 Å-3000 Å. This spectral region shortward of the atmospheric transmission limit at ∿ 3000 Å

de Jager/Nieuwenhuijzen (eds.), Image Processing Techniques in Astronomy. 103-108.

has increased in importance for astronomical studies because of the possibility of using instrumentation carried in space vehicles.

The most direct way to combine the high sensitivity of an image intensifier with the data storage capability of photographic film is to use fibre-optic coupling with the film placed in close contact with the output of the image tube. Visible to ultraviolet wavelength conversion is obtained by the use of a thin layer of a fluorescent material on the input window of the intensifier. This type of image detector system is discussed in the second section of the paper. The use of fibre-optic coupling can be avoided if the primary photoelectrons are recorded directly by placing an electron-sensitive photographic film inside the image tube. This electronographic technique is discussed in the third section of the present paper with particular reference to an internal grating ultraviolet spectrograph system.

2. IMAGE INTENSIFIERS WITH PHOTOGRAPHIC RECORDING

Image intensifiers can be classified into three different categories by the method used for electron-optical focusing, which can be either magnetic, electrostatic or proximity (ie dependent on close spacing of the co-planar photocathode and anode surfaces). Electrostatic intensifiers were selected for the studies described in this paper because they require only a high voltage power supply without an extreme stability requirement, while magnetically focused detectors require both a stabilised high voltage and also a uniform magnetic field. Proximity focused devices with sufficient spatial resolution for this application have only recently become available.

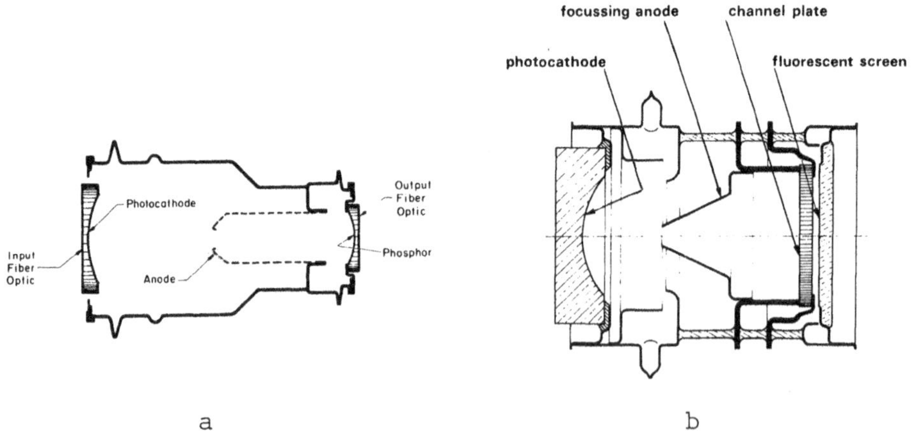

Fig. 1. Diagrams of (a) WL30677 and (b) XX1330 Image Intensifiers

Two different types of electrostatic image intensifier have been used in the present study, the Westinghouse (USA) Type WL30677 and the Mullard (UK) Type XX1330, schematic diagrams of which are shown in Figure 1. Both devices use an electrostatically focused inverting image stage which requires a spherically figured photocathode surface. To provide a plane input surface, a plano-concave fibre-optic faceplate is used to couple the input image to the photocathode. In the single stage WL30677 the photoelectrons are imaged directly onto the phosphor which is coated on the inside surface of the output fibre-optic plate, but in the XX1330 these primary photoelectrons are focused on a microchannel plate electron multiplier which provides considerable gain ($\sim 10^5$) before the output electrons are proximity focused onto a plane phosphor-coated fibre-optic. The effective input and output window diameters of the WL30677 are 40 mm and 25 mm respectively and the corresponding values for the XX1330 are 50 mm and 40 mm.

Both of these intensifiers have S25 photocathodes but the necessary use of glass fibre-optic faceplates eliminates the possibility of using windows with ultraviolet transmission. Ultraviolet sensitive proximity-focused image converters are available and could be coupled to the input fibre-optic of the intensifier. However, for the present application it was decided to use a fluorescent wavelength converter coated on the intensifier input faceplate because this very simple procedure gave better spatial resolution and extended the far ultraviolet response well shortward of 1050 Å, the transmission limit of available image tube window materials. The wavelength converter selected was tetraphenyl butadiene (TPB) which has a high conversion efficiency over an extended ultraviolet wavelength range and produces a blue fluorescence spectrum well matched to the peak sensitivity of the S25 photocathode. A vacuum-evaporated coating of ~ 1 mg cm^{-2} thickness produces no significant degradation of the image intensifier spatial resolution, having a modulation transfer function (MTF) of more than 90% at 10 ℓp mm^{-1} (Burton and Powell 1973), while the WL30677 has a MTF of 70% at 10 ℓp mm^{-1}.

To utilise the available sensitive area of these detectors, the two-dimensional format spectrum of a cross-dispersed echelle system is more efficient than a single dispersion spectrum. A compact echelle spectrograph using the WL30677 intensifier with a TPB wavelength converter has been developed for ultraviolet observations from rocket vehicles (Burton et al. 1971). Figure 2 shows the spectrum of an electrodeless microwave source recorded on Kodak 103aD film using this spectrograph. The instrument covers the wavelength range 1200 Å-3000 Å and provides a spectral resolution of 0.1 Å at 1200 Å ($\lambda/d\lambda \simeq 10^4$). At the detector, this spectral resolution element corresponds to ~ 50 μm which is approximately one pixel at 70% modulation.

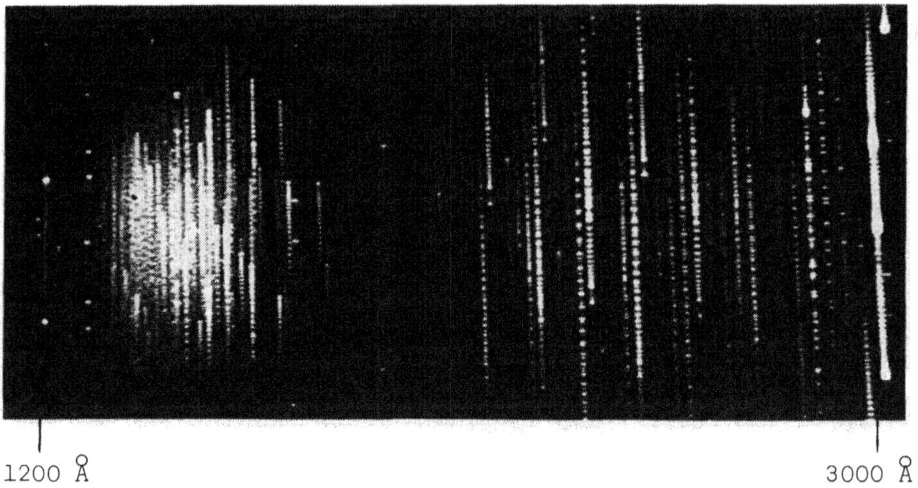

Fig. 2. Cross-dispersed echelle format spectrum recorded using the TPB-coated WL30677 with Kodak 103aD film

The Mullard XX1330 microchannel plate (MCP) image intensifier is now being evaluated as an alternative detector for this spectrograph with significantly increased gain. By varying the voltage across the MCP between 0.7 kV and 1.1 kV the gain of the detector can be varied from $\sim 10^3$ to $\sim 10^5$, while the MTF is about 40% at 10 $\ell p\ mm^{-1}$. Tests have been carried out by forming images of resolution test patterns on the tube input surface and recording the output on photographic film placed in contact with the fibre-optic output surface (see Figure 1b). The resulting images show that the performance of the intensifier changes significantly as the gain and illumination levels are changed. Figure 3 shows three images recorded under different operating conditions.

In Figure 3a the illumination level is relatively high and the gain is low. This image shows a hexagonal multifibre shading pattern caused by the non-uniform gain of multichannel bundles in the MCP which have different surface conductivities. In Figure 3b the gain is higher and the MCP uniformity is better, but a square mesh pattern is recorded which results from variations in the optical transmission of the output fibre-optic where the square multifibre bundles are bonded together. Figure 3c shows an image recorded using optimum high gain conditions which minimise these effects and enable individual photon events to be detected. When a detector is operated in this way the spatial resolution in the image is limited by photon statistical noise and the sensitivity is essentially that of the photocathode surface.

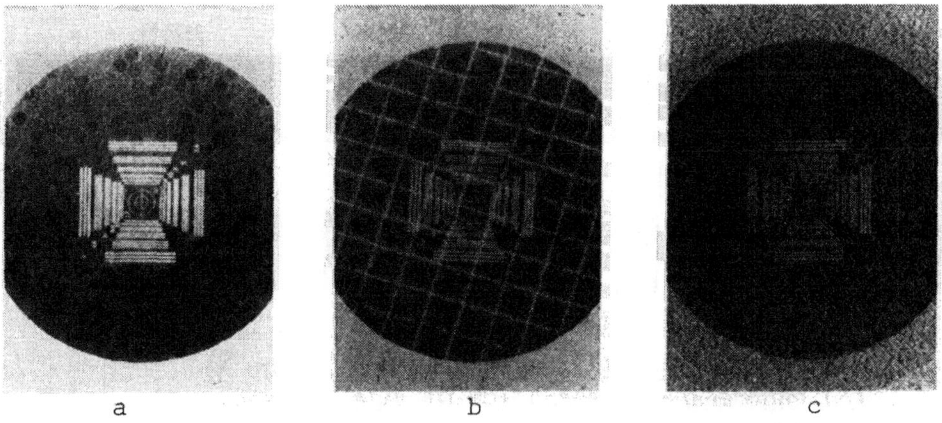

Fig. 3. Test patterns recorded using the XX1330 Image Intensifier

3. ELECTRONOGRAPHIC DETECTORS

In the electronographic process photoelectrons are accelerated directly into the recording film, thus avoiding the problems associated with phosphor-coated fibre-optic windows discussed above. A simple magnetically focused electronographic image tube with a LiF input window coated with a semitransparent CsI photocathode is being evaluated as a possible alternative for the intensifiers described in the previous section. An internal grating electronographic UV spectrograph of the type described by Carruthers (1972) has also been developed for spectroscopy in the wavelength range 900 Å-2000 Å. Figure 4 shows the spectrograph design which is based on a 40 cm concave grating forming a spectrum on an opaque CsI photocathode. The photoelectrons are accelerated by a 20 kV field and magnetically focused to form an image of the spectrum on electron-sensitive film (Kodak 088-03). The energy of these accelerated electrons is sufficient to ensure that each primary photoelectron has a high probability of producing a developable photographic grain and the effective quantum efficiency is therefore determined by the photocathode response and the transmission of the optical system.

Fig. 4. Ultraviolet Electronographic Spectrograph

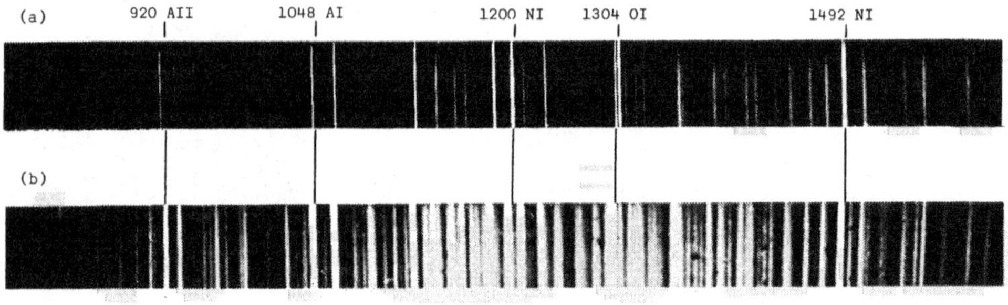

Fig. 5 Comparison ultraviolet spectra (900 Å-1600 Å)
(a) Photographic: Kodak 101-01 film (30 sec)
(b) Electronographic: Kodak 088-03 film (5 sec).

To estimate the relative sensitivity of photographic and electronographic processes a comparison exposure was made with Kodak 101-01 photographic film placed at the photocathode position. Figure 5 shows spectra recorded by the two techniques, demonstrating clearly the gain in sensitivity provided by electronography. Although the photographic exposure is larger by a factor of six, the electronographic spectrum is comparable at 1600 Å and significantly better exposed at 900 Å. This variation follows from the quantum efficiency of the opaque CsI photocathode which increases from $\sim$ 15% at 1600 Å to $\sim$ 60% at 900 Å, while a typical photographic DQE value would be $\sim$ 1% in this wavelength region. An electronographic detector of this type would be very suitable for astronomical instrumentation on the Shuttle-borne 'Spacelab' vehicle.

ACKNOWLEDGEMENTS

The work described in this paper includes important contributions made by several of my colleagues, particularly Mr A T Hatter and Mr R A Hardcastle. Image tube performance data has been provided by Mullard Ltd and Westinghouse Inc. This paper is published with the permission of the Director of the Appleton Laboratory.

REFERENCES

Burton, W.M. and Powell, B.A: 1973 Appl. Opt. 12, 87.
Burton, W.M., Reay, N.K., Shenton, D.B. and Wilson, R: New Techniques in Space Astronomy (IAU Symposium No. 41), Labuhn and Lust Eds. (Reidel, Holland 1971) p. 304-312.
Carruthers, G.C: 1972, Adv. Electronics and Electron Physics, 33B, 895.

MESURE DES CONTOURS DE RAIES PAR LES TECHNIQUES DE TELEVISION

Ch. FEHRENBACH et A. VIN

Observatoire de Haute Provence, Saint-Michel l'Observatoire, France.

At the Haute Provence Observatory, we have been developing a television detection system, TV Raies, which will eventually be available for the observation of spectra in real time, at the telescope.

It is our aim to record faint spectra by the technique of photon counting. Such systems are expected to be used with the 3.60 meter telescopes with which France is currently associated.

We decided, as a test of the technique, to begin with a system for the measurement of line profiles in the spectra of stars with extended atmospheres such as those observed by Mrs Hermann, Mrs Doazan, and their colleagues.

An analog system has already been built and we hope to start the construction of a counting system towards the end of this year.

Observations have been made at the Coudé focus of the1.52 meter telescope. The televison camera acted as the detector in the Echelec (Fig. 1) spectrograph. When used in the

de Jager/Nieuwenhuijzen (eds.), Image Processing Techniques in Astronomy. 109-116.

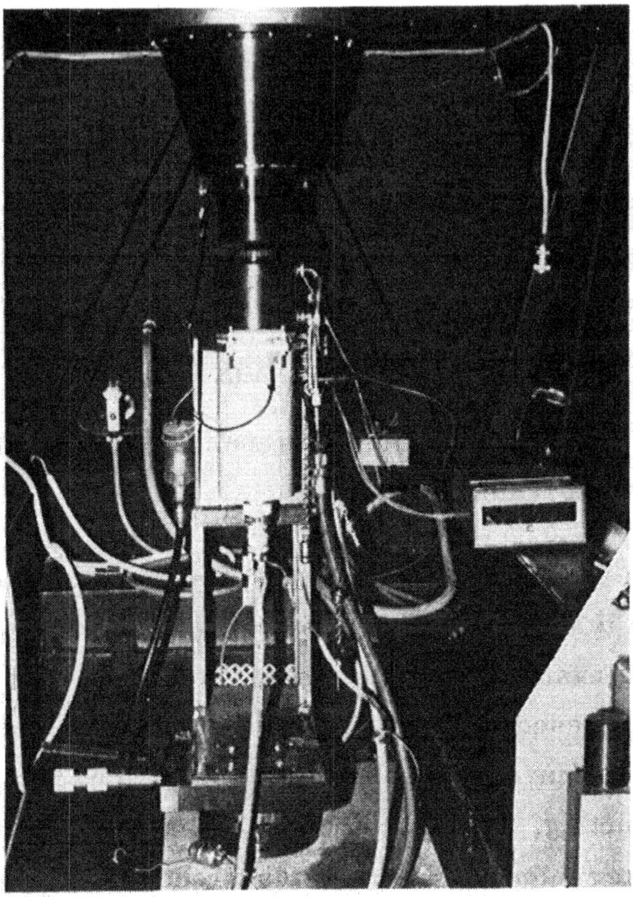

Figure 1. - Caméra TV à l'Echelec

echelle configuration, the dispersion was 4 Å/mm, and 100 Å/mm with the normal grating. The camera operates at an f ratio of 2.6, and the 40 microns resolution of the detector are equivalent to two arc-second on the sky.

The image, on the television monitor, is obtained as follows (Fig 2) :

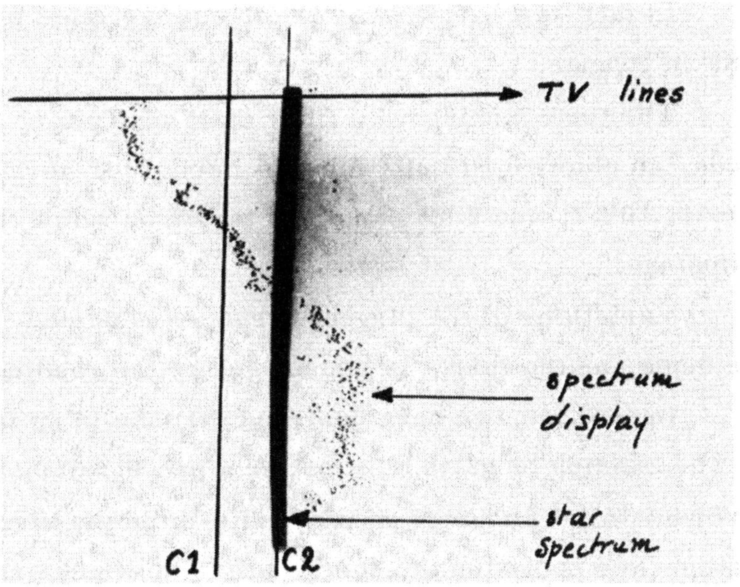

Figure 2. - Scan TV

The spectral dispersion is normal to the TV scan lines. Consequently, one line corresponds to a single spectral element.

Information is sampled twice on each line : One, C1, gives a measure of the sky spectrum, the other, C2, is the stellar spectrum itself. The position of C1 and C2 can be moved anywhere across the screen..

The whole system (scan, integration time, and so on) is controlled by the data acquisition electronics. While a standard scan rate can be used, slower rates are also possible, which, as we point out below, allows the use of detectors of very different properties.

For the first experiments, we have taken an unmodified camera and operated it at the standard scan rate of 50 frames/second with interlacing, each frame containing half of the image.

In fact, we have used a Nocticon tube made by Thomson-CSF in France.

The tube consists of a fiber optic face plate, S 20 photocathode, an electrostatically focused intensifier-section which operates at 10 kV, and a silicon diode target on which the electrons impinge.

Variations of the target charge are detected in the vidicon mode, as the target is discharged by the reading beam.

In addition, we have operated the tube in an integration mode. The reading beam is blanked for a preset number of frames in order to let a charge pattern build-up in the target. The integration time is limited by charge diffusion which gradually degrades the image, and thermal noise in the target. The latter effect is very serious at room temperature and, inconsequence, the tube is refrigerated using a freon pump.

A temperature of about - 25°C is achieved in the region of the target, which allows useful integration times up to 5 seconds.

The video signal is treated in the following way :
It is first filtered (Fig. 3), in order to retain only the higher frequencies of interest, that is to say the crossing of each spectrum, but not the lower background along the line. This tends to extend the available integration time by reducing the level of the thermal background continuum.

The signal is then sampled and integrated (Fig. 4) at the points corresponding to C1 and C2, over an interval which can be varied from 100 ns to 2 µs. The width of a star spectrum corresponds to about 1 µs.

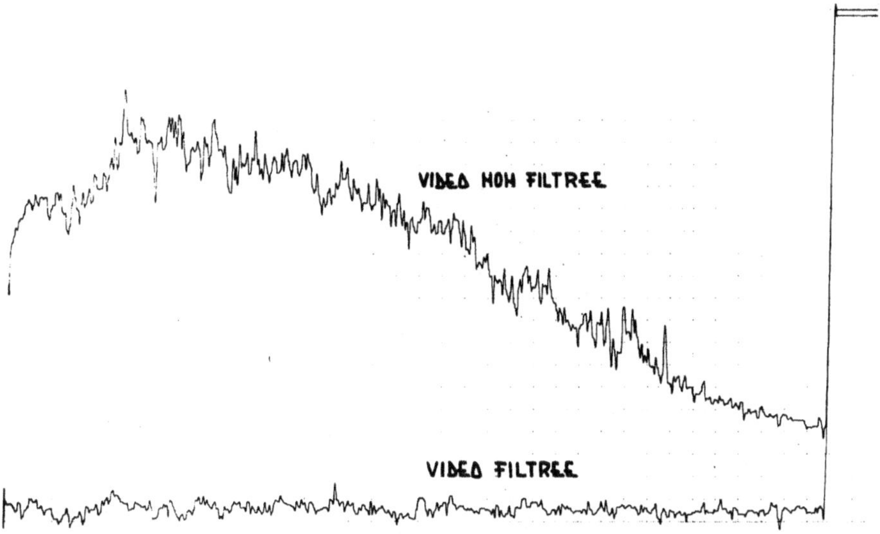

Figure 3. - Video filtrée et non filtrée

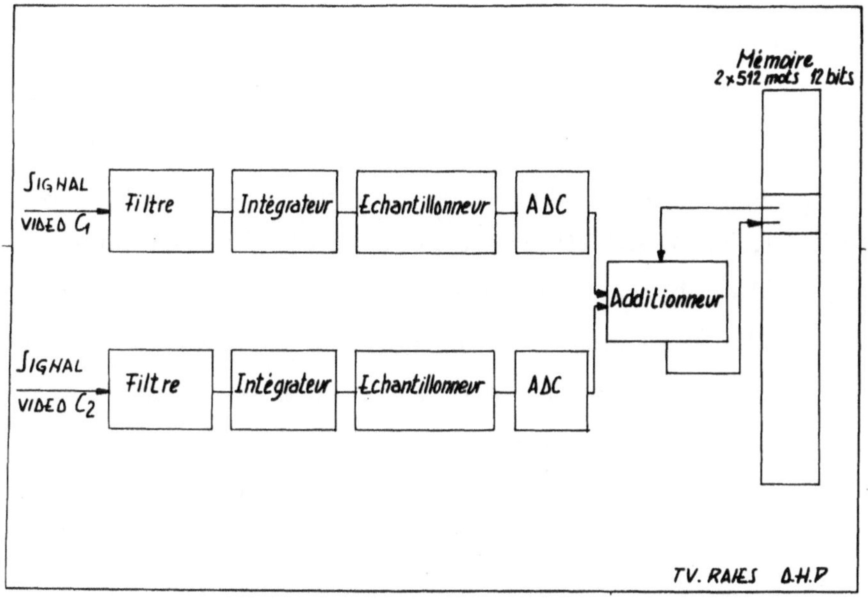

Figure 4. - Block diagram

The output voltages of the sample and hold modules are digitized to twelve bits by an ADC. There are two parts of the memory, one corresponding to C1, and the other to C2. There are 512 locations for each.

The value of the memory is transmitted to a delayed pulse generator, such that the corresponding TV lines displays a bright point whose delay corresponds to this value. As a result, the observer can watch the development of the spectrum and can stop the exposure at an appropriate signal level.

The first observations were made without the TV system coupled to a computer, the video signal was filtered, the sample time was only 100 ns, that is to say that we picked up, unfortunately, only the tenth of the useful signal.

There are some of the spectra we have obtained, this one, for example, of the star 32 Ursae Majoris, in 45 seconds, without any beam blanking, ... and this one of the star ε Ursae Majoris in 50 seconds. (Fig. 5-6)

Later, this year, we plan to interface the system to our computer Télémécanique T 1600, and we expect to correct for dark background, night sky spectrum, and variations of sensitivity. This will allow us, also, to observe for longer times, while substracting the baseline at each integration.

We shall use also the same tube in a slow scan mode, which allows an improvement in signal to noise ratio by lowering the effective amplifier noise.

For the observation of very weak spectra, we plan to incorporate an image intensifier, in order to detect individual photon events and the data acquisition would then be modified to a pulse counting mode.

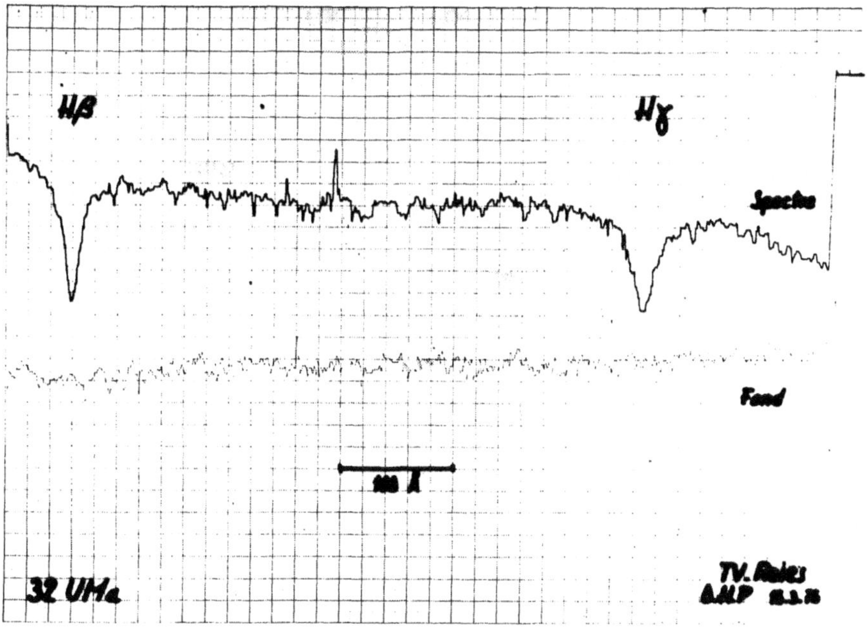

Figure 5.- 32 UMa

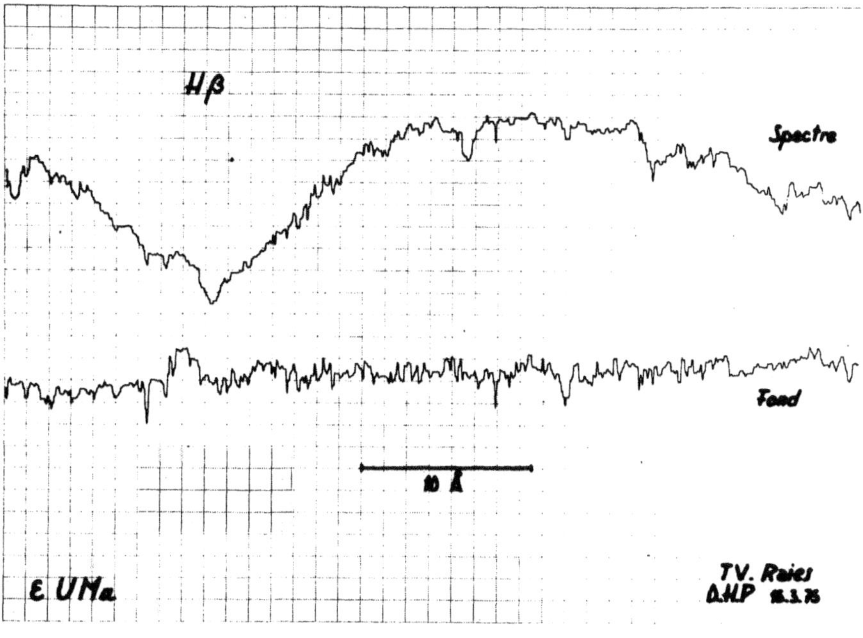

Figure 6.- ε UMa

Finally, we shall soon connect to the same system a Reticon solid state array of 1024 diodes, which is sensitive in the near infrared. This will allow us, in the same way, to observe at the Cassegrain focus of our 1.93 meter telescope, using the Roucass spectrograph, which has been specially treated for light transmission in the region 7000 Å to 11000 Å.

PHOTOMETRY USING ELECTRONOGRAPHIC TECHNIQUES

J. Ring and S.P. Worswick

Astronomy Group, Physics Dept., Imperial College, London S.W.7.

INTRODUCTION

Electronography has two main advantages, namely a linear response and high storage which leads to a fairly substantial dynamic range.

With the growing application of computer techniques to the processing of observational data it is becoming increasingly important to understand each stage of any detection process which produces the required numerical information. Thus for those processes, such as electronography, where an emulsion is used as an intermediate store, the measuring which produces the digital data must be considered as an integral part of the detection system.

This paper deals with the photometric performance and limitations of the specific combination of a Spectracon image tube, used with Ilford L4 nuclear track emulsion, with a Joyce-Loebl MkIII CS microdensitometer which produces digital output on magnetic tape. The system has been applied to broad and narrow-band surface photometry of extended objects and it is with these types of observations in mind that the evaluation has been made.

EMULSION

Studies of the various emulsions used for electronography have shown that, except for the very faintest of sources, Ilford L4 produces the best performance in terms of linearity and noise (Cohen and Kahan (1972), Cohen (1972)).

However, for emulsion noise to determine the photometric accuracy that can be obtained, one has to assume both a uniform

de Jager/Nieuwenhuijzen (eds.), Image Processing Techniques in Astronomy. 117-124.

surface-response in a detector and a small noise contribution from the digital read-out of exposures. In practice these assumptions may not be valid and the extent to which the system can be made to conform to these conditions will determine the photometric accuracy.

The main investigations that are required are into the performance of the microdensitometer and its effect on the measurements of the electronograph and into the removal of changes in sensitivity across the detector surface which will limit the accuracy with which extended objects can be mapped.

THE MICRODENSITOMETER

Electronographic data impose rigorous conditions on the performance of any measuring machine. The requirements are not only the accurate representation of spatial information, which is necessary for any high resolution emulsion, but also an accurate measurement of densities that are much higher than those that are useful in photographic exposures. The primary areas of importance are:

1) Accuracy of spatial sampling
2) Linearity
3) Noise levels at high density

The Joyce-Loebl microdensitometer used for this work is a two-dimensional scanner with a fairly slow data rate in the region of 10 to 20 samples/sec. The data is produced in a format that is compatible with the CDC computers on which it is processed. The density resolution is dependent on the grey wedge attenuator used in the reference beam, the full range of the wedge being divided to 1700 levels.

ACCURACY OF SPATIAL SAMPLING

For mapping and photometry it is necessary to have an accurate representation of the spatial distribution of the emission of any object.

Positional information used for the mapping of images scanned on the Joyce-Loebl is based on the assumption of equal spacing between samples. Stepping motors and lead screws provide the X and Y motions of the table and have a minimum step length of 5μm. No evidence has been found for large jumps in table position; however a cumulative error amounting to about 6μm in a 25mm stroke is present. A certain amount of backlash, which is compensated during automatic scanning, exists between the sample table and lead screw. Therefore accurate relative positions are obtained from computer produced plots of scans rather than by using the step-counters on the motors.

LINEARITY

The linear response of the electronographic process requires that the microdensitometer itself be highly linear. Measurements made on Ilford L4 with the Joyce-Loebl yield a linear density exposure relationship up to densities in excess of 5D (Ables and Kron (1967)).

A problem with all microdensitometers, and probably the major source of non-linearity in the Joyce-Loebl, is the low assessment of the true density due to transmission-averaging across the measuring slit. Since electronography exhibits a linear density-versus-exposure response it is necessary to obtain a measure of the true density for this feature to be of any use. By taking a simple model of a linearly varying density gradient across the measuring aperture it is possible to do some order of magnitude calculations on the change in density across the slit that is tolerable for a given percentage error in density measurements.

For a given image content, that is the largest density gradient likely to be found in the sample, these figures can be used to give a guide to the maximum size of measuring aperture that can be used.

MICRODENSITOMETER NOISE

Having considered the systematic errors of the microdensitometer it is necessary to deal with the random noise arising from the photometric system of the machine.

Measurements of system noise have been made by holding neutral density filters fixed in the beam while the table was scanned in the normal manner. At high transmission the noise levels are set by the digitising electronics whereas at low transmission it is the photon shot noise of the system which predominates (Ring and Worswick (1974)).

REMOVAL OF DETECTOR NON-UNIFORMITIES

In electronographic detectors the photocathode usually exhibits variations in surface sensitivity to a greater or lesser extent and the removal of the effects of these features governs the accuracy with which the incoming fluxes can be measured. The applications for which we employ electronography are biased towards photometry of extended objects. Thus interpolation across the object using the night sky background as a uniform light source is not the best method of correction. Therefore it was decided to develop a means whereby two exposures could be matched and divided.

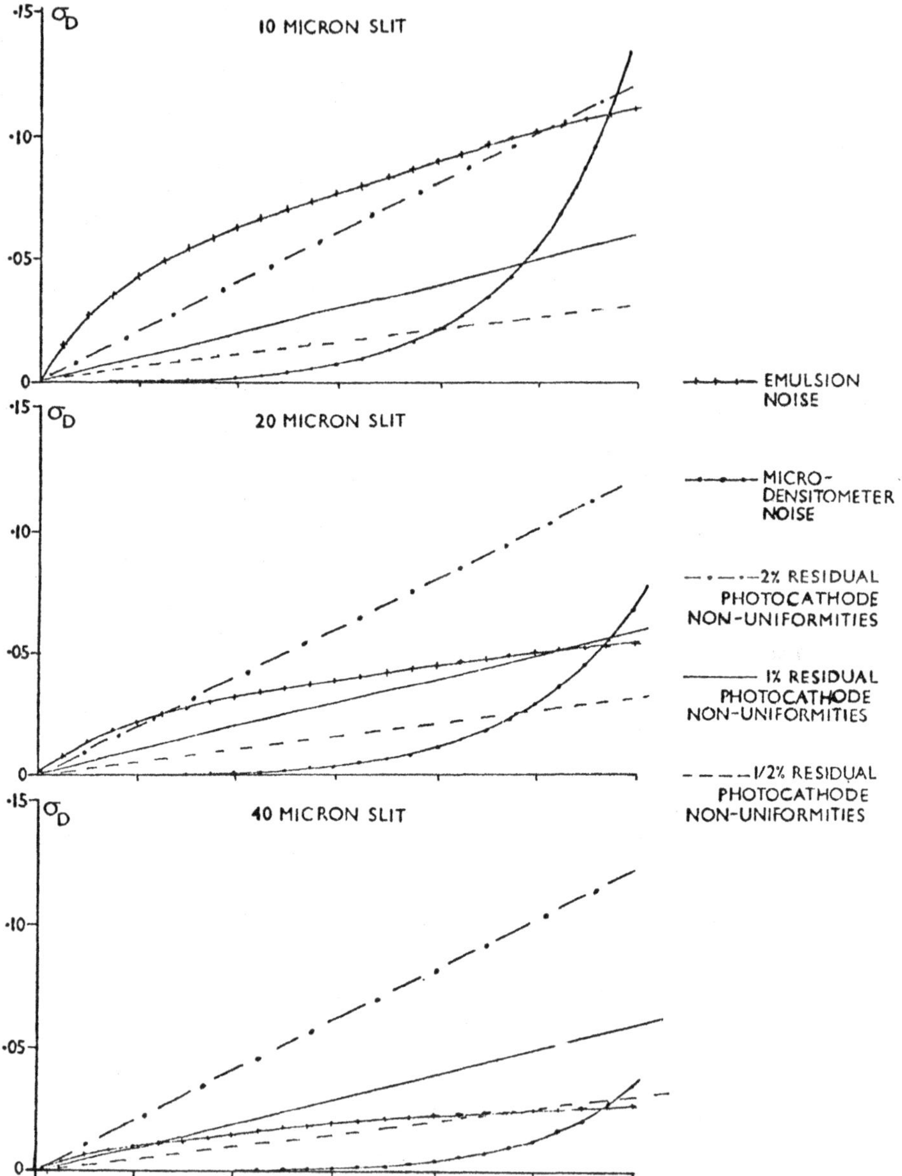

Figure 1: Sources of Limitations to Photometry

A tube exhibiting a fairly wide range of defects including a step in sensitivity due to a change in thickness of the mica window was used to take several laboratory exposures. The same areas of photocathode were located on each exposure using dead-spots, as fiducial marks. Scans of each exposure were made and the final data matched and divided in the computer.

Comparing noise in both the calibrated data and the original scans one finds in the corrected exposure the noise is a factor of $\sqrt{2}$ greater than the emulsion noise of the originals. With exposures taken to different background densities it is found that this regime holds until the noise level on the corrected data reaches the region of 1% to 2%. From then on the improvement in signal-to-noise ratio one would expect from the higher density in the emulsion is not reached. These figures have been obtained from fairly large areas of photocathode a few mm across. If a closer examination is made and those areas clear of dead-spots are selected then the noise levels obtained again would seem to be determined by the emulsion grain statistics (Worswick (1975)). Thus for areas corresponding to typical image diameters it would seem that, at present, the response can be made sensibly uniform to the region of 1% to 2%.

LIMITATION TO PHOTOMETRIC ACCURACY

Figure 1 shows a summary of the sources of noise that have to be considered as possible limitations to photometric accuracy. Each graph represents a different aperture (10,20,40μm). It can be seen that for the case of emulsion and microdensitometer noise alone the density at which the machine noise exceeds that of the emulsion is the same for all apertures. This would seem reasonable since the improvement in machine noise due to increased photon flux is matched by the corresponding drop in emulsion noise due to grain statistics. The effects of 1%, 2% and ½% residual photocathode non-uniformities are shown in the straight lines. These set another value of limiting density above which one will not sensibly gain an improvement in signal-to-noise ratio over substantial areas of photocathode.

Taking 1% as the best accuracy one can reasonably hope to obtain across a fairly large area of photocathode, it is possible to calculate the emulsion density required for 1% emulsion noise for different sizes of analysing aperture. For slits smaller than 20μm the densities required are higher than the limiting density set by the microdensitometer and therefore it is the machine noise which will set the limit to photometry on resolution elements smaller than 20μm. For apertures larger than this, photometry is emulsion noise limited until the 1% residual photocathode non-uniformities are reached.

The uniformity limitations set by the photocathode mean that features 4 to 5 magnitudes down on the sky background should be discernible. This is important in broad-band photometry where one is trying to distinguish faint features against a high sky background.

It should be noted that the accuracy of measurement of relative fluxes across an object is set by the dynamic range of the electronograph and although each individual measurement may be accurate to the order of a few percent the smallest measurable flux may be only a few tenths of a percent of the peak flux from the object. For example, in the case of broad-band photometry where the sky background has been taken to a density (D_B) corresponding to 1% emulsion noise for the required resolution element to be used, then the smallest useful density fluctuation that can be measured is $0.01D_B$. If microdensitometer noise is not to play a role then the upper density which can be usefully measured is D_L, the limiting density set by the microdensitometer. Thus the dynamic range (R) of the exposure will be given.

$$R = \frac{D_L - D_B}{0.01D_B}$$

Taking a 40μm resolution element the dynamic range is about 350:1. In fact if one lowers the tolerance on accuracy the dynamic range can be increased. For example, for 2% accuracy on a 40μm aperture the dynamic range becomes 800:1. The limiting dynamic range of the process is set by the smallest and largest density that the microdensitometer can usefully measure and is in the region of three orders of magnitude.

Thus it would seem that the extra effort required to reduce the effects of photocathode non-uniformities is still recompensed by the linearity and dynamic range of the process.

COMPUTER PROCESSING

The computer processing developed for use on the scans is undertaken using CDC 6400 and 6600 computers. Thus the lack of flexibility in the measuring machine is to some extent countered by the large amount of computer power that is available.

When the measuring system was first set up three main routines were produced - a translation program to take the data from the microdensitometer tapes and format them into an archive file which could be read using FORTRAN programs, a display program which plotted out large arrays using a fairly small amount of computer power and a contouring routine.

Since then the number of routines has proliferated, some being specific to certain observing programs, others being of more general use. All the programs will read the standard archive format but otherwise the subsequent processing is left very much to the individual user. With the data stored on a magnetic tape it is possible to produce both photometric as well as morphological information from the same scan. For example, a program has been developed to assess the general background level of an exposure and the noise on it; this is then used as a guide to setting the outer contour of the object under study. Figure 2 shows NGC 6543, a planetary nebula, with an outer contour 2 standard deviations above the background. Incidentally, it is at about 0.2% of the peak surface brightness of the nebula.

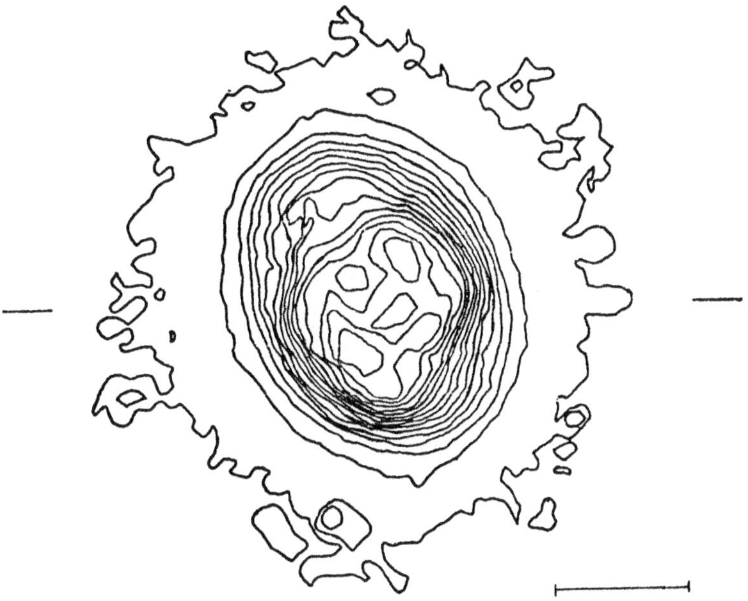

Figure 2: Hβ Contours of NGC 6543.

Until recently the graphical output has been restricted to CALCOMP incremental and MICROFILM plotters. However, a TEKTRONIX visual display unit which is run on-line to the CDC 6400 has become available and many of the routines are now being modified so that graphical output may be displayed and changed before plotting. In this way it is hoped that we will in the future only get hard copy of the specific data required and all preliminary viewing of scans will be undertaken using some form of interactive program.

REFERENCES

1. H.D. Ables and G.E. Kron, Pub.A.S.P., 79, 423, 1967.
2. M. Cohen and E. Kahan, Advances in Electronics and Electron Physics, 33A, 53, 1972.
3. M. Cohen, Ph.D Thesis, University of London, 1972.
4. J. Ring and S.P. Worswick, Presented at Sixth Symposium on Photoelectronic Imaging Devices, London. To be published in Advances in Electronics and Electron Physics, 1974.
5. S.P. Worswick, Ph.D Thesis, University of London, 1975.

INTEGRATING T.V. SYSTEM DEVELOPMENT AT CAMBRIDGE*

C.D. Mackay

Institute of Astronomy, University of Cambridge, Cambridge, England

The development of an integrating television system was started at Cambridge at the beginning of 1974, and is now nearing completion. It is intended to use the camera for accurate two - dimensional multi-colour photometry of galaxies and nebulae. The camera uses an SIT-vidicon and is a development of the work of Westphal at Caltech [1]. It differs in that the control of the camera and the reduction of the observations is handled by a small computer. The use of a computer gives a great deal of flexibility in operation, enabling the geometric (electro-optical) distortion in the camera tube to be corrected for with accuracy. It allows a number of housekeeping-type checks to be made on the operation of the camera (such as camera temperature, signal line noise levels and EHT supply levels) which would make any hand-wired controller extremely complex. The data are recorded on magnetic tape in digital form. A medium-persistence oscilloscope is used to display the data from the camera. A core image is held packed in bytes which provides a continuously refreshed image at about four frames per second fully interlaced. Images or parts of images may be displayed directly from the camera (when it is used for focussing or field finding) or from images stored on magnetic tape. Routines are provided which allow all or part of a frame to be displayed as a grey scall or in contour form with the option of compressing the dynamic range of the data for ease of viewing. One-dimensional profiles may also be displayed. Images may be added, subtracted, multiplied or divided by another image or images. As this is all

* This work has been supported in part by the Royal Society and the Radcliffe Trustees.

de Jager/Nieuwenhuijzen (eds.), Image Processing Techniques in Astronomy. 125-126.

done on magnetic tape it can be a little slow, but two images can be added to produce a third image in about 60 seconds. The reduction and display of the data may proceed while an exposure is in progress. The entire operating, reduction and display programme is core-resident and includes real-time multitasking facilities within 16 kilobytes occupied by the programmes in the Nova used. A further 16 kilobites is used to store the core image of the data used for display, as well as providing data buffers for the magnetic tape drivers. The whole system is therefore extremely compact and economical while still allowing nearly all of the reduction of two-dimensional images to be done at the telescope.

1. Westphal, J.A., 1973, "Astronomical Observations with Television-Type Sensors", Vancouver.

SEC VIDICON SYSTEM FOR A BALLOON ULTRAVIOLET STELLAR SPECTROMETER

T.M. Kamperman

Space Research Laboratory, Beneluxlaan 21, Utrecht, Netherlands

1. INTRODUCTION
2. SYSTEM GOALS AND CONSIDERATIONS
3. DETECTOR SYSTEM
4. TESTING HARDWARE

1. INTRODUCTION

In 1972 the Space Research Laboratory proposed a continuation of her high resolution stellar spectroscopy research program by means of a telescope-spectrograph-detector combination as a balloon borne payload. It appeared that on a national basis the costs for such a program were prohibitive. Looking for a collaboration partner we found the group of the Astrophysics section of the Johnson Space Center of NASA at Houston under supervision of Dr. Kondo willing to cooperate. They have been flying with a large 4 m diameter balloon platform and a 40 cm diameter telescope with its own fine-pointing system in combination with a Ebert Fastie spectrograph and an Image dissector tube (Kondo et al., 1972).
Figures 1 and 2 are showing the balloon platform and the telescope respectively.

de Jager/Nieuwenhuijzen (eds.), Image Processing Techniques in Astronomy. 127-134.

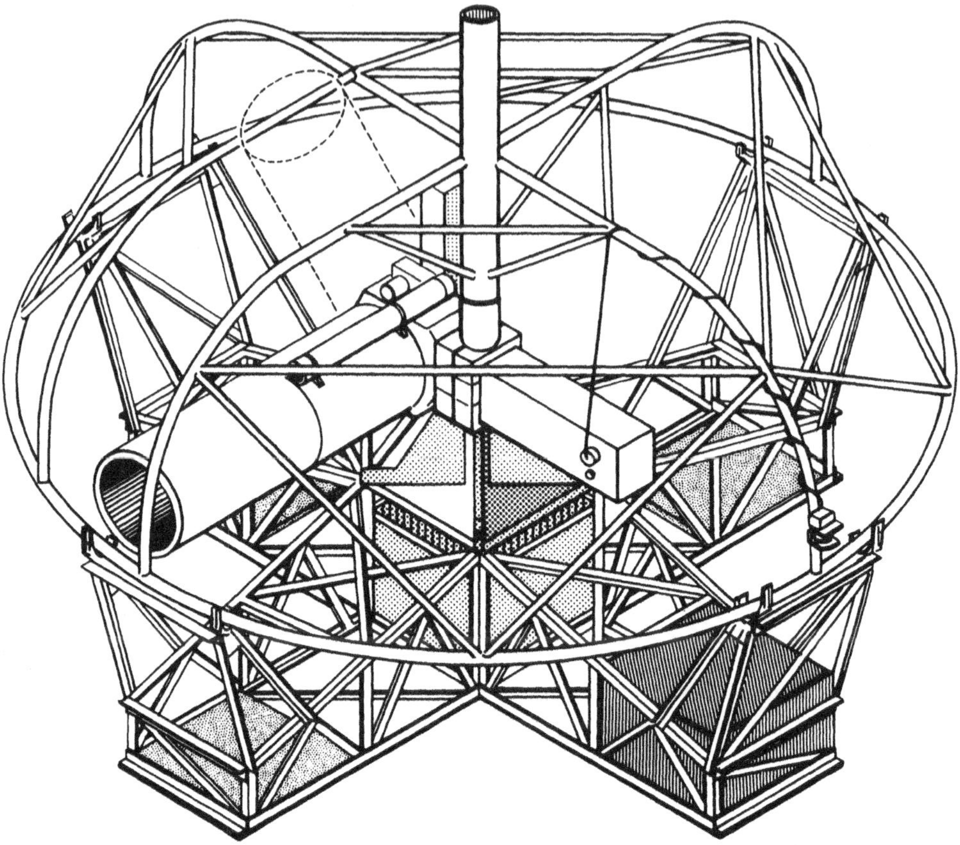

Fig. 1. BUSS platform (NASA- JSC).

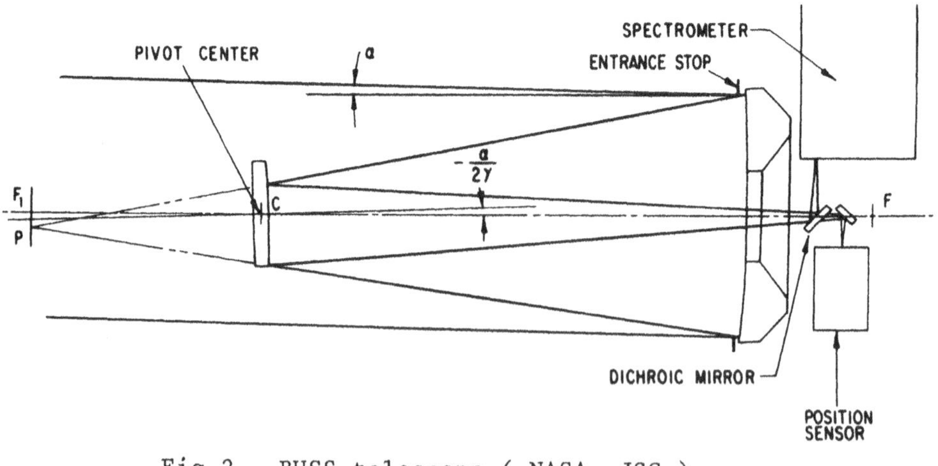

Fig.2. BUSS telescope (NASA- JSC).

2. SYSTEM GOALS AND CONSIDERATIONS

The Image dissector tube used by NASA has the same disadvantages for spectrometry as our S-59 UV Spectrophotometer (de Jager et al., 1974) in the European TD1-A satellite: relatively long integration times are required as the system lacks an integrating element. In the collaboration program we chose for an integrating television-type detection system behind an echelle-type spectrograph.
The goals for the system currently in development at our laboratory are the following:
- Wavelength region: 200 - 340 nm. The short wavelength limit is imposed by the remaining atmosphere at float altitude of 40 km.
- A spectral resolution of 3×10^4 to enable studies of the dynamic properties of stellar atmospheres, abundancies of rare elements, their isotopes, and the interstellar medium.
- Observation of $m_v = 5$ stars (of type B_o) with 3% noise in 15 minutes.
- Compatibility with already existing hardware.

The main optical components for the spectrograph (figure 3) are:
- a dichroic mirror, reflecting the full UV region and transmitting the visible light for startracking purposes,
- an off-axis paraboloidal mirror, which can be shifted to correct the focus by means of telecommand and provides a collimated beam to the echelle grating,
- an echelle grating with 79 grooves/mm,
- a fused silica wedge, reflecting from its backside, providing the cross dispersion and
- a catadioptric imaging system in front of the television camera tube.

The dispersion of the spectrograph is between 2.1 nm/cm and 1.2 nm/cm at wavelengths of 340 nm and 200 nm respectively.
The spectrograph has been optimized for equal contribution of
- pointing errors and image quality of the telescope
- degradation as a result of aberrations in the optical system
- degradation as a result of the limited resolution of the detector
- degradation of the quality of the spectra from structural tolerances.

As an example: the first contribution mentioned was kept small by chosing a relatively high focal ratio of the collimation mirror, resulting in wide beams inside the spectrograph, a large echelle grating and especially a high focal ratio for the catadioptric imaging system.

The contract for the design and fabrication of the spectrograph itself was given to the Institute of Applied Physics TPD at Delft, Netherlands.

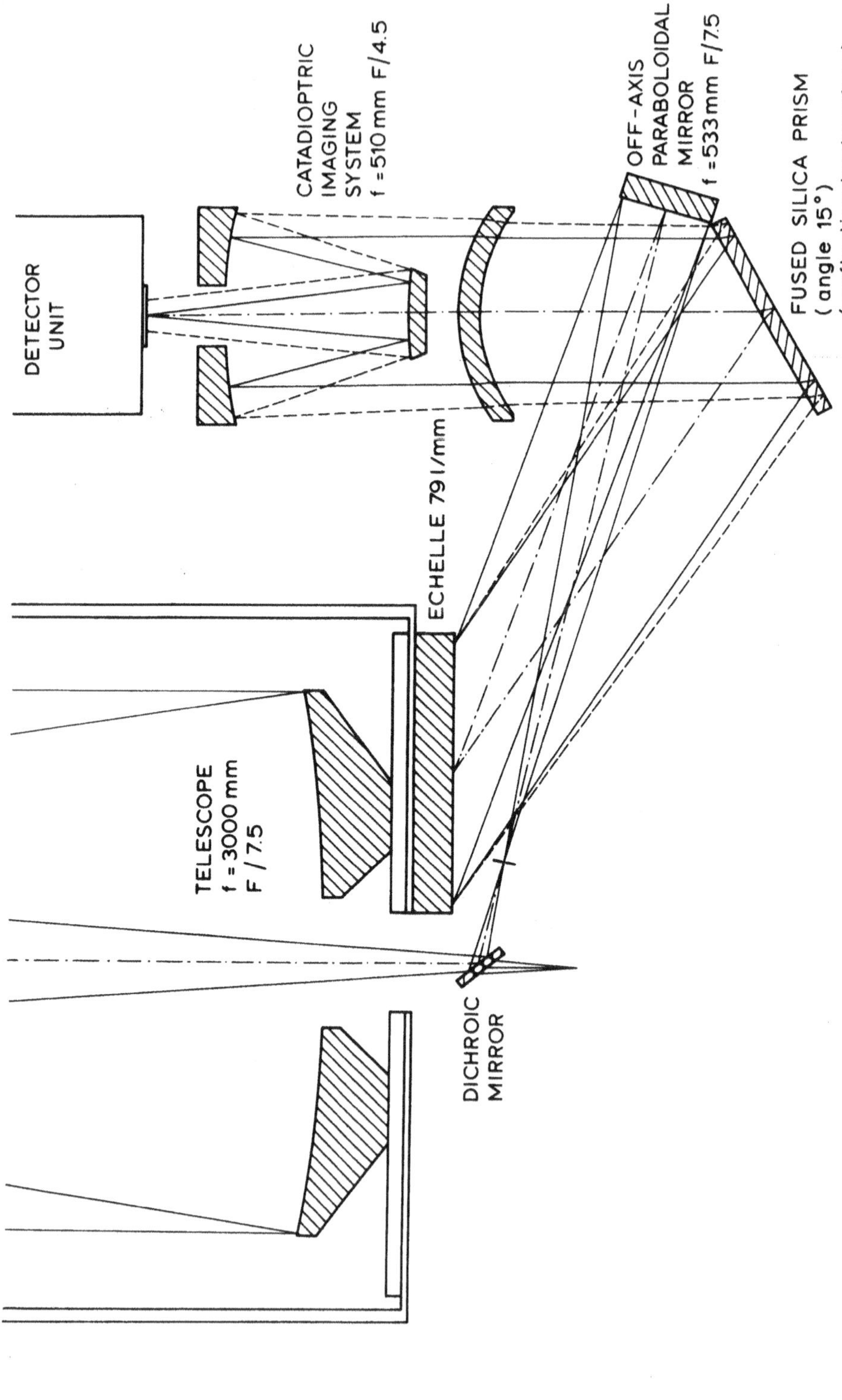

Fig. 3. Optical lay-out spectrograph.

3. DETECTOR SYSTEM

Our requirements in the choice of a detector were:

- Two dimensional device to detect the echellogram.
- Long-term integration capability and single frame read-out to limit the telemetry bandwidth.
- UV sensitivity with quantum efficiency in the order of 10%.
- Noise had to be limited as much as possible by photon statistics.
- High spatial resolution.

We selected a magnetically focussed SEC camera tube (Westinghouse WX31718) with a 25 x 25 mm target and a CsTe photocathode.

As we did expect problems in operating the tube with the required high voltages at float altitude at a pressure around 2 Torr, we decided to put the camera and its high-voltage converters in a pressurrized container as was done in the Stratoscope project (Flory et al., 1966).
The corrector lens used in the catadioptric imaging system appeared to be a good entrance window for this container.
We are scanning the SEC-vidicon with 1024 lines, each digitized in 1024 picture elements. As we chose the television-lines more or less parallel to the echellogram orders, the spectral information is in the direction of the lines. To prevent saturation, the orders are chosen as wide as possible, just not overlapping. Doing this, each order is covered by about 8 television-lines. This might seem overdoing it, but some lines may be lost due to telemetry interference, and we only have to digitize the video-signal per pixel into 8 bits before we put them on telemetry as we are averaging information from several lines any way.

With the 25 μm squared pixels used, we can accumulate 150 primary photo-electrons per pixel in the linear range, giving a total of about 1600 per resolution element in the spectrum.
As a result of this a S/N-ratio of 30 can be achieved easily, even with illumination levels below the optimum.

Owing to the secondary emission in the target the total number of electrons per pixel is about 12000. As we hope to keep the read-out noise of the preamplifier below 300 electrons (at a bandwidth of 24 KHz), this contribution to the noise is lower than the statistical noise.

Further noise is introduced by thermal electron emission of the photocathode material. In our case, using CsTe with its high work-function, this contribution will be negligible.

As outlined above, the spectral dispersion is in the direction of the scan lines. In order to achieve a proper wavelength calibration, this requires an extremely stable horizontal sweep. We chose for

Fig. 4. Electronics mounted in pressurized container.

Fig. 5. Detector components.

Fig. 6. Testequipment for detector test and calibration.

a deflection current generated by an analogue ramp of which the slope is controlled by a D/A converter running in parallel. This leads to a system with the linearity and stability of a D/A converter, still having a ramp function with equal dwell times for all pixels.
The accuracy is 1 pixel over the total sweep in a large temperature range.

We optimized the bandwidth for minimum read-out noise. The resulting sweep-time appeared to be too short to be compatible with the available PCM bandwidth. We solved this by storing the data of each scanline in a 1024 x 8 bits RAM before putting it on telemetry at a lower rate. The read-out beam in the tube remains switched off during that time.
Figures 4 and 5 are showing some of the detector hardware.
Operating the tube in a slow scan sequential mode has the disadvantage that the optimalization of the tube parameters takes a lot of time. Besides, we suffered quite a while of beam landing problems on the target. Not withstanding this we still aim for a first joined flight with NASA next fall.

4. TESTING HARDWARE

Figure 6 shows the laboratory set up for testing the detector system and for optimalization and calibration of camera tubes.
In the past we always made hard-wired test and check-out equipment for our experiments. In this case we first used software controlled testequipment. We did not use a real computer as that would inevitably require a software specialist, but an easily programmable desk-top calculator (HP9830). We designed several interfaces to this instrument and it appeared to be no problem to simulate telemetry and command subsystems and to do housekeeping monitoring under software control.
The actual television-frame is stored on 9-track tape and batch-processes at the University computer center.

REFERENCES

Kondo, Y., Giuli, R.T., Modisette, J.L. and Rydgren, A.E.: 1972, Astrophy. Journ. 176, 153.

Jager, C. de, Hoekstra, R., Hucht, K.A. van der, Kamperman, T.M., Lamers, H.J., Hammerschlag, A., Werner, W., and Emming, J.G.: 1974, Astroph. and Space Science 26, 207.

Flory, L.E., Pike, W.S., Morgan, J.M. and Boyer, L.A.: 1966, Adv. in electr. and electron phys. 22B, 885.

PROBLEMS OF OVERSAMPLING WITH SEC VIDICON TELEVISION SYSTEMS

E. J. Devinney

University of South Flordia

D. Fischel and D. Klinglesmith

Goddard Space Flight Center

Introduction

Many TV sensors, such as the SEC Vidicon, involve a charge distribution image which is scanned by a reading electron beam, providing the signal. In a TV system which is digitally controlled, the electron beam is stepped, then pulsed on, to provide some form of sampling of the charge image. An important practical question is the nature of this destructive sampling process when the sample intervals become smaller than the beam width, a condition termed oversampling. The electron beam shape is assumed to be Gaussian with standard deviation σ. We will examine the one-dimensional case first, and give an example of the two-dimensional case.

1. Procedure

The one dimensional case is applicable to a single line, or to a raster of lines separated by a distance greater than the beam width (e.g., 3σ for a Gaussian beam).

The first sampling of the charge distribution by the beam affects the charge located directly under the beam centroid, but for all subsequent measures, the effective region of discharge lies ahead of the beam's address. (Figure 1.) For the purpose of being able to define a point-spread function, or an

de Jager/Nieuwenhuijzen (eds.), Image Processing Techniques in Astronomy. 135-139.

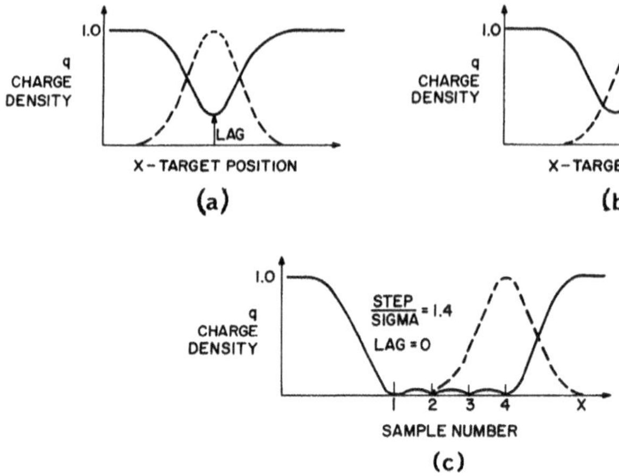

Figure 1. One Dimensional Readout 1a; first position of the beam, solid line shows the charge left after readout. 1b, second position of beam, solid line shows the charge distribution before readout. The dashed line indicated the read beam. 1c, the fourth position of the beam after readout.

optical transfer function (OTF), for this process, let the first measurement at x_0 be written

$$M(x_0) = \int_{-\infty}^{\infty} q_0(x) \exp\,[-(x-x_0)^2/2\sigma^2]dy \qquad 1$$

for the initial charge distribution $q_0(x)$. The remaining charge is

$$q_1(x) = q_0(x)\ (1 - \exp\,[-(x-x_0)^2/2\sigma^2]). \qquad 2$$

The second measure gives, for a step of Δx,

$$M(x + \Delta x) = \int_{-\infty}^{\infty} q_1(x) \exp\,[-(x-x_0-\Delta x)^2/2\sigma^2]dx \qquad 3$$

with remaining charge

$$q_2(x) = q_0(x)(1 - \exp\,[-(x-x_0)^2/2\sigma^2]) \\ (1 - \exp\,[-x-x_0-\Delta x)^2/2\sigma^2]). \qquad 4$$

Let $y = x_0 + n\Delta x$, where n is the sample number. Then, after n samples, the coefficient of q_0 will become

$$P(y-x) = (\exp[-(x-y)^2/2\sigma^2]) \times (1 - \exp[-(x-y)^2/2\sigma^2]) \times \ldots \\ \ldots \times (1 - \exp[-(x-y+n\Delta x)^2/2\sigma^2]). \quad 5$$

For any sample point y we can now write

$$M(y) = \int_{-\infty}^{\infty} P(y-x)\, q_0(x)dx$$

i.e., the process takes the form of a convolution with P being an equivalent non-destructive read beam. P is the point-spread function and its Fourier Transform is the Optical Transfer Function. For the step sizes of interest here, it is sufficient to stop in Equation 5 when n = 3. The shape of the effective convolving beam was computed for a number of ratios of sample interval to standard deviation of the Gaussian beam. What needs to be emphasized here is the shape of the effective non-destructive beam is a function of the sample interval. This is a unique property of destructive readout for non-square beams. Table 1 conveniently summarizes some properties of the beam computed for step sizes of 1, $\sqrt{2}$, 2, $2\sqrt{2}$ and 3 times σ.

Table 1

$\Delta x/\sigma$	$x/\Delta x$	x^2/σ^2	$W/\sqrt{2\pi}\sigma$
1	0.931	0.567	1.048
$\sqrt{2}$	0.283	0.621	0.883
2	0.068	0.603	0.853
$2\sqrt{2}$	0.0072	0.732	0.921
3	0.0043	0.767	0.936

The quantity $x/\Delta x$ is the shift of the centroid of the effective discharge region from the beam's address in units of step size. This would, of course, be zero for a non-destructive read. The quantity $\overline{x^2}/\sigma^2$ is the mean-squared half-width of the beam in units of the variance of the Gaussian (non-destructive) beam. W is the equivalent width, or the width of a rectangle with the same height as the beam.

Not noted in Table 1 is the decreasing peak amplitude, (the peak does not occur at $x = \overline{x}$), nor is there any measure of the skewness of the beam. All this information can be graphically seen in Figure 2 for the two dimensional case, which shows the iso-contours of the point spread function for $\Delta x = \Delta y = 2\sqrt{2}\,\sigma$ on the left. The effects of the previous line can be seen at the

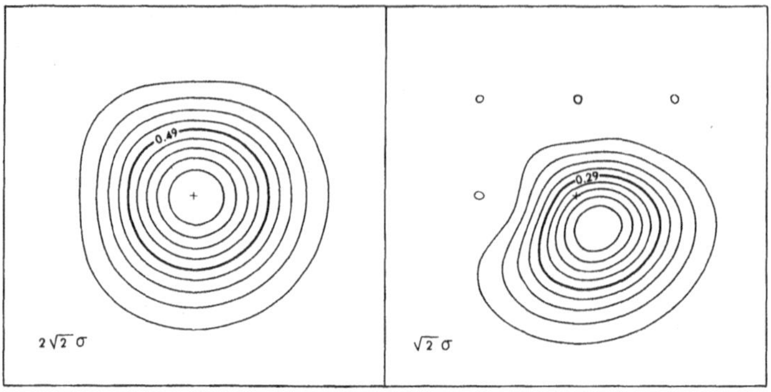

Figure 2. Iso-contours of the beam readout of a uniform field for stepsize equal to $2\sqrt{2}\sigma$ (left) and $\sqrt{2}\sigma$ (right).

top where the contours have been moved down, and likewise for the previous sample in the current line, on the left, where the contours have been shifted to the right. These effects are more pronounced in the small step case in Figure 2, right. The plus (+) sign marks the address point of the beam, showing the shift imposed on the point spread function. It is also clear that the peak amplitude decreases. In the Fourier domain, it is clear that there would be a significant imaginary component in the OTF. For the one dimensional case we have the MTF curves of Figure 3. The angular range of the phase transfer function (PTF) for the selected step sizes is given at top. Also shown are the sample frequencies for the same intervals. Tick marks on the curves mark their equivalent width in frequency space. The arrow on the 1σ curve marks the frequency at which the PTF discontinuously changes sign at 90 degrees. Relative to a non-destructive Gaussian beam, we see that as the sample interval decreases, the low frequency response drops, but the high frequency response is greater than that for a non-destructive Gaussian. One sees that there is some response even out near the sample frequency in each case. Thus, the temptation arises to utilize a format which crams as much onto the detector as possible. One could reduce the scale of the image by a factor of two, for example, and then step twice as fine to recover the same spatial information. However, when one carves up the device into more photometers, they rapidly lose their independence. In any real device, reduced channel capacity will also occur; i.e., a reduction in dynamic range sets in. It should also be noted that in a real SEC camera, the size of the beam grows with distance from the center, so that if one were working on the middle curve of

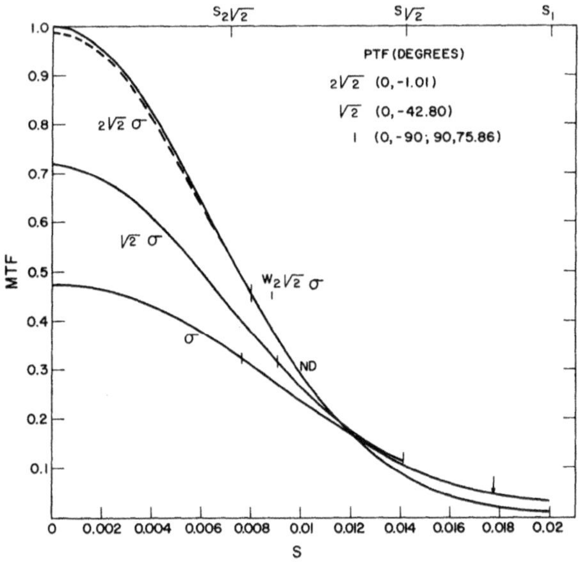

Figure 3. OTF for various stepsize to beam diameter ratio.

Figure 3, in the tube center, then as the beam progressed outward the MTF moves through a series of curves downward in the figure. Thus, real images from an SEC are not isoplanatic, the convolving function depends on position in a major way when oversampling.

To appreciate the sampling situation here it could be compared to the typical sampling conditions usually found, which involve three characteristic lengths: the sampling aperture or data window, a; the sample interval, Δx; and the Nyquist interval $1/(2\ S_n)$, S_n being the Nyquist frequency. Whereas the general conditions for good sampling should be $a \ll \Delta x \ll 1/(2\ S_n)$, here we have $a \sim \Delta x < 1/(2\ S_n)$.

SOME EXPERIMENTS WITH AN SEC IMAGE TUBE FOR HIGH PRECISION MULTI-COLOUR PHOTOMETRY OF GALACTIC CLUSTERS

A. Blecha and P. Bartholdi

Observatoire de Genève , CH-1290 Sauverny , Suisse

ABSTRACT. The possibility of high precision multi-colour stellar photometry with an SEC vidicon camera is studied on the galactic cluster NGC 581. In particular, the best exploitation of calibration information is sought. The correlation length is determined and used for data compression. Several calibration techniques are discussed and the best one is selected and tested. A related problem of zero level is examined. In order to reduce the number of necessary calibration measurements, a separation of the SEC target non linearity and the photocathode quantum efficiency local dependency is carried out. Certain incompatibilities in calibrated data are found. Nevertheless, the magnitude of some stars already measured in the Geneva system is presented, and sources of error are discussed.

1. INTRODUCTION

Although very sensitive television type camera with extraordinary integration capability and practicaly photon noise limitation exists since many years very few use of them for precise photometry of point sources have been reported. The possibilities of such tubes seems particularly interesting for objects of relatively high density of sources, like galactic and globular clusters, where many stars could be measured simultaneously with much better spatial resolution than with single channel photometry. The present work is a first attempt in this promissing direction, and the relatively poor results obtained should be taken as indication for further research and encouragement for those interested in this field, as no fundamental difficulties where encountered.

de Jager/Nieuwenhuijzen (eds.), Image Processing Techniques in Astronomy. 141-155.

2. CAMERA DESCRIPTION AND PARAMETERS.

The camera used is described in detail by Lowrance et al. (1972). Some properties are listed in table 1 of Crane (1973). A schematic view of SEC vidicon tube is given in Deutschman (1972).

3. UNIFORMITY AND LINEARITY.

It is evident that both photocathode and SEC target are not uniform. Further-more, the efficiency of a photocathode will be colour-dependent, and the target response will present some non-linear behaviour, especially near saturation value. In general, it will be impossible to separate the photo-cathode and target contribution to global gain, and so a calibration measure-ment for each colour will be necessary in multicolour photometry.

4. DATA FORMAT.

The SEC camera used is scanned line by line, and a total of 600 lines per picture is recorded analogicaly on magnetic tape. An off-line digitiser produces a 1024 points record for each line. So the picture is divided into 1024 x 600 pixels (Crane 1971).

5. CALIBRATION.

Because of the spatial, colour and intensity dependency of the camera gain, a series of calibration files is necessary for each colour to remove the target non-linearity. It is made for "white colour". The camera is exposed to uniform illumination (Crane 1973) during an increasing period of time (from 30 to 360 seconds). Due to the good integration camera possibilities (more than 8 hours), it is equivalent to illumination intensity variation. Further, two files for G and V1 Geneva photometric system filters (Rufener 1971) are recorded in order to test the validity of the main calibration files for coloured files interpretation. We thus have a signal exposure from 30 to 360 seconds at each point of the camera field. In most points, saturation start gains importance only at 240 to 300 sec exposure, so we have enough points to reconstruct at each point a calibration curve (figure 5.1). Before initiating the calibration data handling, we have to examine data compression possibilities. It is not very convenient to handle continuously 15 files of 10^6 words each.

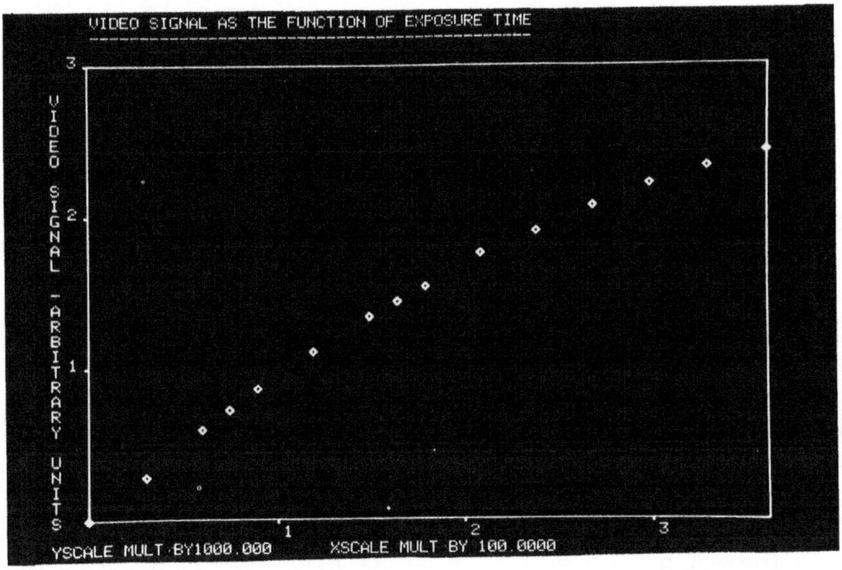

Figure 5.1

6. CORRELATION LENGTH.

Due to the rather arbitrary choice of element size (it was chosen so as to optimise the readout SNR), some correlation between adjacent elements may be introduced during readout of the tube. Let $\delta\overline{N}$ be the rms deviation of data from the mean in a [k x k] square, and $\overline{\delta\overline{N}}$ the mean of this rms deviation in the many [k x k] squares through some fixed region of the data. If there is no correlation, the $\overline{\delta\overline{N}}$ value will be independent of k and this value is about $\sqrt{N}$, where N is the number of photoelectrons per element. Following figure 6.1, we choose a reasonable size of elements.

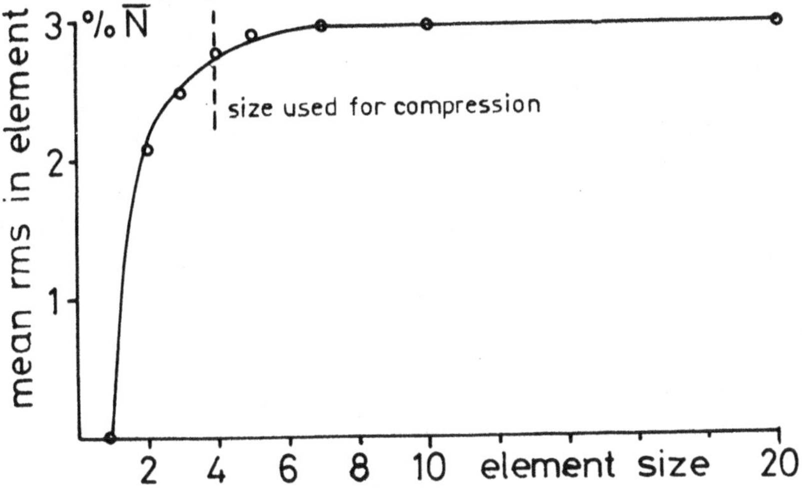

Figure 6.1

k = 4 (4 x 4 primary elements) as a compromise between data losing and data compression. Thus the first step in data handling is a data reduction, replacing an original 1024 x 600 points grid by 256 x 150 points of means over a 4 x 4 original elements region.

7. CALIBRATION PROCEDURE .

We tested three methods : linear, quadratic interpolation, and an all-points parabolic fit. In order to test these three procedures, we used an extra 75 seconds uniform file unused in the calibration procedures. We calculated the rms in the central region of this extra file calibrated by all three procedures. The overall parabolic fit is clearely the best one.

procedure	rms (75 sec)
no calibration	53.13
linear interpolation	28.67
quadratic interpolation	36.18
parabolic fit	24.86

There is also a physical motivation for our choice, in the sense that in the expected dependency of the form

$$S = A\,(1 - \exp\,(B.I)\,) \qquad (7.1)$$

where s is the electrical signal, I the illumination, A and B position dependent coefficients, the exponential can be replaced by the approximation (Abramowitz 1965)

$$\exp(\,x\,) = 1 + A\,x + B\,x^2 + e\,(\,x\,) \qquad (7.2)$$

where $A = -0.9664$, $B = 0.3536$ for $0 < x < \ln 2$ $|e(x)| < 3.10^{-3}$

As B.I is generally within the limits, we fit the data rather on 7.1 by means of parabolic approximation, wich is very fast on small computers.

8. SELF-CONSISTENCY TEST OF CALIBRATION .

The selected calibration procedure has a great advantage in comparison with those using an interpolation procedure because of the facility for the self-consistency tests. Generally 15 points are used for two coefficients' fit, and so each point enters into the fit with only limited weight.

Calibration files after calibration procedure

T_e exposure time T_m mean time

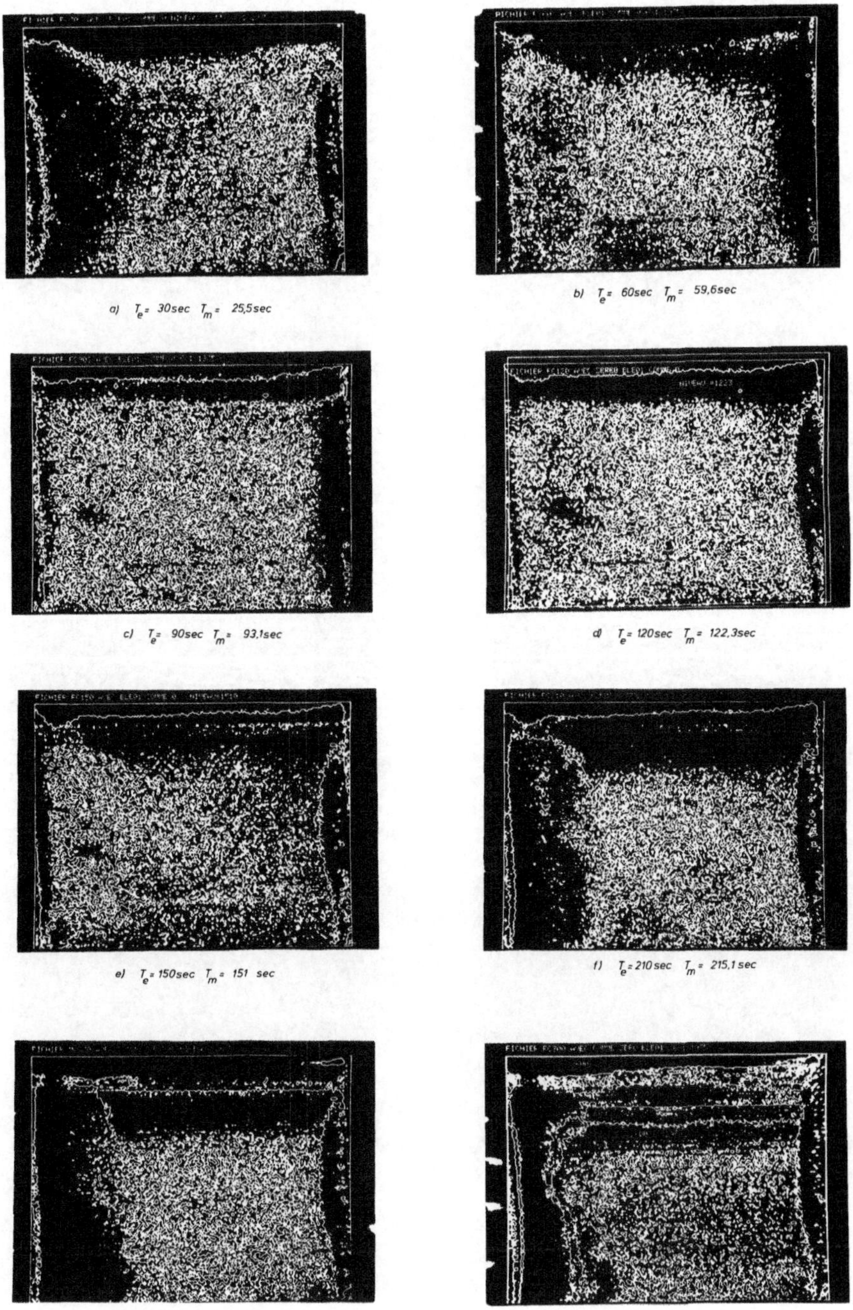

a) T_e = 30sec T_m = 25,5sec

b) T_e = 60sec T_m = 59,6sec

c) T_e = 90sec T_m = 93,1sec

d) T_e = 120sec T_m = 122,3sec

e) T_e = 150sec T_m = 151 sec

f) T_e = 210sec T_m = 215,1sec

g) T_e = 270sec T_m = 268,5sec

h) T_e = 300 sec T_m = 295 sec

Figure 8.1

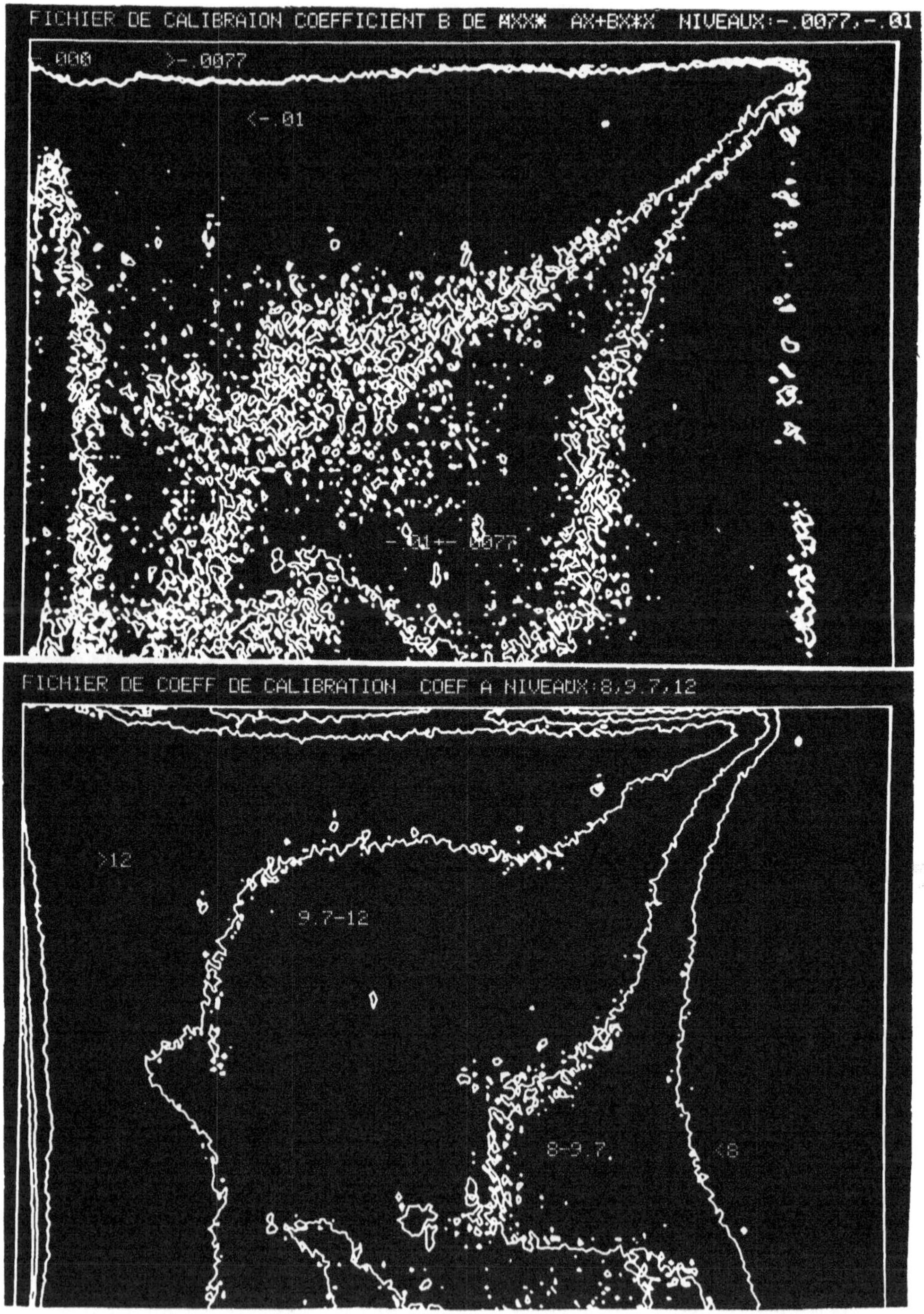
FICHIER DE CALIBRAION COEFFICIENT B DE AXX* AX+BX*X NIVEAUX:-.0077,-.01
-.000 >-.0077
<-.01
-.01+-.0077
FICHIER DE COEFF DE CALIBRATION COEF A NIVEAUX:8,9.7,12
>12
9.7-12
8-9.7
<8

Figure 8.2

Thus we can use the fitted curves to linearise calibration files and examine the efficiency of the procedure. The linearised one is almost filed with the random noise patterns of one isolevel curve only. We used this test systematically on most of the files (fig. 8.1). Starting with the 210 sec exposure we can see local saturation on the upper edge of the picture.

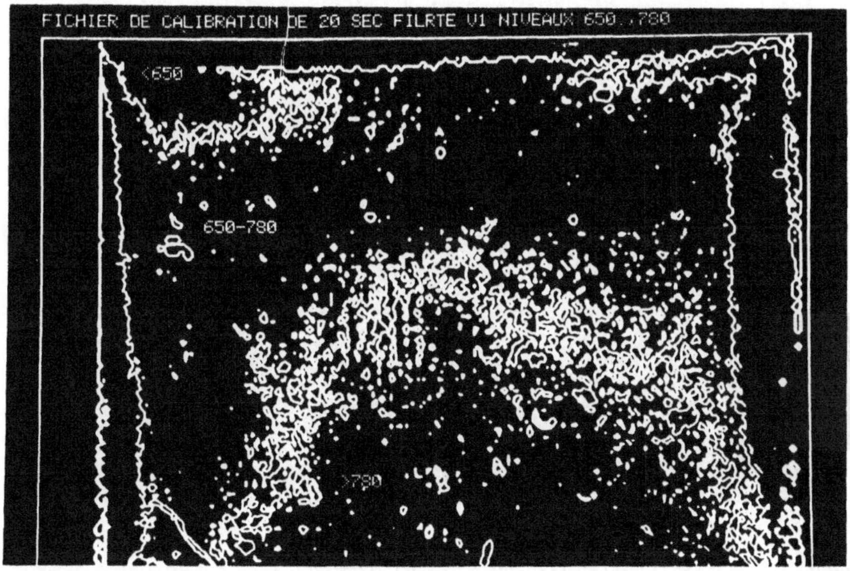

Figure 8.3

We notice that some part of the picture at the boundary cannot be used and that care must be taken to account for the local saturation effect wich starts at the place of the highest camera gain. Figure 8.2 displays the linear and quadratic coefficients of the parabolic approximation. Figure 8.3 shows G and V1 coloured calibration files linearised by means of the same coefficients extracted from the white light calibration files. The linearisation is not completely successful, and so a special procedure must be used for the calibration of coloured measurements.

9. ZERO LEVEL.

Zero level must be subtracted from all files treated (calibration files included). The stability of this zero level used for subtraction was tested.

Comparison of zero level of the calibration and data file.

Level 20, 40, 60, correspond to the 2, 4, and 6 sec exposure.

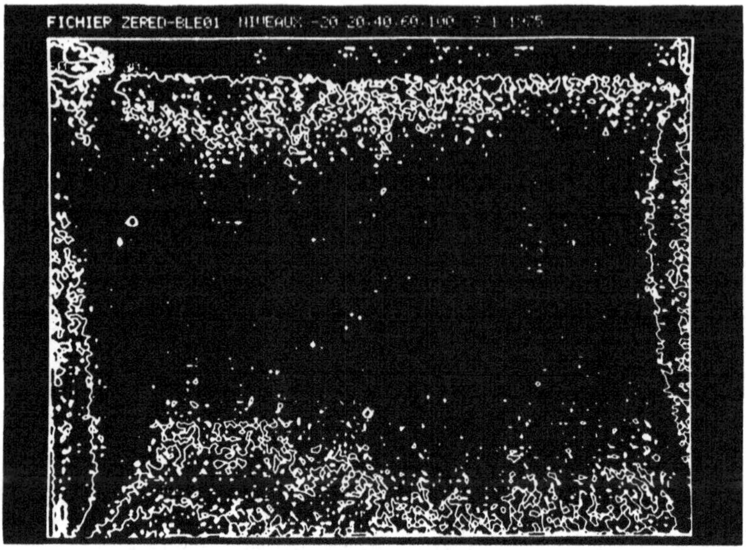

Random noise patterns at mean value 1,6 corresponding to .16 sec exposure.

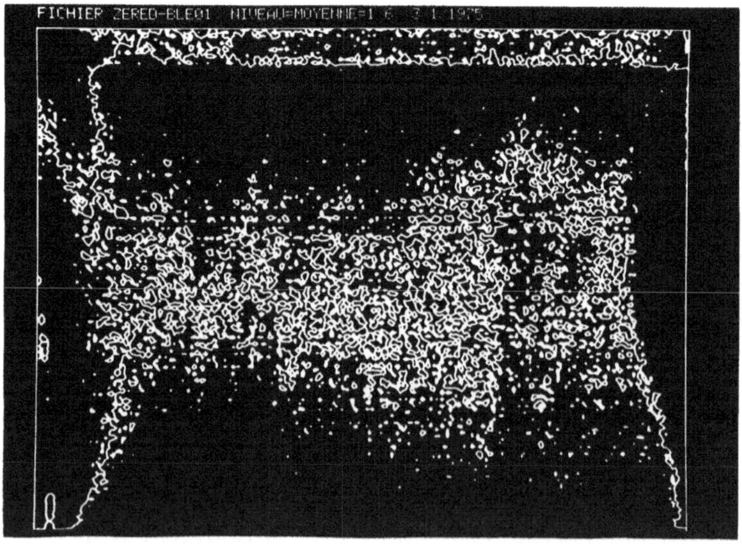

Figure 10.1

This test can serve also as general tube stability criterion if its variation is within some background size limit. Unfortunately, we have to subtract this level from the original points without the possibility of subtracting some mean value, as it varies from point to point. This procedure also increase the noise of the resulting file. Furthermore, two files of zero level made before and after the calibration measurement are slightly different (fig. 10.1). So we subtracted the first one from the calibration files and used the second one as zero level for the measurement files.

10. COLOURED DATA CALIBRATION.

We can express the signal at place x, y as :

$$S_i = T(x, y, Q_i . I) \qquad 10.1$$

where i is a colour indice. Indeed Q_i is also x, y dependent. We have a calibration file for white light, which means that at each point x, y we know the function

$$S = S(x, y, I) \qquad 10.2$$

For any coloured filter c, 10.1 can be expressed as

$$S_c = T(x, y, Q_w . \frac{Q_c}{Q_w} I) \qquad 10.3$$

Replacing $I.Q_c.Q_w^{-1}$ by I', we obtain

$$S_c = T(x, y, Q_w . I') = S(x, y, I')$$

and so

$$f_{cw} = I' . I^{-1}$$

I' is an input flow calculated by means of the "white" calibration data, and I is a known input calibration c coloured flow. As we are interested in I regardless of any common multiplication factor, we can assume I=I, and so f_{cw}=I'. This means that we do not need to repeat all calibration measurement for each colour, and that only one colour calibration file supplies information about photocathode efficiency. In the absence of coloured calibration files we can use the sky background level as constant level calibration file. Unfortunately, as this value is generally only a few times more than a noise rms value, this method can introduce important errors.

11. DATA HANDLING.

Off-line data processing was carried out on the Geneva Observatory HP 2100 computer equipped with 2 Mbytes removal cartridge disks and magnetic tape. Files were disk resident and some 40 interactiv programs were used via the operator console for performing different file treatment following the schematic flow chart 11.1. At this stage, a graphic mode display seems to be a necessary tool for a quick general look at intermediate data.

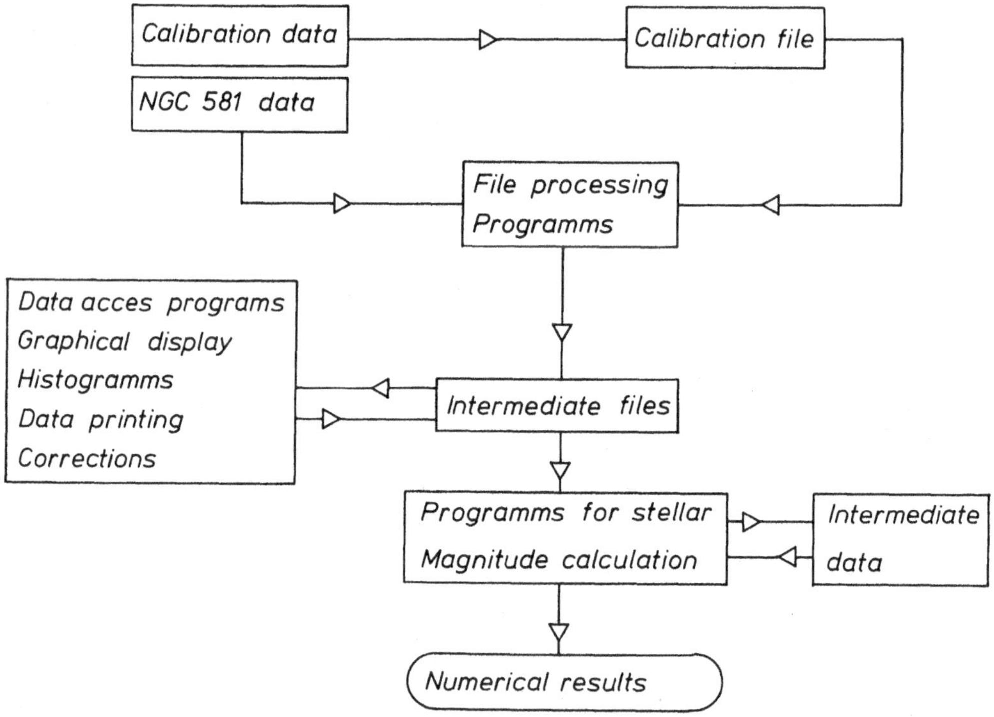

Figure 11.1

12. MAGNITUDE CALCULATION.

As the purpose of this work was to test the possibility of stellar photometry with SEC vidicon tube, the astronomical data collected are those of the galactic cluster NGC 581. As we have colour calibration files for V1 and G filters only, we used the sky background to correct the other colours. A two-dimensional gaussian profile fit and line by line one-dimensional

Gaussian fit on luminosity data of 10 and 11 magnitude stars.

Dotted line joins the data points as the solid line is a twodimensional gaussian model.

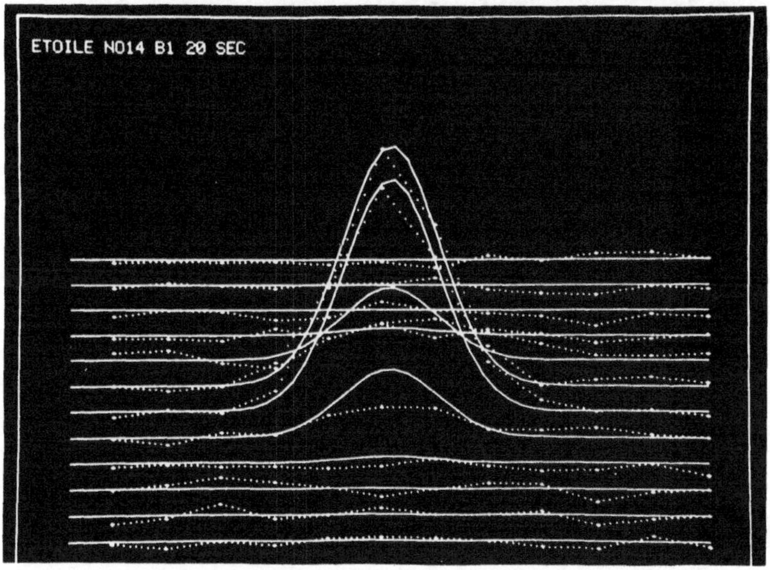

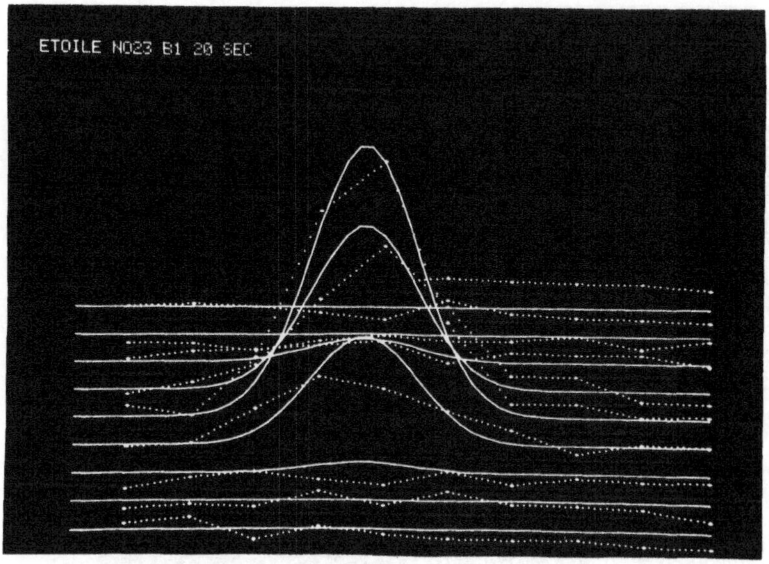

Figure 12.1

LINE BY LINE SCANNING OF THE CALIBRATED FILE

10 SEC EXPOSURE WITH B FILTRE

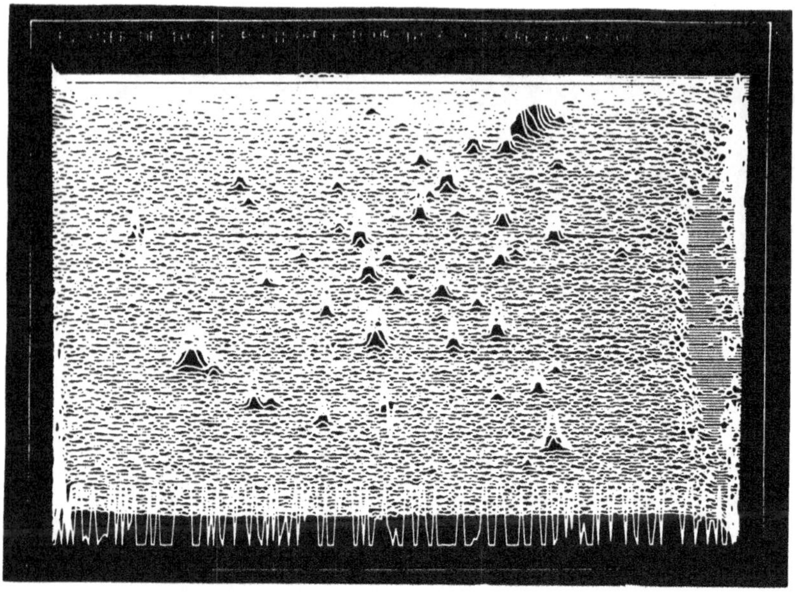

Isolevel contours near to the backround level

20 sec exposure with a V filtre

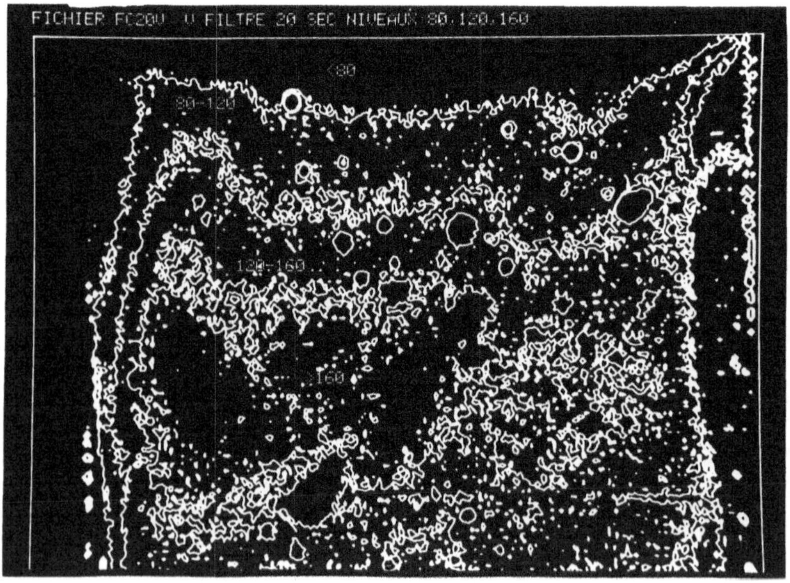

Figure 12.2

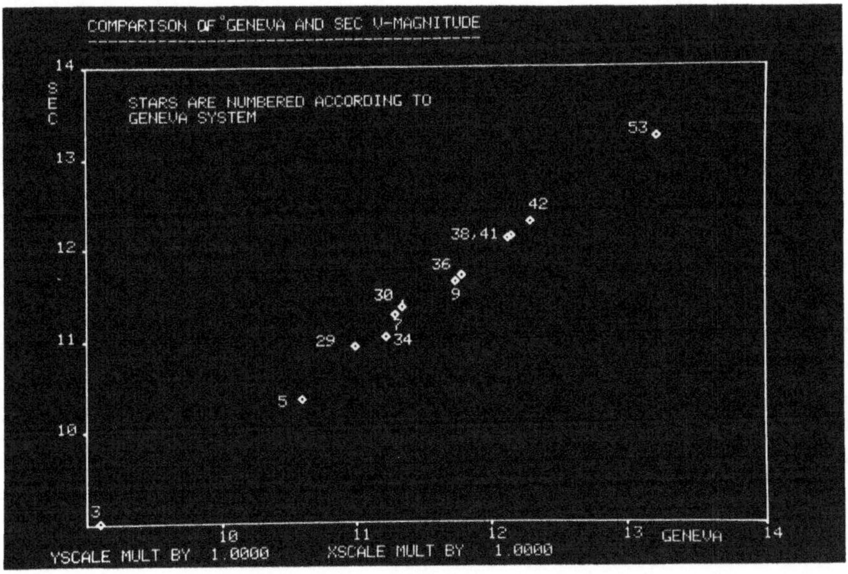

Figure 12.3

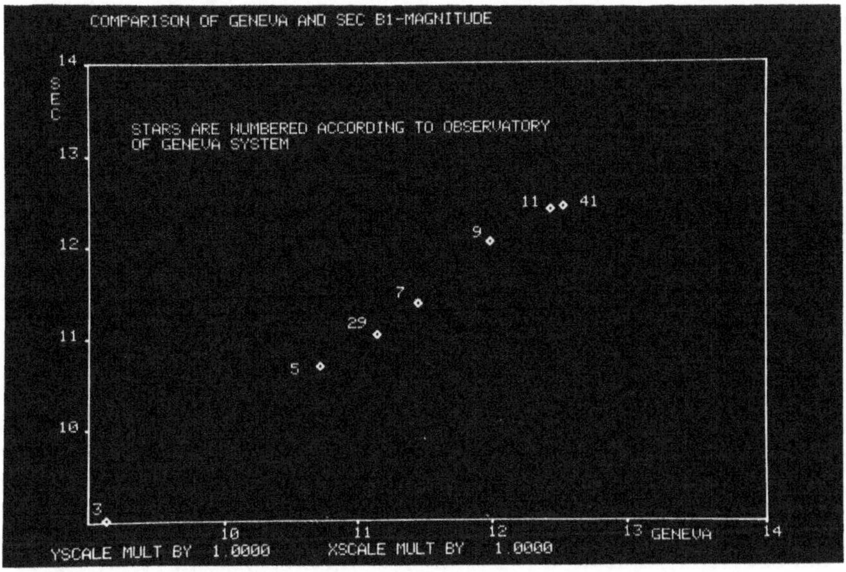

Figure 12.4

gaussian fit are compared with direct summing the flux above the backgroun Finally, the two-dimensional fit was selected as the best. A typical fit of stars of magnitude 9 and 11 is shown in figure 12.1. The solid line represent the calculated profile, while the calibrated data points are connected by a dotted line. Figure 12.2 shows a typical display of the whole calibrated field. Line by line display (a) is convenient for a control of star recognition. On the other hand, an isolevel display (b) is more sensitiv to background gradient. Theoretically, a homogenous background should be expected according to the calibration file in the corresponding colour. Unfortunately, even in the case of the coloured calibration file available, as for V1 filter, a strong local gradient in the background is present. It may be explained only by zero level uncertainty or data inconsistency. Nevertheless, we carried out a series of photometric estimations of stellar magnitudes in NGC 581. We compared them to the value of the Geneva catalogue as given by Rufener (1971). (figures 12.3 and 12.4) If V and B1 measurements can be considered as compatible with the system inherent noise (essentially photon noise), the U magnitude are completely beyond admissible limits. It can be pointed out that the large dispersion of experimental points is due rather to data incoherency than to some lack of information about photocathode efficiency.

13. CONCLUSION.

We must leave unanswered the question wether an SEC camera can be used for high precision multicolour photometry. Nevertheless, we can state that perhaps the most important part of this job lies in calibration measurements. Care must be taken to test stability and reproductibility of calibration measurements. The hypothesis of the separation of photocathode (linear) and target (non-linear) effects should first be verified experimentally. The possibility of the suppression of zero level signal introducing a supplementary noise has to be studied. The handling of important amounts of data necessary in such experiments is possible even on a small processing unit providing there is a random access memory of adequate capacity, as in the case of the present study.

14. ACKNOWLEDGMENT.

We wish to thank Dr John Lowrance, Paul Zucchino and Philippe Crane from Princeton Observatory, who provided the camera, complete support during the observation, and then the data in computer readable format.

REFERENCES .

Abramowitz, A. and Stegun, A. : 1965, Handbook of Mathematical Function, Dover Publ. Inc.
Crane, Ph. : 1971, BAAS, 3 , 399
: 1973, Astronomical Observations with Television-type Sensors, Vancouver.
Deutschman, W.A. : 1972, PASP 84 , 123.
Rufener, F. : 1971, Astron. and Astrophys. Suppl. 3 , 181
and Maeder, A.: 1971, Astron. and Astrophys. Suppl. 4 , 43

P A R T 3

PROCESSING HARDWARE

FOUR YEARS EXPERIENCE IN APMS STAR PLATE PROCESSING - RESULTS AND FUTURE PLANS*

James S. Newcomb

Control Data Corporation

1. INTRODUCTION

In the previous session of this conference interesting and significant advances in techniques of astronomical data acquisition with television, diode array, and electronographic techniques with the possibility of data storage from these sensors directly on computer compatible magnetic tape, core, or disk storage have been presented. This paper is about a measurement system for data that has been stored only on photographic film - the measurement of stellar proper motions. More than eight years of experience in digitizing the data stored on photographic plates has developed in our group at Control Data Corporation a deep and abiding respect for the vast amount of information--wanted and unwanted, good and bad--that is stored on each photographic star plate! The photographic plate, with all of its non-linearity of response, granularity, and noise, is still an outstandingly compact, practical data storage device.

In this paper we will review the operating principles of the measuring and data recording systems. Then the computer operated data reduction system will be described, with emphasis on the image evaluation criteria. The effectiveness of the image reconstruction and evaluation procedures will be presented by comparison of the image validity measurements made by the measuring system with the evaluation of the same images by Dr. Luyten using his own blink microscope. Future plans for measurement of the high stellar density Milky Way star plates will be presented, as well as calibration and qualification measurements for problems other than proper motion.

de Jager/Nieuwenhuijzen (eds.), Image Processing Techniques in Astronomy. 159-170.

2. OVERVIEW OF THE SYSTEM

The purpose of the automation of the stellar proper motion measurements is to discover, evaluate, and measure the proper motion of candidate star images on pairs of photographic star plates of identical areas of the sky. A symbolic diagram of the measuring system is shown in Figure 1. Data collection is accomplished by the 48" Schmidt telescope at the Hale Observatories on Palomar Mountain. The plates, covering 6.5 square degrees of sky and are nominally 355.6 mm square. The images on the star plate are digitized by a scanning microdensitometer using the light from a Helium-Neon gas laser. Control of the microdensitometer and arranging the measurements in tape and computer-compatible form is accomplished by a special purpose computer built by our group for this purpose. The resulting magnetic tapes are placed in the expanded version of the CDC 6600 large scale computer installed at the University of Minnesota for the data reduction and image and motion evaluation. The resulting data on proper motion candidates and their measurements is presented to Dr. Luyten in the form of computer printout. Those motion candidates passing his visual inspection on the blink microscope are published; future plans include entry of both the motion star data and the data for the fixed stars from which the motion was measured into the data bank of the National Aeronautics and Space Administration at Greenbelt, Maryland, U. S. A.

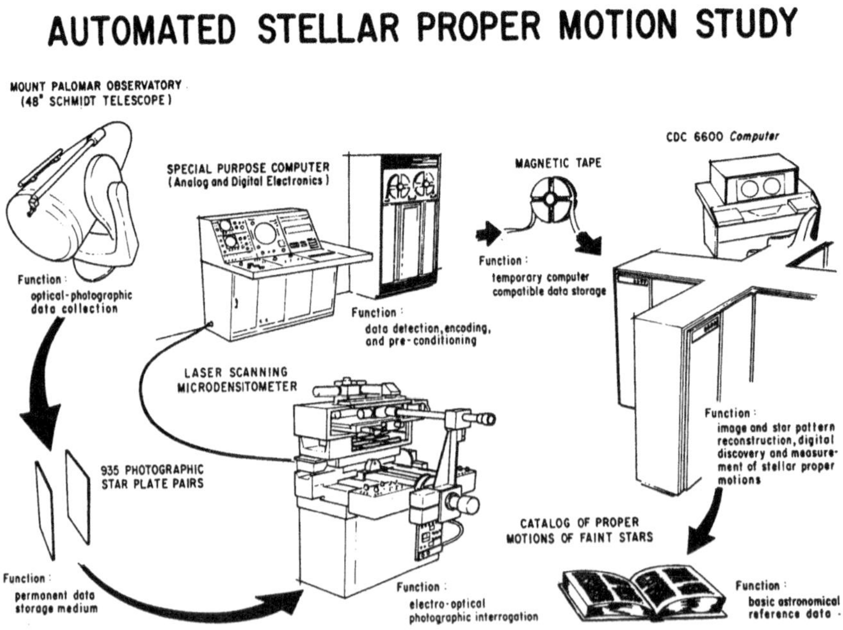

Figure 1.

More detailed descriptions of the measuring systèm and the data reduction scheme have been given earlier; Newcomb (1971) gives a system description and accuracy measurements from the first year of experience in scanning plates in "production" quantities; Luyten (1974) gives a practical survey of the results after three years of measuring in production quantities, and develops a cost figure of about $1 per published motion, and compares this value with the cost of other systems for the measurement of proper motion. The three papers given at the Colloquium No. 7 of the International Astronomical Union held at Control Data Corporation in April, 1970--Luyten (1970), LaBonte (1970), and Newcomb (1970)--give the most detailed description of the hardware and software, as well as preliminary measurement accuracy data.

3. THE SCANNER AND IMAGE RECONSTRUCTION

The pair of star plates are interrogated by rapidly moving spot scanning each plate and a reference reticle simultaneously. The spot moves 12 mm in the y direction as the star plates are moved under the spot. The motion of the spot is coordinated with the x motion of the carriage so that each scan is 5 microns from the previous scan. The orientation of the pair of plates in relationship to the scan is shown in Figure 2. The presence

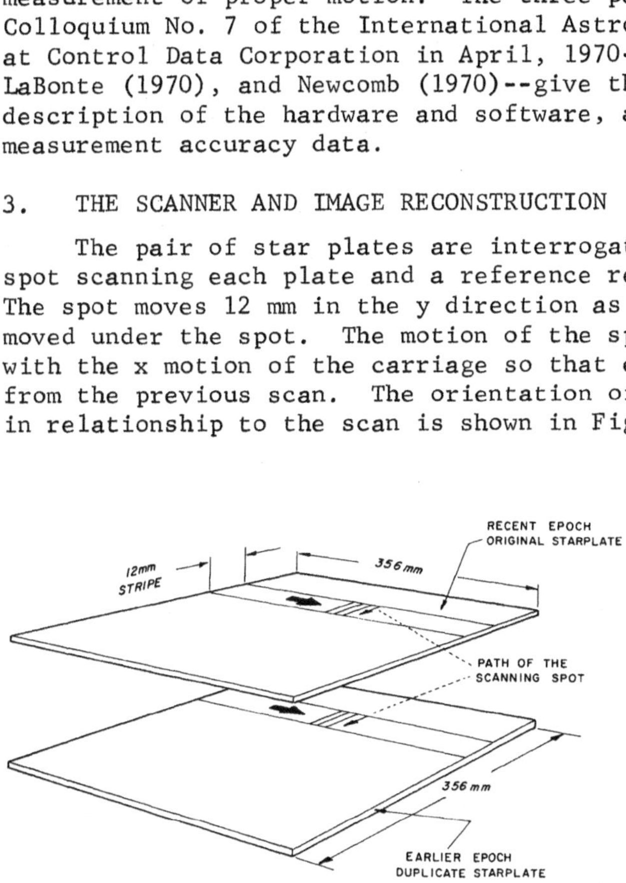

Figure 2. Schematic of the Scanning "Stripe" and the Star Plates

of a star image is detected by the decrease of light collected by the collection optics-photomultiplier system on the opposite side of the star plate from the scanning beam. The decrease of light intensity at which the scanner detects an image, or the "image threshold" is manually set for each plate based on a trial run of a small region of the plate. The setting of the threshold is a very important part of the measurement procedure; if the threshold is set too high, faint star images may be missed; if it is set too low, the proportion of unwanted images to star images make the processing time (and cost) per faint star image too high. The method of digitizing the images can be seen with the aid of Figures 3 and 4. In Figure 3, the path of the spot is shown traversing two images; in Figure 4, the information that is sent to the magnetic tape at each of the image boundaries is shown adjacent to the point of the scan at which the information is generated.

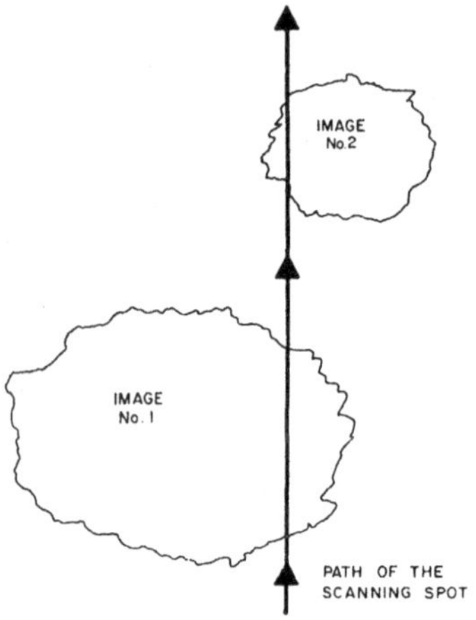

Figure 3. A Single Scan of Two Images

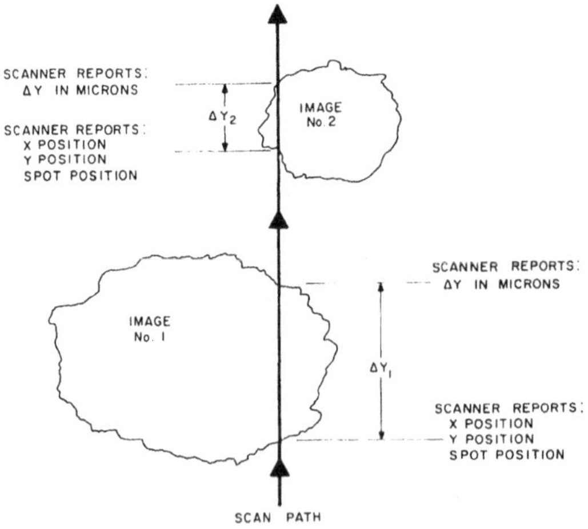

Figure 4. Information Transfers Triggered by a Scan Spot Encountering a Star Image

The first task of data reduction by the computer is the reconstruction of the image from the digitized data. Reconstruction of the two sample images shown in Figures 3 and 4 is shown as the solid lines in Figure 5; the dotted lines of the actual images are shown around the scanned lines for comparison.

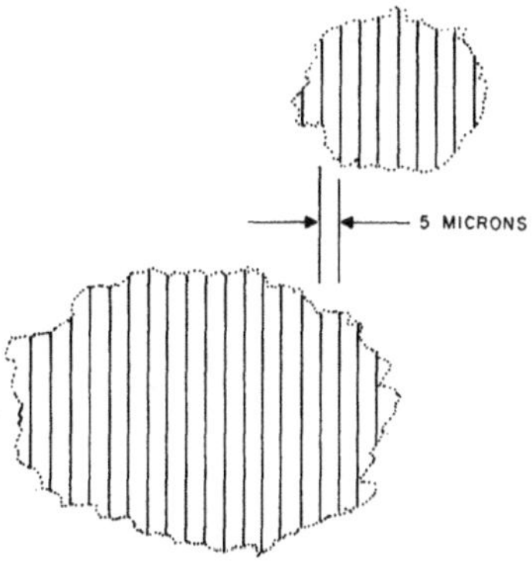

Figure 5. Reconstruction of the Images of Figures 3 and 4 by the Data Reduction Program

4. IMAGE EVALUATION TECHNIQUES

During and after the reconstruction process the assembled images are tested to show whether or not they possess the requisite shape to qualify as a star image. An image is rejected as a star image if:

(1) It consists of less than four or five transits.

(2) The longest transit is not approximately equal to the total number of transits times the scan spacing. This criterion is especially useful in rejecting scratches on the emulsion.

(3) The longest transit is longer than 200 microns. This indicates that the image size is outside the chosen range of 20 to 200 micron image diameters. Images of 200 microns and larger are m_{red} = 12 or less, and have already been documented; images smaller than 20 microns consist of relatively few photographic grains; this effectively accentuates their non-circularity and makes them difficult to distinguish from non-star images.

Reconstructed images passing the above tests are next examined to establish how closely they approach the ideal isolated round star image. This closeness of approach is expressed as a "quality index" and is computed so that the smaller the quality index, the better the image. Reconstructed images having excessively large quality indices for either or both are immediately rejected.

The horizontal symmetry quality index is based on the variation of the center of each transit about the average value of the transit center; the larger the deviation, the less circular the image is, and the farther its shape deviates from the ideal circular image. The horizontal symmetry for the larger of the two sample star images is shown in Figure 6.

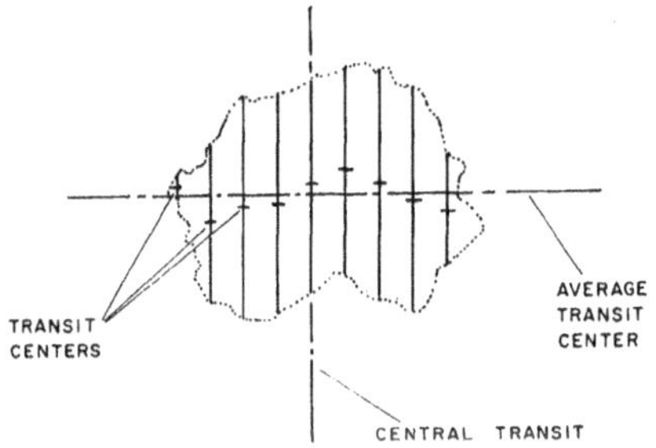

Figure 6. The First Quality Index - Horizontal Symmetry

The circular symmetry index is essentially a logical extension of the horizontal symmetry test. The center of the image is considered to be the center of a circle whose circumference is the best possible fit to all of the transit end points. The circular symmetry quality index is based on the deviation of the circular center from the center of the average transit computed above for the horizontal symmetry.

In addition to the quality indices based on the relationship between the actual image an ideal circular image, comparison between the size of the image on the recent (original) star plate and the approximately corresponding image on the older (duplicate) plate has been found to be a significant measure of whether the two images are or are not formed from the same star.

The discovery of motion star candidates and matching of "fixed" stars is done in small (12 by 12 mm) regions; the resulting transform expression for each region is compared with the transforms of the surrounding region. The transform quality index indicates the magnitude of the deviation of the transform in the region from which the star image originated.

Experience with large candidate motions has indicated:

(1) If it is a valid motion, it is worthy of detailed examination.

(2) The possibility of image mismatch increases rapidly with the size of the motion.

For these reasons, a quality index associated with large candidate motions is used to call attention to them.

The regions in the immediate vicinity of the image of a bright star are characterized by an apparent increase in the fog level which contaminates the images of faint stars; a quality index noting the proximity of an image to a bright star image indicates the possible cause of non-circularity of the image.

A given image can be matched to more than one other image as a motion candidate; the multiple match quality index expresses that a condition.

These quality indices represent numerical methods of expressing judgements on the validity of the images and candidate motions that reflect the judgements of the visual images by Dr. Luyten as he has observed images at the blink microscope. The final quality index presented with the computer printout is a linear sum of the individual quality indices weighted by factors reflecting four years' experience in production processing of over 600 pairs of star plates, a quarter of a million measured proper motions, and 170,000,000 measured fixed stars.

The final criterion of the validity of the quality index is the extent to which the motions having acceptable quality indices are accepted by Dr. Luyten and the motions of doubtful validity are rejected under visual inspection. That the quality indices are indeed valid criteria for evaluation of candidate motions can be seen in Table I, which summarizes the results of detailed visual examination of five pairs of star plates for motion candidates having motions greater than 0".09 and less than 0".18 annual motion.

An additional example brought to my attention by Dr. Luyten, Plate 558 with center at 13:36 R. A. and +6° Dec. was recently (March, 1975) visually checked on the blink microscope; the measuring system indicated 1979 motions of quality index of 24 or less; of these candidate motions, 1953 motions were accepted and only 26 rejected, or an acceptance of 98.7% of the motions.

TABLE I

Candidate Motion Evaluation for Five Pairs of Star Plates in the Range of 0".09 to 0".18 Annual Motion

Evaluation	Quality Index	Number of Motions with this Index Range	Number of Motions Rejected	Percent of Total Rejected
Acceptable	$Q < 24$	1514	27	1.78%
Doubtful	$24 < Q < 30$	527	235	44.5%
Unacceptable	$Q > 30$	62	50	80.6%

Thus the function of the automation--doing the measuring, and calling to the attention of the astronomer acceptable, doubtful, and borderline motion candidates--has been successfully implemented, and the effectiveness of the astronomer has been increased by requiring of him judgement of the difficult, non-routine motion candidates, and wasting little of his time on routine and obvious either acceptable or completely unacceptable motion candidates.

5. RESULTS OF THE MEASUREMENTS

The basic aim of the stellar proper motion survey of the National Geographic - Mount Palomar plates compared to the later epoch plates taken principally by Dr. Luyten was to provide a statistically significant body of measurements of faint stars. The measurements of the faint stars have necessarily been of two classes: faint stars whose positions during the 13 year epoch difference between plates did not change measurably - the "fixed" stars, and the faint stars having proper motions with respect to the "fixed" stars. Of the proper motion stars either hand-blinked by Dr. Luyten or measured by the automated measuring system and verified by visual inspection by Dr. Luyten, some 30,949 motions have been published. Most of the listings are part of a series published by the University of Minnesota, Minneapolis, Minnesota under the title "Proper Motion Survey with the 48-inch Schmidt Telescope". The listings under this title are shown below in Table II.

In addition to the publications in this series, Dr. Luyten has written a monograph entitled "The South Galactic Pole", also published by the University of Minnesota (April, 1973) containing data for 6878 stars in an area of 2925 square degrees around the South Galactic Pole.

TABLE II

Listings of Proper Motions in the Series Entitled "Proper Motion Survey with the 48-Inch Schmidt Telescope"

Series Number	Title	Date
XXVIII	Faint Proper Motion Stars near the South Galactic Pole	December, 1971
XXX	Proper Motions for 1357 Faint Stars	March, 1972
XXXI	Proper Motions for 2520 Faint Stars	June, 1972
XXXII	Faint Proper Motion Stars near the South Galactic Pole	September, 1972
XXXIII	Proper Motions for 3478 Faint Stars	November, 1972
XXXIV	Proper Motions for 4452 Faint Stars	July, 1973
XXXVI	Proper Motions for 6955 Faint Stars	February, 1974
XXXVII	Proper Motions for 4483 Faint Stars	July, 1974

In addition to listings of for their proper motions, Dr. Luyten has prepared publications concentrating particularly in stars of low luminosity, using measurements of the automated system, verified by visual inspection, and his own hand blinked measurements. The earliest listing which included measurements by the automated system was "The Stars of Low Luminosity" published by the University of Minnesota in March, 1970. A more recent listing appears in the Proceedings of the National Academy of Science 71: 4813, 1974, and lists 122 stars of less than 1/10,000 the bolometric luminosity of the sun. Of the stars listed, 119 were found by the automated system and verified by Dr. Luyten using the blink microscope.

The publications above present only data for proper motion; data on the fixed stars used to discover and measure the proper motions is considered to be a separate problem in the information retrieval of an even larger body of data than the proper motion data - numbers of fixed stars in each star plate pair in 10 micron wide classifications from 40 to 200 microns diameter. Present plans are for the deposition of these measurements in the computer data bank of the National Aeronautical and Space Administration in Greenbelt, Maryland sometime after January 1, 1976. Requests for this information should be addressed to:

Dr. Jaylee Mead, Code 671
Goddard Space Flight Center
National Aeronautics and Space Administration
Greenbelt, Maryland 20771 U.S.A.

The present state of the scanning of the pairs of plates for the proper motion survey is that all of the star plates outside the Milky Way have been scanned; only the dense plates are left. These plates represent a challenge in that one of the very important properties used in image evaluation of the other plates is necessarily missing: the concept of the ideal star image as a round, isolated object. In the Milky Way region plates, it is a very common situation to find that the spaces between star images are smaller than the star images themselves. Thus, much of the image sifting and quality indices concepts will have to be revised to accept non-circular images as star images. A fringe benefit from successful implementation of this concept will be that close double stars can then be detected on any star plate; the bulging shape of a close double star has hitherto been automatically rejected by the system. An additional improvement that Milky Way star plates will impose on the software will be modification of all software to improve efficiency of execution, since processing time will be increased markedly for these plates.

6. PREPARATIONS FOR THE FUTURE USE OF THE SYSTEM

The future of the scanning system depends very markedly on use of the system by other astronomers and a cordial invitation is extended to all astronomers to consider the capabilities of the system as a general purpose measuring system for the digitizing of photographic plates.

In general, the system is capable of digitizing any photographic emulsion at any chosen single density level without any changes in the existing hardware. Digitization at multiple density levels could be accomplished by multiple scans at different thresholds or by changing the hardware to give multiple density level information in a single scan. For single level scans, then the data acquisition could be accomplished without changes in hardware. The software presently in use is Fortran IV, with all of the inherent flexibility available.

Recent inquiries about the measurement capabilities of the system have enabled us to measure the capability of the system for evaluating cm type distances between two points on the same plate, in contrast to the present requirements which are micron distances between two images on two different plates. The National Oceanic and Atmospheric Administration presented us with a calibrated glass plate with 100 micron diameter dots on 5 cm centers covering a 20 by 20 cm square. The results of measurements of the location of the dot centers when the grid was placed in each of the four possible orientations in the scanner showed that a scale difference of less than one micron per 10^6 microns between axes and a non-orthogonality skew of less than 0.2 seconds of arc. The internal precision of the multi-orientation measurements of the grid showed an unbiased estimate of 1.6 microns one sigma for circular error. In addition, the APMS system measurements exhibit a one sigma variation of about 1.2 microns in each axis compared to an independent manual comparator measurement of the same plate made by NOAA. These results indicate that the relatively long distance measurements made between points on the same plate show the same accuracy that the short inter-plate measurements for proper motion have been indicating all along.

In addition, measurements of a small area of a star plate taken by the Schmidt telescope of the European Southern Observatory have recently been completed. Measurements of 292 star images seen in multiple orientations of the plate in the scanner in the diameter range of 40 to 200 microns have shown a one sigma variation in the unbiased estimate of position to be 0.97 microns in each axis in plate coordinates and a one sigma variation of 0.88 microns in the y axis and 1.06 microns in the x axis in scanner coordinates. The diameter measurements showed a one sigma variation of 1.32 microns.

7. CONCLUSIONS

The measuring system developed for the measurement of proper motions from the plates taken at the 48 inch Schmidt telescope has shown that the automation of the discovery and measurement of proper motion can speed up the documentation of a statistically significant body of astronomical data. The time has come to offer to the rest of the astronomical community the opportunity to discover, measure, and document other significant bodies of astronomical data; Control Data Corporation welcomes this opportunity to serve astronomy, and looks forward to meeting the challenges of their data measurement and reduction problems.

REFERENCES

1. LaBonte, A. E., "Automated Proper Survey: Data Reduction and Initial System Performance," Proceedings of the International Astronomical Union Colloquium No. 7, University of Minnesota, Minneapolis, MN, December, 1970, pp. 26-47.

2. Luyten, W. J., "The Performance of the Automated - Computerized Plate Scanner and Measuring Machine," Proceedings of the International Astronomical Union Colloquium No. 7, University of Minnesota, Minneapolis, MN, December, 1970, pp. 48-52.

3. Luyten, W. J., "Stellar Motion Survey by Automation," SCIENCE, 26 July, 1974, pp. 351-352.

4. Newcomb, J. S., "The Automation of Stellar Proper Motion Measurement - Method and Measurement Accuracies," Mittelungen der Astronomischen Gesellschaft, Nr. 30, (1971), pp. 53-65.

5. Newcomb, J. S., "The Luyten - Control Data Stellar Proper Motion Measuring Machine," Proceedings of the International Astronomical Union Colloquium No. 7, University of Minnesota, Minneapolis, MN, December, 1970, pp. 5-25.

*This work was supported by the National Aeronautics and Space Administration through the University of Minnesota; prime contract NSR 24-005-062.

THE USE OF THE "GALAXY" MACHINE AT THE ROYAL GREENWICH OBSERVATORY

C.A. Murray and W. Nicholson

Royal Greenwich Observatory, Herstmonceux

ABSTRACT. A GALAXY plate measuring machine, similar to that constructed for the Royal Observatory, Edinburgh, has been in use at the Royal Greenwich Observatory, Herstmonceux, since late in 1972. The design and performance of the Edinburgh machine have already been reported; in this paper the current operational procedures at Herstmonceux, including minor modifications to the original design of plateholders, and also the planned enhancement of the system by installation of a NOVA 2/10 Control Computer, are described. An account is given of the main computer programmes which have been developed for data processing, and analysis, and some aspects of the performance of GALAXY in the measurement of astronomical photographs are discussed.

1. INTRODUCTION

The basic concept of the GALAXY measuring system has been fully described by Fellgett (1970). The first such machine was constructed for the Royal Observatory Edinburgh in 1966-69, under the general scientific supervision of Dr. V.C. Reddish. The development and design of the Edinburgh GALAXY has been described by Walker (1971), and an assessment of its performance has been published by Pratt (1971).

In 1969, a contract was placed with Messrs. Paul Coradi Scotland Ltd. for the construction of a second GALAXY machine, for the Royal Greenwich Observatory. As in the case of the first machine, the main engineering work was carried out by Sogenique (Services) Ltd.

de Jager/Nieuwenhuijzen (eds.), Image Processing Techniques in Astronomy. 171-184.

The R.G.O. GALAXY, which was installed in the late summer of 1972, is operated as a national facility for the benefit of U.K. astronomers. Up to the present time, the major use of the machine has been on astrometric programmes, but measurements for several photometric investigations have been undertaken.

GALAXY was originally conceived as a solution to the so-called "Schmidt Problem" (Fellgett 1970) and the basic design has been conditioned to the measurement of Schmidt plates in which the number of stars per unit area is high.

The digitization of the information on a plate is carried out in two stages. First, a search is made, to find the approximate coordinates of all images above a set threshold. These are recorded on an intermediate storage medium, currently paper tape (the "search tape") and are then fed into a tape reader, causing the plate carriage to be driven to each successive address for the accurate centring and measurement of the image diameter. Because of random variations in the positioning of different exposures on the same field, on different plates, it is necessary to have a separate search tape for every plate. The search and measurement operations, which are carried out by two separate scanning systems have been described by Walker (1971).

The full exploitation of the measurements made on a machine such as GALAXY requires an efficient system of computer programmes to handle and analyse the raw data. For example, almost all astronomical investigations involving direct photographs require more than one plate on each field to be measured. It may be desirable to search more than one plate in such a series, in which case the measurements derived from different search tapes will not be in the same order; thus a basic problem is the sorting and collation of measurements made on different plates.

At R.G.O. most of the plates which have been measured have been taken on long focus telescopes, for which, unlike Schmidt plates, the star density is low and the search operation correspondingly less productive. It has been found necessary to develop methods whereby a synthetic "master" search tape is made, entirely off-line from a catalogue or from manual measurements made on another machine, for short measuring lists, or alternatively from a GALAXY search on one or more plates on a field. In all cases a coordinate transformation is required in order to match the search coordinates to the actual position of the star images in GALAXY; this transformation is carried out by means of manual measurement on GALAXY, of three stars which can also be identified on the master search tape.

Most problems also require the collation of external data, such as standard catalogue numbers, photo-electric magnitudes or absolute positions, with the GALAXY measurements of the corresponding stars. Since the only data which are directly output from GALAXY are coordinates and image diameters, stars can only be identified by these parameters, and further auxiliary measurements and computer runs are required in order to insert the external data.

During the past three years a set of computer programmes has been developed at R.G.O. for editing and examining the raw measurements, and for the various auxiliary data processing functions outlined above. These will be described in sections 4 and 5.

Many of the operations are elementary, but have required the use of the main computer. At the present time the GALAXY system at R.G.O. is being enhanced by the installation of a NOVA 2/10 Control Computer System which will enable much of the auxiliary monitoring of measurements and transformation of coordinates to be done in real time during measurement, thus increasing the efficiency of the system. A brief outline of the plan of operation using the NOVA system is given in section 7.

2. SEARCH PROCEDURE

Searching is carried out in multiples of a rectangular unit area which has dimensions $2^{13}\mu$ m in X and $2^{12}\mu$ m in Y. Each unit area is subdivided into unit squares equal in size to the diameter of the scanning spot, $2^{n}\mu$m, where n is set to 3, 4 or 5, according to the size of the faintest images to be detected. The area is scanned linearly in the X direction, and after each line scan the plate is stepped on in Y by one spot diameter; thus there are 2^{12-n} line steps per unit area. Since the search time is dominated by this step time, which amounts to $0^{s}.263$/line, it is important not to set n smaller than necessary. For example, the effective area of a 16 cm x 16 cm plate which can be scanned is about 200 cm^2, or 600 unit areas; this takes between 5 and 6 hours with the largest spot, n = 5, and proportionally longer with the smaller spots.

The number of images detected in a search run depends critically on the threshold set, and on the background fog level on the plate. There is a natural tendency to try to extract the maximum information from a search by setting a low threshold, but this can lead to an unacceptably large number of spurious images. The usual procedure is to choose a sample area of the plate to be searched and to test the detection rate for different threshold settings, knowing, from visual examination, how many real images should be detected. Even this process can produce an excessive number of spurious images if there is a slight increase in background fog over that in the sample area.

One naturally chooses the most suitable plates for a search, the criteria being the length of effective exposure and the uniformity of background. However a particular problem which we have encountered at R.G.O. has been the need to search old plates which have markedly variable background. By using a master search tape only on a good modern plate, in a proper motion investigation there is a real risk that stars with large proper motion, which could be the most interesting, will be missed in the measurement of the old plates.

3. MEASUREMENT PROCEDURE

In the measurement phase, an annular area of the plate is scanned in a circular pattern with radius decreasing from a maximum R_1, in uniform radial "increments" of $\Delta r = 2^{-10} R_1$, either to zero or to a minimum radius $R_2 = 2^{-6} R_1$. Where possible, the latter alternative is chosen, in order to prolong the life of the cathode ray tube. The maximum radius of the scan on the c.r.t. is 30.72 mm and the plate is projected on to it by means of a microscope objective; three optical magnifications are currently available and are chosen according to the range of image sizes which are to be measured. For the three magnifications we have the following maximum image radii, and radial increments:-

Magnification	R_1 (μm)	Δr (μm)
X 30	1024	1
X 75	409.6	0.4
X240	128	0.125

The GALAXY measurement procedure is based on the simultaneous statistical optimisation of estimates of the coordinates of the centre of the image and of a magnitude parameter, in the presence of grain noise which is assumed to be Gaussian.

Let r, θ be polar coordinates of a point relative to the centre of the scan and let $T(r,\theta)$ be a measure of the transmission through the plate at this point, on a scale from 0 to 1. If $P(r,M)$ is the true radial profile of the image, assumed circularly symmetrical, corresponding to a magnitude parameter M, it can be shown (e.g. Fellgett 1970) that the rigorous equations derived from principle of maximum likelihood on the hypothesis that the centre of the image is at r = 0 and that the magnitude parameter is M, are:-

$$I_1 = \int_{R_1}^{R_2} r \frac{\partial P}{\partial r} \int_0^{2\pi} T \cos\theta \, d\theta \, dr = 0$$

$$I_2 = \int_{R_1}^{R_2} r \frac{\partial P}{\partial r} \int_0^{2\pi} T \sin\theta \, d\theta \, dr = 0 \qquad (1)$$

$$I_3 = \int_{R_1}^{R_2} r \frac{\partial P}{\partial M} \int_0^{2\pi} (T-P) \, d\theta \, dr = 0$$

In practice the function P is not known and it is necessary to introduce a simulated profile P'. In the case of GALAXY this is assumed to be linear with constant slope C^{-1}, say. Thus, expressing r in units of the radial increment Δr, $R_1 = 2^{10}$ and, for the profile corresponding to M, we have:-

	P'	$\frac{\partial P'}{\partial r}$	$\frac{\partial P'}{\partial M}$
$2^{10} \geqslant r \geqslant M$	1	0	0
$M \geqslant r \geqslant M-C$	$1+(r-M)C^{-1}$	C^{-1}	$-C^{-1}$
$M-C \geqslant r \geqslant 0$	0	0	0

The integrands in (1) are zero for $r > M$ and $r < M-C$. Further, since the star profile is only approximately linear over part of the radius, the range of integration is reduced by two parameters D_1, D_2 to the range $M-D_1 \geqslant r \geqslant M-D_2$ for $M \geqslant D_2 + R_2$ (saturated images) and $M-D_1 \geqslant r \geqslant R_2$ for $M < D_2 + R_2$. Provision is made for changing the profile function at $M = D_2 + R_2$ to one with constant radius and decreasing slope, as described by Walker (1971) but this system has not been used at R.G.O.

The solution of equations (1) proceeds iteratively, starting with a preset value $M = M_1$ and the initial integrations run from $M_1 - D_1$ to R_2. M is changed, first in steps of 2^4 until I_3 changes sign, and then in steps of 1 until, after a short time lag a second sign reversal occurs; at this stage a solution is assumed to have been obtained, and the X,Y coordinates of the plate carriage (to 1μm) and the value of M are punched on to paper tape. At each step in the solution the image is re-centred on the scan according to error signals generated by I_1, I_2.

If an image is too large for the scan or smaller than some minimum M which depends on the setting of the high voltage supply, or if, in attempting a measurement, the plate is driven by more than $2^8\mu m$ from the search coordinates, a "reject" code is punched on to the output paper tape.

The speed of measurement depends on the value set for M and on the distance between images; it can be as fast as fifteen images per minute.

Binary switches enable the operator to set the following parameters before the start of a measuring run:-

$$2^{-4} M_1 \leqslant 2^6 - 1$$
$$D_1 \leqslant 2^{10} - 1$$
$$D_2 \leqslant 2^{10} - 1$$
$$S \leqslant 2^{11} - 1$$

where S is the slope parameter which is related to C by

$$SC = 2^{20-n}$$

and n can be set to an integer from 0 to 4

The image profile T and the cumulative integral

$$I_{3,r} = \int_{M-D_1}^{r} r \int_{0}^{2\pi} (T-P')\, d\theta\, dr$$

are displayed on an oscilloscope, as shown in Fig.1. Before starting measurement, the operator selects the profile parameters empirically from examination of the shape of $I_{3,r}$ on typical images, the aim being to set it as straight as possible.

The geometry of the simulated profile function and associated parameters corresponding to a saturated image are shown in Fig.2. Clearly we must have $D_2 \leqslant C$. Also since the integrations run from $M-D_1$ to R_2 on small images we must have $D_1 \leqslant M-R_2$; thus in practice, if very small images are likely to be encountered, as will be inevitable in the case of a GALAXY produced search tape, we can never set D_1 larger than about 2^6. This means that, even in crowded fields, it is not possible to exclude much of the outer wing of an image.

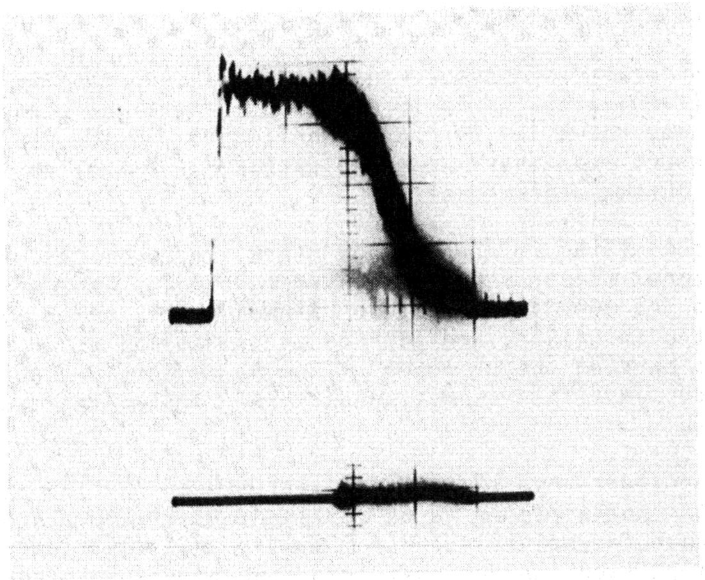

Fig. 1

(a) Transmission T (b) $I_{3,r}$

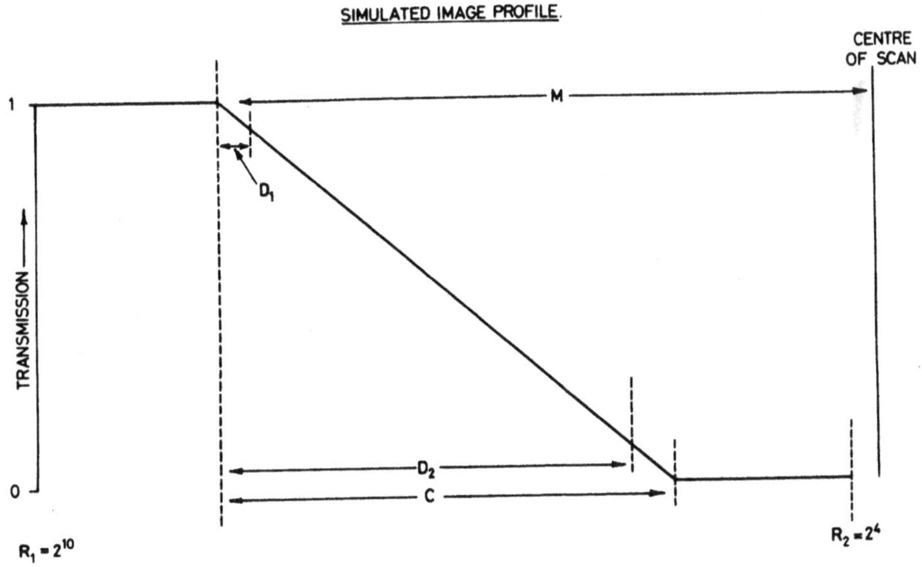

Fig. 2

4. PREPARATION FOR MEASUREMENT

Before any plate is measured on GALAXY, it is necessary to select a few (~20) stars suitably distributed over the plate and in magnitude, known as the "reference loop". These stars are measured about once every hour and provide a control on machine drift during measurement.

Coordinates of the reference loop stars are generally obtained from manual measurement on another machine. They must be transformed to the coordinate systems of each plate, as it is actually located in GALAXY, and punched in the format of a search tape; this is carried out by means of manual measurement on GALAXY of three identified stars, usually the first three, in the reference loop.

The plateholders have been fitted with three locating pads and springs, to enable plates to be replaced after removal to an accuracy adequate for acquisition (~ 50μm). This modification has greatly simplified the somewhat laborious procedure of determining these "three star" transformations for each plate.

During a GALAXY search, the coordinates of the reference loop are automatically reproduced on to the search tape after every 400 images detected in the search, and at the end of the search.

For some investigations the whole measuring list, including reference loops and programme stars, is obtained from manual measurement, or even from a catalogue; this list must likewise be transformed by means of "three stars" to the system of each plate.

5. POST MEASUREMENT DATA PROCESSING AND ANALYSIS

The raw measures of X,Y,M for reference loops and programme stars are output on to paper tape. Before any astronomical analysis can be performed, the data are processed by three computer programmes.

Programme 1 transfers the measures to magnetic tape; any measures with "reject" codes (see section 3), or parity errors, are eliminated. A fourth data word is created, which contains the serial number of the image in the measuring run. The numbers of images in ranges of 10 in M-value are listed on the line printer and also, as a check on data transfer, the total numbers of each binary bit in X,Y,M are given.

Programme 2 transfers the data from magnetic tape to disc storage. The coordinates for each measurement of the reference loop stars are transformed orthogonally to those of the preceding measurement using four constants (scale, orientation and zero points in X,Y); any measures which deviate by more than a tolerance of 2μm or 3 standard deviations, whichever is the larger, are rejected.

Any drift in the coordinates or in M is applied linearly to successive measurements of programme stars between two sets of reference loop measurements. Drifts larger than about 3μm in X or Y, or 5 in M, occur in about 5 per cent of measuring runs and are considered unacceptable; when this happens it is usually near the beginning of a run, even though measurement is not started until after a warming up period of between 15-30 minutes. The coordinates and M values, corrected for drifts, are transformed to the system of the first measurement of the reference loop. To avoid rounding-off errors, guard figures are introduced, thus X,Y are recorded to 0.1μm and M to 0.1 of a radial increment.

Programme stars are sorted in order of the X measurements and any duplicates, within a chosen tolerance (assumed to be 30μm by default) in both X and Y, are rejected, only the latest measurement in these cases being retained.

It sometimes happens that a single plate contains more than one exposure on a field. If the search tape has been created from a master search tape, all the images corresponding to one exposure will be consecutive on the output tape. Alternatively, if the search tape has been created from a GALAXY search of the multiple-image plate itself, the images will have been measured essentially in random order. In either case, Programme 2 determines iteratively the 4-constant linear orthogonal transformation from one image to another starting from information in the reference loop, using an initial tolerance of 5μm, and eliminating measures which have residuals in excess of 5μm or 3σ whichever is the larger. Images are then matched within a tolerance which can be set appropriate to the circumstances of the plate.

Fig.3 shows a histogram of M-values of images, measured on one exposure, of a two-exposure plate. The characteristic "double hump" arises because in this case the search tape was derived from a GALAXY search using a threshold below the plate limit. The sharp peak at around M=100 represents grain noise; real star images start at about M=150. This is confirmed by examination of the residuals $\Delta X, \Delta Y$ between the transformed coordinates of corresponding images. Fig.3 shows the relative numbers for which $\Delta S = (\Delta X^2 + \Delta Y^2)^{1/2}$ is less than or greater than 20μm. The high proportion of $\Delta S > 20\mu$m (shaded area) at low M-values indicates again that measurements have been made on random grain clumps.

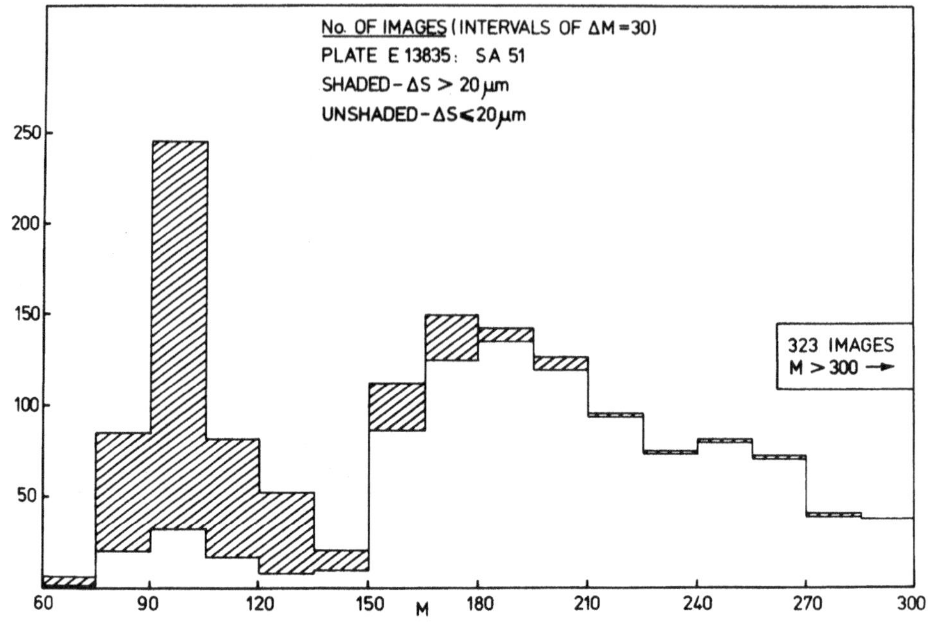

Fig.3

Data from all plates measured on a field are accumulated on to magnetic tape by successive use of Programme 3. The general philosophy underlying the merging procedure is similar to that used in matching images in Programme 2, except that matching is performed after independent 3-constant linear transformations have been applied to each coordinate, since the plates may well have been taken on different telescopes under a wide variety of circumstances.

All measurements for each star are stored together in integer format, in as many consecutive records as are necessary, dummy "measures" with negative M-values being set where a corresponding image was missed. The stored coordinates are all transformed to the coordinate system of the first set of measures on the tape by means of 4-constant linear orthogonal transformations, the elements of which are also stored on the tape.

The first eight words for each star contain "housekeeping" data and identifying information such as typical coordinates (generally identical to the first measurement) which are used in all merging and matching operations. Two of the words are reserved for astrometric and photometric standard star numbers. These can be inserted by Programme 3 by merging a synthetic "measurement" run, in which the relevant standard numbers have been inserted in either the M-value or serial number words, on a magnetic tape in the format of the output from Programme 1, which must first be transferred to disc by Programme 2.

Subsequent analysis depends on the nature of the astronomical investigation, but a set of general programmes has been developed which carry out the basic reduction. The final output is a magnetic tape in which all the measured coordinates on all plates on a field have been reduced to heliocentric tangential coordinates (unit $0''.01$), referred to some arbitrary plate centre, by applying corrections for refraction and aberration but excluding parallax. Provisional positions, and, if the time-base is sufficiently long (>0.5 year), proper motions, are derived from a linear solution of the astrometric data. Calibration of photometric plates is carried out in the usual way by means of standard stars and the resulting magnitudes and colours are also stored on the tape. Circumstances relating to each plate, including the epoch and time of exposure, the equatorial direction cosines of the observer's zenith, components of the heliocentric position and velocity of the observer and the matrix for converting the listed tangential coordinates to equatorial direction cosines, are stored in a set of records at the beginning of the tape.

6. MEASUREMENT PERFORMANCE

6.1. Coordinates

The coordinates of a star image as measured in GALAXY depend on the image size; this is the analogue of "magnitude equation" which is familiar to astrometrists in the bisection of images. The effect is illustrated in Fig.4, which shows comparisons between measures made of the same plate in four orientations differing successively by 90°. The measures X_iY_i ($i = 0, 1, 2, 3$) were made using magnification X30 and have all been transformed to a common system. Figs.4(a), (b), (c) and (d) show residuals in X_2-X_0, Y_2-Y_0, Y_3-Y_1 and X_3-X_1 respectively. The plots have been generated on the computer line printer; abscissae are quantized in units of 10 radial increments and ordinates in μm. The numbers indicate the number of images falling in each plot point.

By taking the average of measurements made in opposite orientations, a repeatability of better than 1μm (s.d.) is achieved.

The magnitude effect is reduced, in proportion, in measurements made using the higher magnifications. The cause is ascribed to residual asymmetry in the scanning system, possibly due to irreducible internal reflections from elements surrounding the scanning train. Such accidental reflections will have the effect of changing the level of background illumination on the plate, thus affecting the scanning equilibrium position; the size of the effect will vary with the position of the scanning beam, which depends on the image diameter.

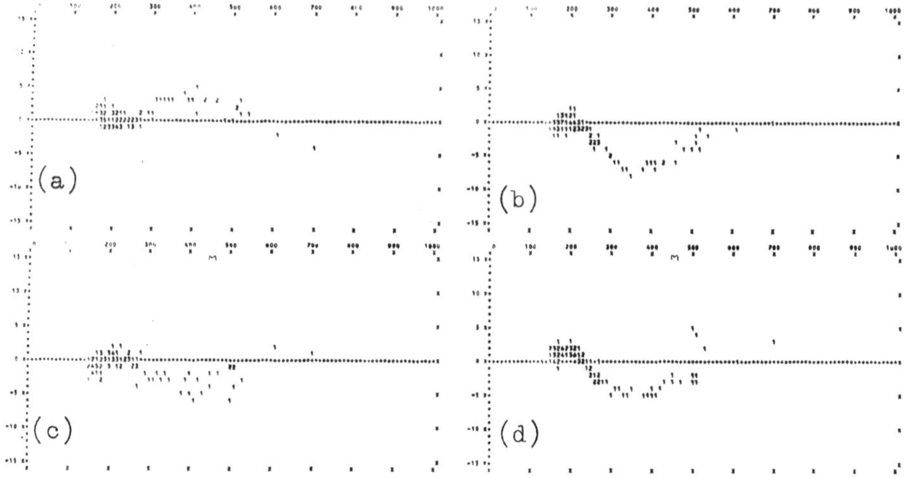
(a)
(b)
(c)
(d)

Fig.4

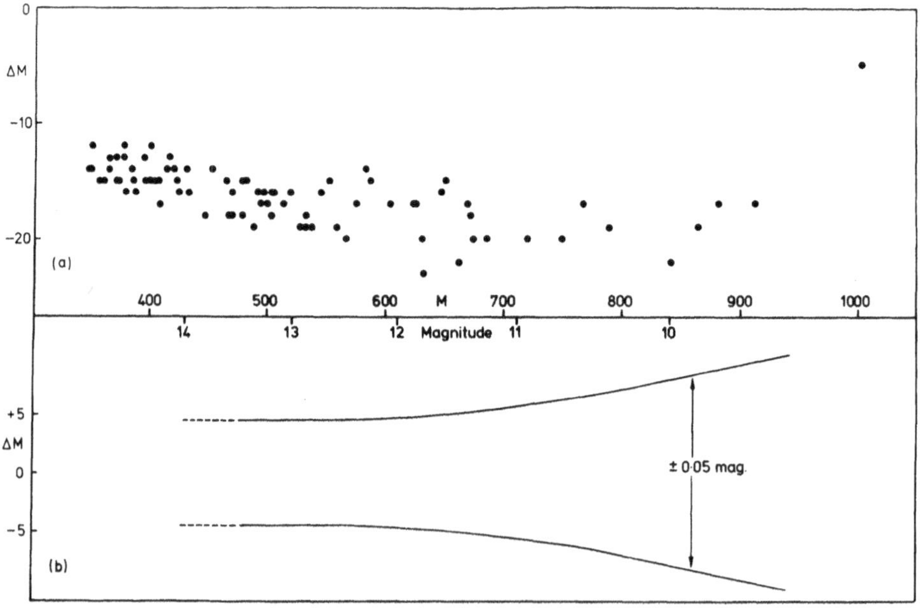
0
ΔM
-10
-20
(a)
400
500
600
M
700
800
900
1000
14
13
12
Magnitude
11
10
+5
ΔM
0
-5
± 0·05 mag.
(b)

Fig.5

In order to eliminate the magnitude equation from measurements made for precise astrometry, it has been the practice at R.G.O. to measure each plate in two opposite orientations. To facilitate this, the original design of the plateholders has been modified by the addition of three further ball-ended screws, which makes them reversible with respect to the machine carriage.

6.2. Magnitudes.

The scale of M-values in relation to magnitude, derived from repeated measures of the same plate depends not only on the profile parameters, but also on the illumination of the plate which is controlled by the setting of the high voltage supply. Measurements made without any change in the plate illumination give virtually identical M values for the same profile parameters. However, random variations in the level of illumination are inevitable when a plate is remeasured after an interval of time. This is illustrated in Fig.5, in which two independent measurements of a blue plate on the Pleiades are compared. The top diagram shows the difference of M-values, ΔM, plotted as a function of M. It will be seen that while the shapes of the corresponding calibration curves will differ, the random scatter is quite small. The lower diagram shows the extent of the scatter for ± 0.05 mags, and, along the bottom scale, the magnitudes corresponding to the plotted M-values; this scale has been determined from photo-electric standards from Johnson and Mitchell (1958).

7. FUTURE ENHANCEMENT.

There have been two main causes of lost output in the system during the past two years, failure of the paper tape input/output equipment, which accounted for about 75 per cent of "down" time, and unacceptable drifts in the datum, referred to in section 5. Both of these should be considerably reduced by the introduction of a mini-computer into the system.

A Data General NOVA 2/10 control computer with magnetic disc and tape drive has been purchased and will be installed during the summer of 1975.

The mini-computer will be used not only to replace the present paper tape equipment, but also to supervise the operation of the machine itself. Ultimately it is intended that other measuring machines (e.g. the Zeiss ASCORECORD and the Coradograph) will be included in the system, thus facilitating the flow of auxiliary data derived from these machines, such as coordinates of reference loop stars, into GALAXY itself.

In the search phase, coordinates will either be stored directly on the disc or on tape. In the measurement phase there will be several innovations. Transformation of search coordinates from a master list will be determined and applied at run time, thus obviating the necessity of providing, off line, search tapes for each individual plate. Furthermore, during the course of measurement the stability of the datum can be monitored and, if any large drifts occur, the measures affected can be rejected and then repeated, or the run abandoned.

Subsequent handling of measures will be much simplified by the inclusion of meaningful serial numbers with the search coordinates. The matching of measurements of the same star on different plates will then become a simple matter of correlating the star numbers, instead of the current laborious process of transforming, sorting and merging on positional coordinates.

It is hoped that the addition of the mini-computer will at least double the rate of output of useful data.

8. ACKNOWLEDGEMENTS

The successful operation of GALAXY depends on many individuals to all of whom we are most grateful. Mr. K.A. Long and Mr. M. Fisher are responsible for maintenance of the machine, and Mr. G.S. Walker of Computer Advisory Services, Heriot-Watt University, has given much advice on the operation of GALAXY as well as carrying out the regular servicing of the machine. Several former and present members of the staff of the Astrometry Division have been involved in developing programmes and operating the machine; the operating and data handling team currently consists of Mr. G.M. Harvey, Mr. D.L. King and Mrs. S. Stone. Exploitation of the GALAXY measurements would not be possible without the assistance of the staff of the Computer Department.

The project for installing the NOVA computer was planned by Dr. J.S. Beale, and the software development is being carried out by Mrs. D.E. Hobden.

Our thanks are also due to Mr. Khao Muanwong and Mr. P.M. Corben for useful discussions, to Mrs. M. Everest and Mr. D. Calvert for preparation of diagrams, and to Mrs. E. Howard for preparing the "camera-ready" typescript of this paper.

REFERENCE

Fellgett, P.B.: 1970, Optics Technology. **2**, 61
Johnson, H.L. and Mitchell, R.I.: 1958, Astrophys. J. **128**, 31
Pratt, N.M.: 1971, Publ. Roy. Obs. Edinburgh. **8**, 109
Walker, G.S.: 1971, Publ. Roy. Obs. Edinburgh. **8**, 103

THE FULLY AUTOMATIC PROGRAM-CONTROLLED MICRODENSITOMETER-COMPARATOR OF THE ASTRONOMICAL INSTITUTE AT UTRECHT

J.R.W. Heintze°), R.L. Porteous°°) and W. Brandie°°)

°) Astronomical Institute, University of Utrecht, Utrecht, The Netherlands
°°) Faul-Coradi Scotland Ltd

ABSTRACT. A fully automatic program-controlled microdensitometer-comparator has been built by Faul-Coradi Scotland Limited for the Astronomical Institute at Utrecht. This instrument fulfils the stringent requirements for accuracy in the measurement of both position and density considered necessary for the reduction of stellar spectrograms. It features air-bearing transport of the spectrogram in one direction and of the measuring optics in the orthogonal direction; moiré-fringe position measurement and automatic focussing. Measuring sequences are fully automatic and under computer control.

The spatial resolution of the machine (½ μm) permits, in principle, its use in the reduction of electronographic plates. With this aim, development work continues on the optical system in order to improve the linearity of measurements at higher densities.

The machine can reduce all types of solar spectra and is also capable of two-dimensional densitometry of extended images on plates up to 25 cm x 10 cm .

Section 1 of this paper gives the historical background of the machine, section 2 describes the mechanical system, section 3 the optics and the automatic focussing, section 4 the control system and section 5 presents some performance data. This performance has proved better than the original specification.

1. INTRODUCTION

In astronomy much information is and will continue to be obtained by photographic means. This holds also for astronomical apectroscopy.

de Jager/Nieuwenhuijzen (eds.), Image Processing Techniques in Astronomy. 185-198.

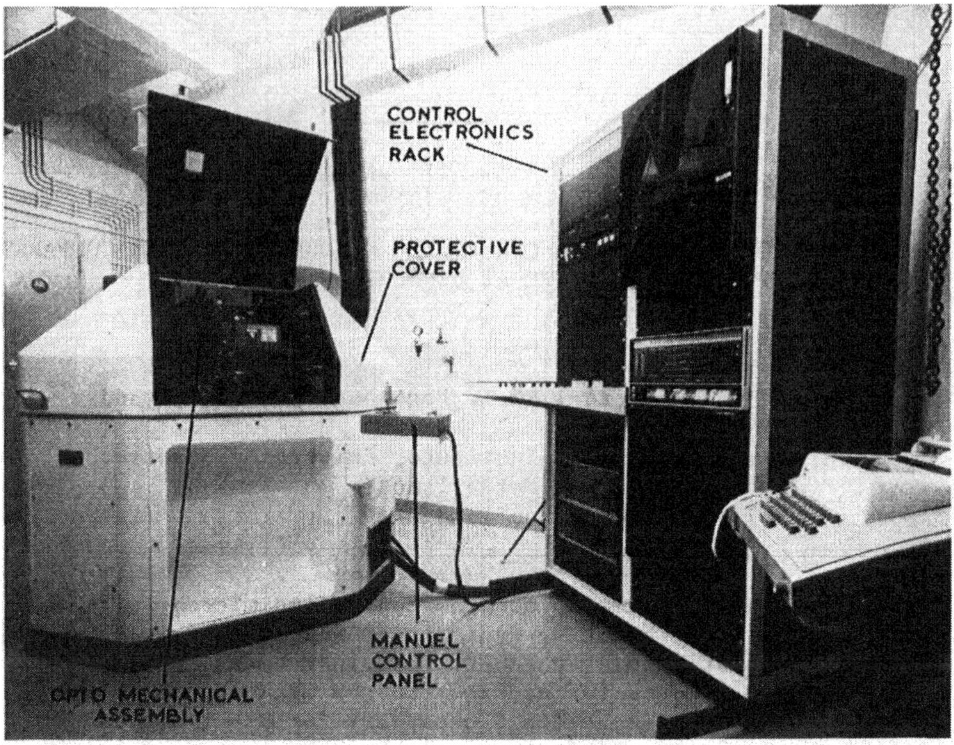

Figure 1. General view of microdensitometer-comparator showing access hatch open for plate loading.

Some very advanced reduction instruments are available for the determination of radial velocities from spectroscopic plates; for example the Grant machine. Quite often, however, radial velocities cannot be found exactly enough by determining the position of the central depths of observed absorption lines: sometimes the continuum of the spectrum changes rapidly with wavelength causing the observed central depths to be shifted with respect to the true line centre; sometimes the line centre is formed in surrounding gas masses and the radial velocity of the star has to be determined from the centre of the line wings. In both cases the continuum has first to be rectified.

Therefore for radial velocity work one should like to have a combination of a microdensitometer and a comparator, which can provide plots of spectrum intensity (if necessary noise filtered) against wavelength, on which the continuum can be drawn and rectified and from which consequently "rectified" line profiles can be obtained, allowing for better radial velocity determinations for line-centra as well as for line-wings.

In 1963 J.R.W. Heintze and H. Nieuwenhuijzen of the Utrecht

observatory started to explore the possibilities in this direction. The design study and the manufacturing was given to Faul-Coradi Scotland (1969-1971). The machine, of which Figure 1 gives a general view, was handed over to Utrecht in May 1972. At Faul-Coradi the work of design and assembly was undertaken by teams led by R.L. Porteous and W. Brandie. At the Astronomical Institute Utrecht the following staffmembers were involved in preparing the specifications: drs. L.D. de Feiter, T. de Groot, J.R.W. Heintze, J.T.J.A. Lafleur, H. Nieuwenhuijzen, H. van de Stadt, A.B. Underhill and C. Zwaan. Mr. G. van Gelder operates the machine, while among others Mr. J. Rosenbaum and Mr. R. Rutten are working on software techniques.

The emphasis of this machine lies on accuracy of position and density and not on speed. In the most accurate mode the plate is stationary during density measurement. In this mode the time necessary for stepping, settling and density measurement is about 40·m sec, this time being almost independent of step size for steps between 1 and 30 μm . The resulting speed-range is 0.025 - 0.75 mm s^{-1}. "On the fly", 100 density-measurements are completed per second, giving a speed-range of 0.1 - 3 mm s^{-1}.

Two special features, the automatic focussing facility and the programmed, computer-controlled plate drive, substantially reduce the total time required for plate measurement.

The philosophy applied in designing the instrument has been outlined elsewhere (Parks et al., 1971). The machine closely resembles the system described in that paper, but includes additions and modifications indicated below.

2. BASIC TWO-AXIS POSITIONING MACHINE

The main body of the instrument is a massive casting in Meehanite, octagonal in plan form, on which two carriages ride independently on orthogonal ways. The carriages are both supported and steered by linear air bearings, their movements being controlled by Inland d.c. servomotors driving ball screws. The positional information is derived from moiré fringe transducers.

Figure 2 is a three-quarter rear view of the assembly in which can be seen, straddling the spectrogram carriage, the upper half of the carriage which supports the optical projector. Below the machine bed a similar gantry, integral with the upper section, carries the efflux optics

The various constraints on the carriage motions are so disposed as to reduce to a minimum positional errors resulting from roll, pitch and yaw. The quality of the resulting motion is illustrated in Figure 3. Over the full stroke of the spectrogram carriage, the path traced out upon it by the optical axis departs from a true straight line by no more than 0.3 μm in 250 mm. This

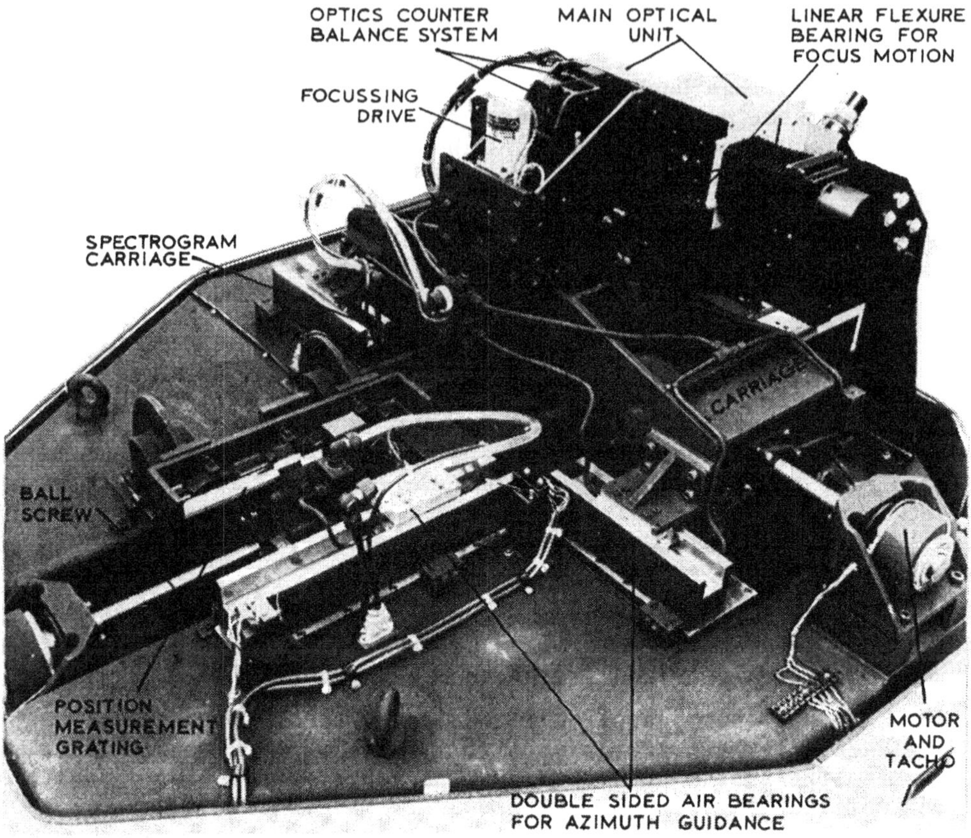

Figure 2. Three-quarter rear view of opto-mechanical assembly with protective cover removed.

is measured with an optical flat, 250 mm long and about 25 mm square, and an electronic dial gauge of high sensitivity and low operating forces. The flat is aligned approximately parallel to the carriage motion and a record is made of the response of the gauge to a selected face, the gauge being clamped to the machine bed. Rotating the flat 180^0 about its long axis, the measurement is repeated on the same face of the flat. The mean of the two traces represents the line traced on the carriage by the optical axis.

A simple method of temperature control has proved adequate since to some extent systems making use of glass gratings for position measurement are self-compensating for temperature variations. Each measurement loop consists of two sections of metal, namely machine bed and carriage, and two of glass, namely grating and spectrogram. With appropriate constraints on the latter, changes in length due to temperature fluctuations, which would

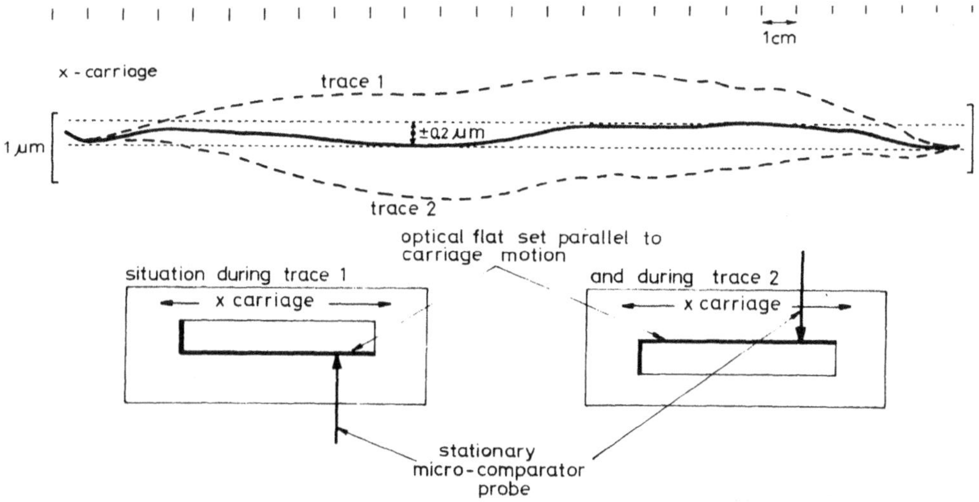

Figure 3. Straightness of motion. Mean of traces 1 and 2 depicts path traversed by plate carriage with respect to optical axis. Straightness of x-axis is within ± 0.2 μm. Ripple due to motion of lead screw is ± 0.1 μm. Total maximum deviation from straight line is ± 0.3 μm.

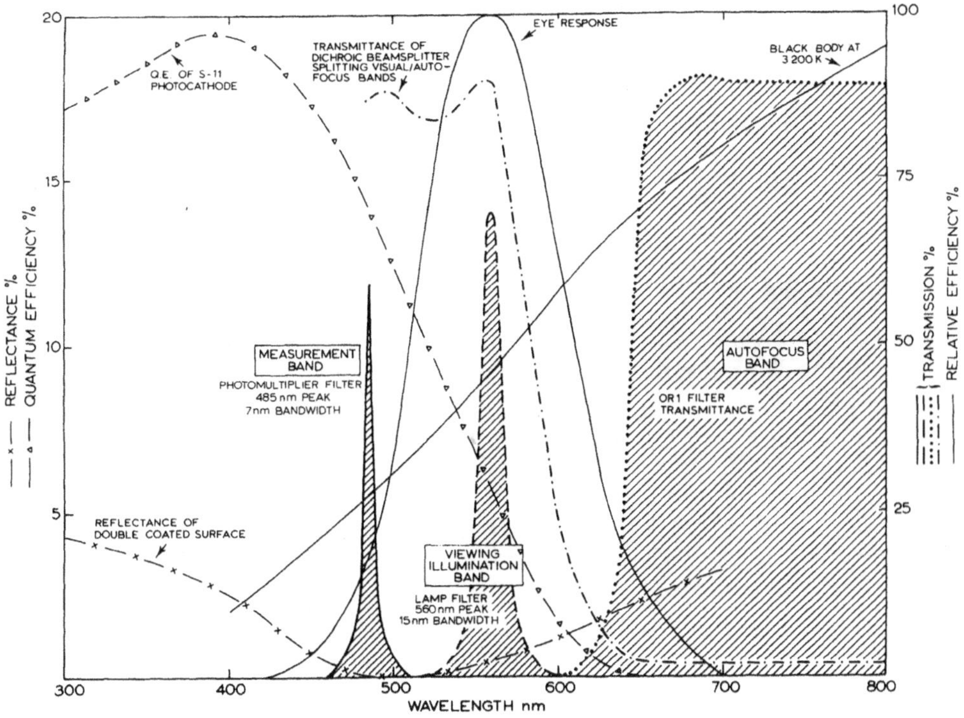

Figure 4. Spectral characteristics of light sources, filters and detectors.

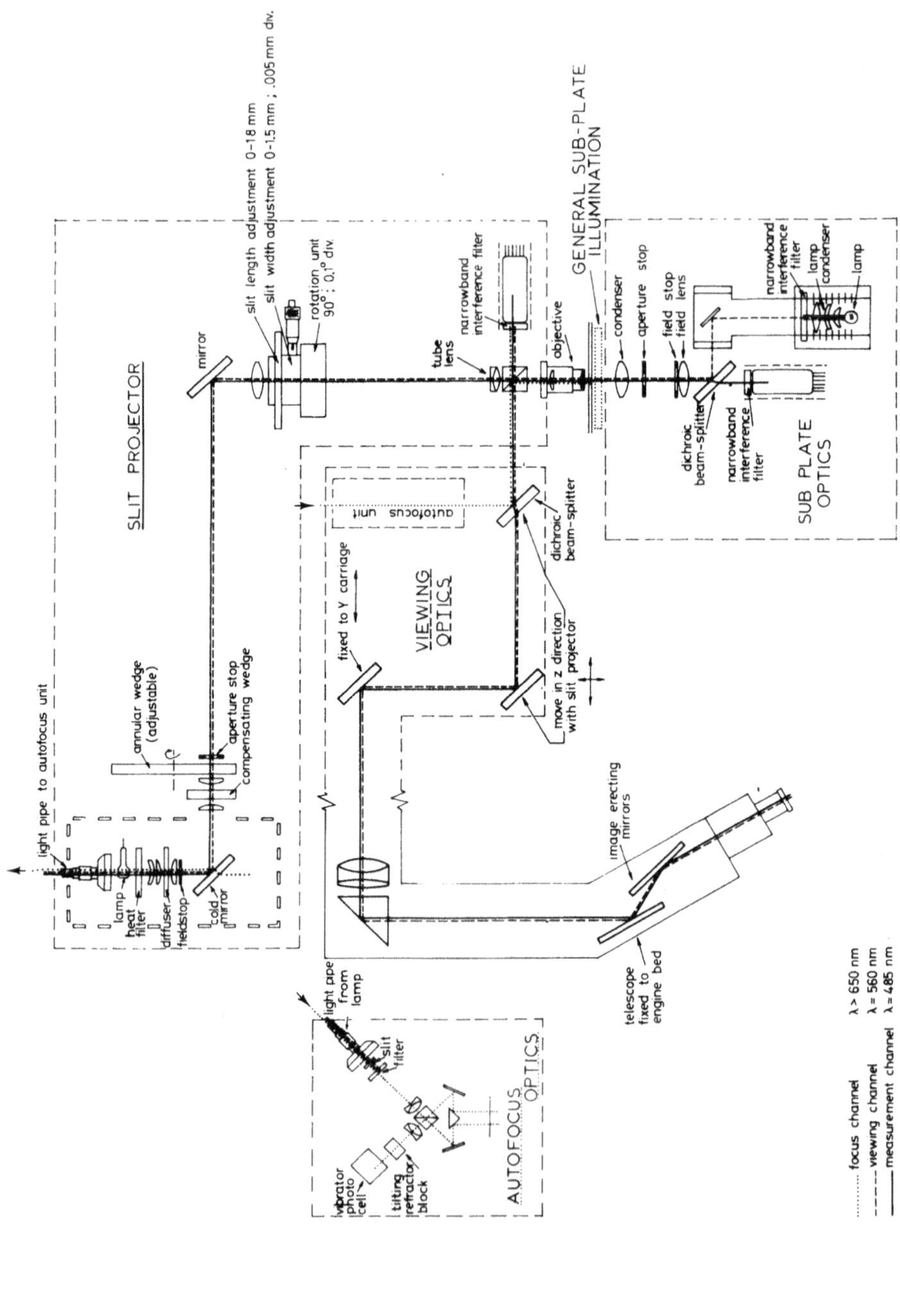

Figure 5. Schematic arrangement of optical channels.

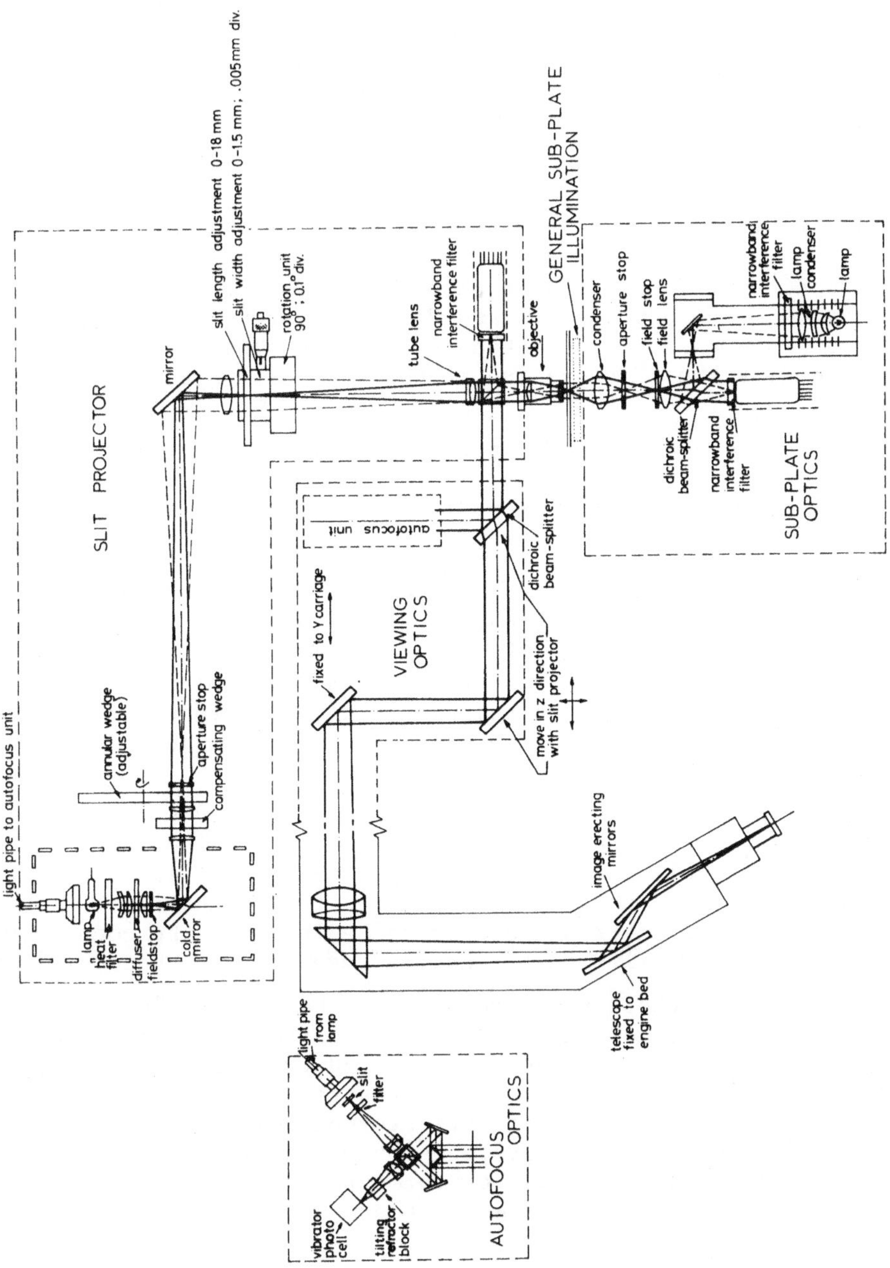

Figure 6. Optical ray paths (schematic)

result in position errors, tend to cancel out. Under these conditions adequate temperature stability is achieved by removing air from the vicinity of all heat sources within the protective cover of the machine, allowing this to be replaced by air from the laboratory which is controlled to within $\pm$ 1^0 C.

Other aspects of machine performance are summarized in section 5.

3. OPTICAL SYSTEM

The optical system consists of three functional parts, the Measurement Optics, Autofocus Optics and Viewing System, operating in independent wavebands as indicated in Figure 4. These functional parts are distributed over five packages of hardware, as shown in Figures 5 and 6.

The main optical unit comprises the Slit Projector which illuminates the required scanning area on the spectrogram, the Autofocus opto-electronic assembly and part of the Viewing System. This unit is mounted on the transverse carriage, moving with it over a horizontal stroke of 100 mm. The entire package can also move vertically, for focussing, guided by compensated parallel leaf-spring hinges. The vertical stroke is 3.25 mm. Mounted in this unit are the objective changer with three objectives and the operator's controls for most of the optical variables - length, width and orientation of the scanning slit, fine focus, reference photomultiplier current (indicated by a meter) and photomultiplier gain matching.

In normal operation the spectrogram is scanned by the impulse response function of the slit projector objective and the transmitted light is collected by a wide aperture condenser in the Sub-Plate package. This sub-assembly is fixed to the underarm of the transverse (optics) carriage and contains, besides the condensing system, the viewing illuminator and the measurement photomultiplier.

The Viewing Telescope is mounted on the bed casting of the machine. It consists of an objective, wide-field eyepiece and appropriate mirrors for image transfer and erection. Light from the substage illuminator, together with back-reflected light from the image of the scanning slit, is reflected out of the slit projector along the same path as the autofocus beams. A dichroic beam splitter diverts the red and infra-red content into the autofocus unit. The remaining blue-green light from the scanning slit and the green light from the field are passed into the viewing system.

Figure 7 shows the main optical assemblies in position on the machine.

The Autofocus sensing unit consists of a small sub-assembly mounted within the main optical unit. A slit, illuminated by an

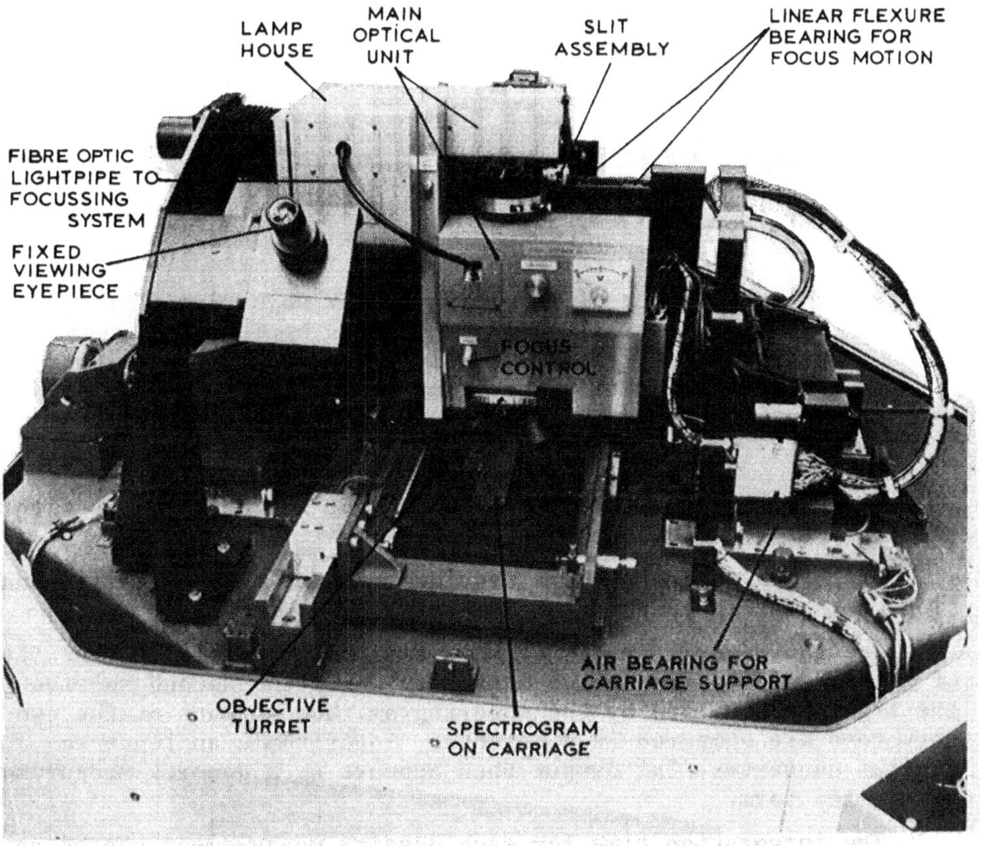

Figure 7. Front view of microdensitometer-comparator with cover removed to reveal the main optical unit.

infra-red beam is mounted at the focus of an infinity-corrected microscope objective (see Figure 6). The rays issuing from the lens are split by a semi-reflector and by means of an arrangement of mirrors are reflected as two similar parallel beams into the slit-projector beam-splitter and thence into the main optical path.

The slit projector objective converges both beams to a common image on the surface of the spectrogram and the reflected light is returned to the autofocus unit. Here the beams are recombined and the slit re-imaged on to a vibrating Schwarz photocell.

A change in height of the spectrogram surface results in a lateral deflection of the slit image and hence in an asymmetry in the output from the detector to the servo-control system. A tilting refractor block, manually controlled by a screw, allows initial fine setting of the slit image on the detector and,by subsequent servo-command, fine setting of the slit projector focus.

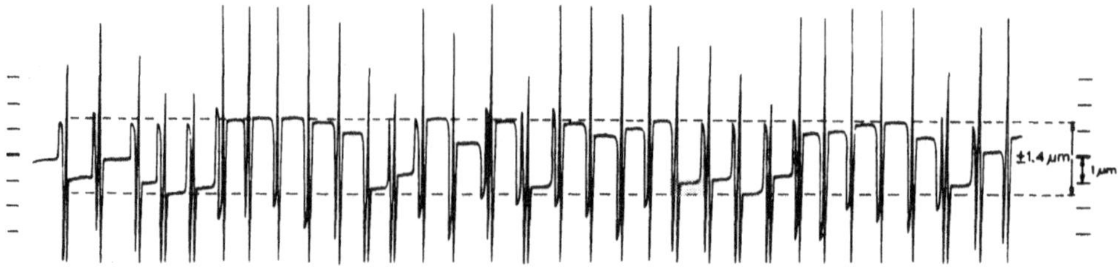

Figure 8. Repeatability of focussing of 25x objective. Trace showing vertical position of main optical unit against time during repeated displacement of this unit. Overall spread of settling position in 32 trials (homing in from both directions) is ± 1.4 μm.

As demonstrated in Figure 8, the error in focussing is within the depth of focus of the objectives.

The fifth optics package provides general sub-plate illumination to assist in spectrogram loading and setting-up.

The photometric system employs two photomultipliers. The first of these monitors the source intensity while the second receives the light transmitted by the spectrogram. The outputs of the two detectors are compared in a patented, logarithmic analogue-to-digital converter. The output thus appears as a density measurement in digital form.

The integration time for each density measurement can be varied from 1 - 4 ms . The uncertainty in the measured density depending as it does on the number of photons sampled, naturally increases with decreasing integration time, with decreasing projected slit area and with increasing measured density. The standard deviation is always less than 1% for projected slit area's of 100 μm^2 or more and exceeds 2% only when measuring densities greater than D = 3 with slits of less than 20 μm^2 .

4. CONTROL SYSTEM

When the control system requirements were initially analyzed it was recognized that the design would have to be flexible, bearing in mind the probable changes in emphasis that would have to be accomodated in the future. Thus what evolved was a self-contained arrangement of modules each of which can be separately optimized or changed with changing conditions:

a) The instrument is controlled by a small computer - PDP 8/I - linked to a small amount of purpose-built hardware.
b) No external data-preparation facilities are necessary. The

system operates in two distinct modes: manual and automatic. Drive tapes generated in manual mode are used as input to automatic mode.

c) Density measurements are made either by stopping before measurement, or by measuring "on-the-fly". Data output from either automatic scanning mode is stored on compatible digital magnetic tape which can if required be read and printed by the system.
d) Density measurement and automatic focussing are, except in a control sense, independent of each other and of the computer.
e) The servo control loops are live even in the absence of the computer, and fail-safe in the absence of a response.
f) The visual displays are driven directly from the computer. Hence position, density or any other required quantity can be monitored on a shared basis.

A manual control box is provided which enables the operator to slew the axis at high or low speed in any combination of axis directions or to make single increment steps. This unit also furnishes the controls which cause drive tapes to be generated. It should be noted that even these tape generation controls do not have hard-wired functions but result in status bits which are analyzed and manipulated by the computer program. Instrument fault conditions are similarly treated.

The computer itself is reasonably loaded but rearrangement or deletion of program modules would free capacity for system extensions. In addition to the teletype and the magnetic tape system, the following facilities are used:

g) Two input/output addresses (maximum possible 64).
h) Approximately 40% of real time.
j) 80% of 4096-word store (12 bit words)
   1) General organization: 287 words
   2) Interrupt organization and control updates: 718 words
   3) Axis control algorithms: 312 words
   4) Associated maths routines: 367 words
   5) Input/output data and operator interlocks: 1270 words
   6) Magnetic tape control and data flow: 314 words.

There are two types of control loop in the densitometer system. The automatic focussing system is a non-linear type 0 holding servo actuated in effective half-micron steps by a stepping-motor, while the machine axis control is a linear velocity servo using a d.c. motor to drive the axis through a leadscrew and ball-nut. The position transducer is of course a moiré fringe grating system.

The details of this latter servo have been more fully covered elsewhere (Parks et al., 1971 and Brandie and Walker, 1972) but it should be emphasized that although it is a half-micron resolution digital servo, finer static repeatability is obtained by subsidiary analogue processing of the moiré fringe signals.

Figure 9 shows the variation in measured position of lines of width of 2.5 μm at the four corners A, B, C and D of a rectangular glass plate. The distances A-B and B-C are 240 mm and 90 mm respectively. Each corner was scanned once in turn with a 3.0 μm slit and the set of four measurements was then repeated another three times. As figure 9 shows the maximum difference in measured position is about 0.35 μm. The mean deviation of these measurements is 0.02 μm. The repeatability is limited essentially by temperature fluctuations. Tests on a similar machine in a more closely-controlled environment have shown that a repeatability of 1 micro-inch (0.025 μm) is within the capability of the system.

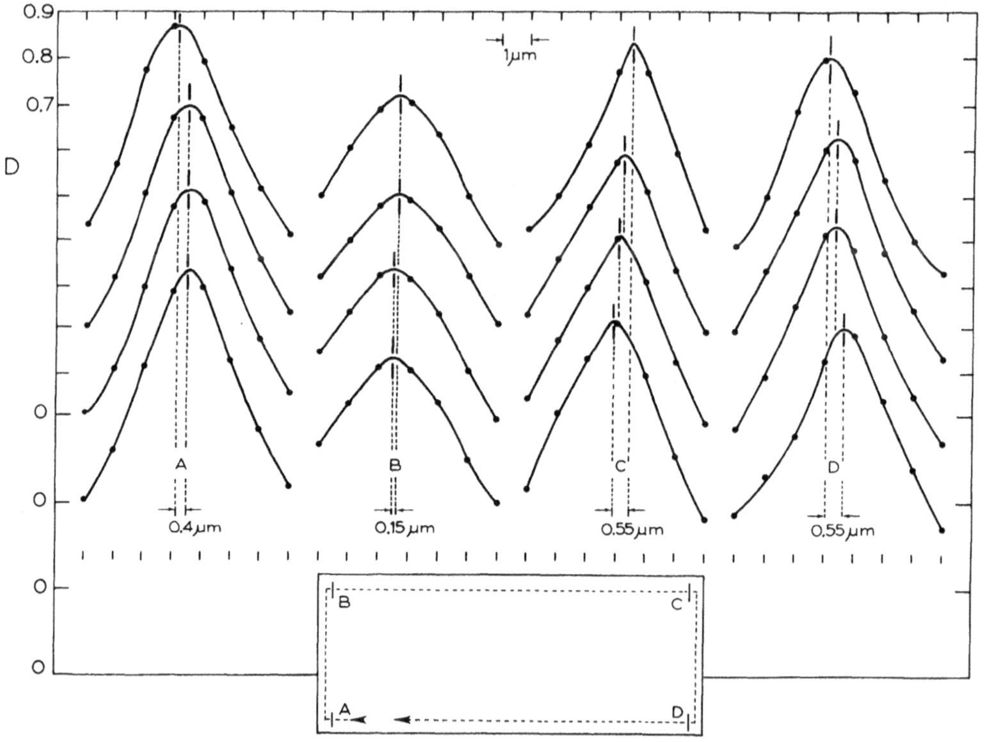

Figure 9. Repeatability of position (See the above text).

5. PERFORMANCE OF THE INSTRUMENT

Mechanical system

Working area	251 cm x 103 cm
Measured deviation from orthogonality of mean carriage motions	1.1 ± 1.2 arc seconds

Mechanical system (continued)

Absolute accuracy of the gratings on each axis at 20^0 C	± 0.8 μm
Minimum step size	0.5 μm
Repeatability of position measurement	± 0.35 μm
Operating speeds:	
fast manual traverse	11 mm/s
slow manual traverse	0.25 mm/s
"on the fly" under computer control	0.1 - 3 mm/s
times for stepping, settling and density measurement at D = 3 for 5 μm steps	39 ms
25 μm steps	40 ms

Optical system

Minimum slit widths and maximum slit lengths	
25 x objective	1 μm; 0.78 mm
10 x objective	2.5 μm; 1.95 mm
4 x objective	6.5 μm; 4.90 mm
Range of focussing movement	3.25 mm
Focussing resolution	0.45 μm
Orthogonality of focus motion to spectrogram plane	within 1.75 minutes of arc
Acceptable slope of spectrogram	1:80 at 3 mm/s
Focus error band of	
25 x objective	± 1.35 μm
10 x objective	± 4 μm
4 x objective	± 19 μm
Range of measured densities	0 - 4.095
Standard deviations in measured densities	See section 3

These tests were made initially on installation in May, 1972 and repeated at the beginning of 1975. No observable changes have occurred.

During this period the only major difficulty has been a defect in the incremental magnetic tape drive which, although trivial in nature, has taken considerable time to resolve.

The machine was delivered with a complete set of control and measurement programs which have been unchanged up to the present. Work in writing reduction programs is currently continuing in collaboration with other institutes especially the Astronomical Institute at Amsterdam. At present the following programs are available:

1) Calculation of power spectrum
2) Noise filtration
3) Production of dispersion curve
4) Determination of line positions (several methods)

5) Plotting intensity against wavelength. Correction factors may be inserted to rectify the continuum.

Currently, some major spectroscopic analyses are being carried out using this instrument. Of the work that has already been published, we would refer to the thesis of Beintema (1975), "An atlas of profiles of interstellar calcium and hydrogen lines", and the thesis of De Boer (1974), "Interstellar absorption lines in the ultraviolet".

REFERENCES

1. D.A. Beintema, Thesis, Groningen, The Netherlands, 1975.
2. W. Brandie and G.S. Walker, NEL Conference on Moiré Fringe Technology, paper No. 7, published by the National Engineering Laboratory, U.K., 1972.
3. K.S. de Boer, Thesis, Groningen, The Netherlands, 1974.
4. R. Parks, R.L. Porteous and W. Brandie, Publ. of the Royal Obs. Edinburgh, **8**, 130, 1971.

THE GRANT MEASURING ENGINE

Edwin W. Dennison

Hale Observatories, Pasadena, California 91125, U.S.A.

GENERAL DESCRIPTION. The Grant Measuring Engine is a highly refined multi-purpose instrument which has a double carriage with one stage mounted on top of the other. Precision lead screws provide both the stage motions and the mechanism for measuring the carriage positions. The design is intrinsically simple and the high accuracy is the result of good detail engineering and construction. The engine design is continuously being improved and this paper describes the most recent performance characteristics.

The basic design has been implemented in a variety of different formats and configurations for making measurements of images on photographic and electronographic materials. The engine can be used as a spectroscopic comparator, a two coordinate stellar measuring engine, a microphotometer and an iris photometer. The measurement settings are made manually with the use of an image scanner. Settings by the use of computer controlled image scans are currently being evaluated. Coordinate and configuration interfaces are designed to permit a wide variety of hardwired and computer controlled data systems.

MECHANICAL DESIGNS

Each stage moves on precision roller bearings which travel on hardened rectangular ways. The guide rollers have vertical axes which are orthogonal to the axes of the load bearing rollers. The ways are made

de Jager/Nieuwenhuijzen (eds.), Image Processing Techniques in Astronomy. 199-208.

straight to within ±1.3 microns (50 microinches) for a 460mm (18 inches) travel. The one mm pitch lead screws are ground from hardened steel and are lapped to give an overall linearity of lead of ±1/2 micron with an even smaller periodic error. The correction bar is straight and can be set to give accurate absolute measurements at any specified ambient temperature. The four-section nut is designed to allow oil to flow through the screw-nut interface area and this oil flow removes small particles of dirt. The stages are spring-loaded to give a backlash of as low as two tenths of a micron.

Engines have been built with stage motions as large as 250×250mm (10×10 inches) and 350×350mm (14×14 inches) engines are being designed. The stage motions have an orthogonality of ±1 arcsecond which results in an error less than ±1.25 microns for a 250mm travel. The total error from all sources is less than 2 microns for any position over a 250×250mm area.

OPTICAL CONFIGURATION

The optical train consists of a regulated D.C. powered quartz-iodine lamp which illuminates the first slit image plane through a suitable condenser lens and heat absorbing filter. The lamp housing is cooled by a fan and is isolated from the base frame by a thick thermal insulator. This construction eliminates detectable thermal drifts due to lamp heating. This image plane can be clear or contain a pair of bilaterally adjustable amber glass slit jaws or an iris diaphragm. The first objective is a fixed focal length lens and the second objective can be either a fixed or a variable focal length lens system. The final focal plane has a viewing screen, the image scanner, a translatable fixed slit, an iris detector assembly or any other image detector system. By the use of a beam splitter up to 2 systems can be simultaneously illuminated in addition to the viewing screen. The photomultiplier detector behind the microphotometer slit is covered by a blue filter which blocks the amber field illumination light. A series of electrically operated masks are used to select the upper, lower or central portion of the slit. This enables the operator to rapidly select either of the comparison spectra or the object spectrum when the engine is used as a spectrum comparator or a microphotometer. For stellar measurements the slit length can be adjusted to optimize discrimination between the background and the stellar image. Fiducial marks on the viewing screen facilitate the slit setting procedure and all slits are rotatable and translatable to permit exact slit alignment. The slit and imaging systems are designed to give one micron resolution in the measuring image plane.

SETTING DEVICES

The Grant Measuring Engines have always used the rotating prism image scanner as a manual setting device instead of a cross hair. This is a refinement of the concept described by Tomkins and Fred (1951). The rotating prism scanner is effective because in reality the images being measured have at least some degree of symmetry. The setting accuracy increases as the image symmetry increases.

The image scanner is a 35mm (1-3/8 inch) square glass prism which is mounted approximately 8cm (3 inches) in front of a slit mounted in the image plane and is rotated at 1800 rpm. Two of the opposing sides of the prism are covered with opaque material. As the prism rotates around its center, the light passing through the clear sides of the prism is displaced in a direction which is parallel to the direction being measured. The result is that the image is scanned over the measuring slit two times for every full rotation of the prism. A circular cam is eccentrically attached to the drive shaft. This cam covers and uncovers a slit which is mounted radially with respect to the rotation axis of the prism drive shaft. An auxiliary lamp illuminates this slit and a silicon detector generates a signal which is proportional to the part of the slit which is uncovered. The cam is phased to give an increasing signal during the first image scan and a decreasing signal during the second image scan. The signal from the image scanning slit is sent to the Y-axis of a CRT display and the signal from the auxiliary slit is sent to the X-axis of the CRT display. The net effect is to cause the first image scan to be displayed from left to right and the second scan is displayed from right to left.

When the measuring stage is moved to a position where the symmetric portion of the stellar or spectral line image is precisely centered, the two scans are superposed on each other. This enables the operator to very accurately determine the center of the image being measured. With small symmetric images this setting is easily reproducible to within one or two tenths

(a)

(b)

(c)

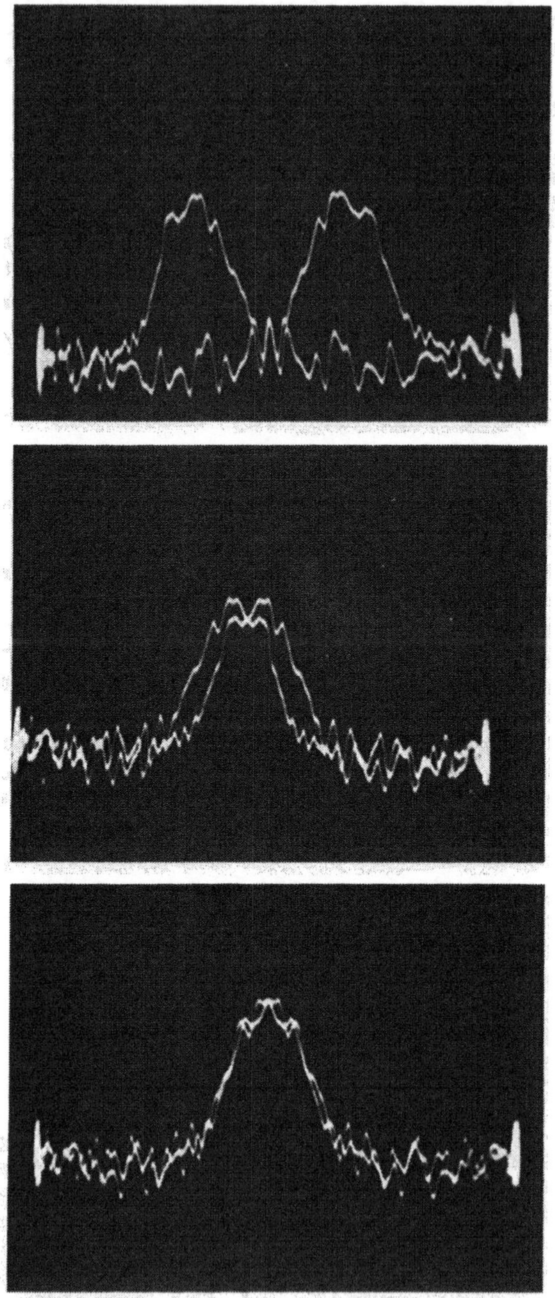

Fig. 1. C.R.T. trace from image scanner.
(a) 72 microns from center, (b) 12 microns from center, and (c) at center.

of a micron. With asymmetric images or blended lines the operator can center on the most symmetric part of the image and ignore image imperfections, asymmetric portions of the image, or unwanted components of a blended line. The accuracy of this setting method far exceeds that of a classical cross hair because it enables the operator to use a large portion of the image to determine the center. This setting device enables an operator to locate an image position with a precision which is greater than the mechanical accuracy of stage motions.

The iris measuring system, currently under development, can be manually set by observing a null meter which indicates when the light passing through the annulus defined by the iris diaphragm on the outside and the stellar image on the inside is equal to a constant reference. The diaphragm motion will also be controlled by a simple servo system which maintains this balance condition. The iris diaphragm will be formed by two orthogonally mounted bi-lateral pairs of blades. Each of these blades has two symmetric edges which are at an angle of 135°. The four blades give an octagonal opening which is the same approximation to a circle as an eight blade standard iris diaphragm. The mechanical linkage has a backlash and uncertainty of less than one part in 10,000.

These manual setting devices are clearly capable of producing very high quality measurements, but with the availability of powerful mini-computers it becomes clear that automatic measurements are possible. The practicality of such systems is currently being investigated. A computer program has already been written which can accurately find the center of a symmetric profile even with the presence of large amounts of noise. The far more difficult problem is to develop a computer program which will selectively reject the image defects and produce measurements which are equivalent to or better than those of a human operator.

STANDARD ELECTRONIC COMPONENTS

The lead screws are driven by a servo motor which is controlled by a digital feedback loop. The axis setting servo system was developed because of the increasing complexity of the mechanical linkages required to accurately position the stages of the large measuring engines. There was also a need for a system which was easy to interface to a computer. A printed circuit drive motor, a D. C. tachometer, and a 20,000 count per turn incremental encoder are the only drive components attached to each lead screw in addition to the usual graduated dial and hand knob. There are no gears or clutches in the drive system and the knob on the dividing head can be used for making settings when all of the power to the drive motor is turned off. The usual manual positioning operation uses a controlling hand wheel and incremental encoder which can be mounted at any location which is convenient for the operator. The control voltage to the drive motor is generated by a 12 bit digital to analog converter which is connected to the least significant 12 bits of a 24 bit reversing counter. For a positive carriage motion the counter is incremented by the hand wheel incremental encoder, a joy stick controlled variable oscillator, a constant power line synchronized oscillator controlled by a front panel selector or a computer interface unit which generates a specified number of steps at a specified rate. The counter is decremented by the encoder on the lead screw. When the counter is at zero, the motor has a dither motion of ± one count, i.e., 1/20 of a micron.

This system provides flexible and accurate control. The hand wheel control gives a tactile response which is remarkably similar to a direct mechanical linkage, but is both simpler and more responsive than the previously used linkage. This is especially true for the cross motion which rides on the lower stage because the spline coupling is now replaced by a few flexible wires. By the use of prescalers on the hand wheel encoder output it is possible to increase the hand wheel sensitivity. For many applications a 5:1 ratio appears to be very effective, i.e., five turns of the hand wheel corresponds to one turn of the lead screw. This type of control preserves all of the advantages of a sensitive manual linkage but also allows for automated control, remote location control, and the use of a variety of different manual controls. The lead screw encoder pulses are used to drive a register which indicates the carriage position with a resolution of 0.1 micron. Linear encoders can also be used to indicate the carriage position, but these devices generally have a lower resolution.

When the engine is being used as a microphotometer the emulsion exposure can be measured by a linear transmission amplifier or a logarithmic converter which gives

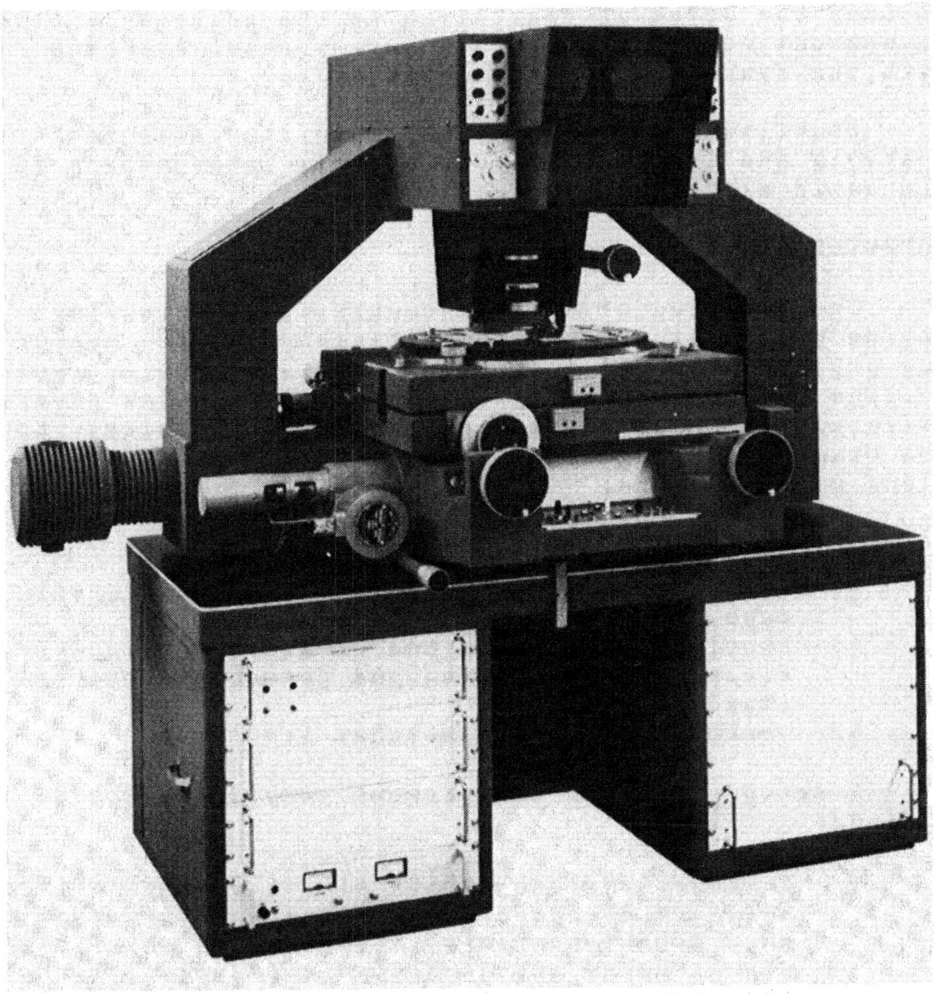

Fig. 2. 250×250 mm (10×10 inches) measuring engine with dual image scanner.

a density output. The recording can be made on a strip chart recorder or processed through an analog to digital converter which sends the data to a computer or hardwired data system. When the strip chart recorder is used the drive is controlled by the adjustable line -synchronized oscillator to insure precise tracking with the synchronous motor chart drive.

Stabilized power supplies, motorized slit mask controls and limit indicators are all incorporated in the basic electronic system.

COMPUTER SYSTEM

The addition of a mini-computer to a measuring engine greatly increases the efficiency of the measuring process. In all instrument related computer applications the computer program can be written for several different levels of sophistication and complexity. For the Grant measuring engines the most elementary functions would include:

1. Data acquisition.
2. Data display on a C.R.T. terminal.
3. Data recording on punched cards and magnetic tape.
4. Provision for entry and recording of identification; information and comments about the image quality.
5. Control of one dimensional scan rate.

A second level of measurement complexity would include:

1. A memory register which stores the approximate position of objects which are manually located on a prototype plate. When subsequent plates are oriented and positioned at a starting point the computer can then automatically move the carriages to bring each new image to be measured within the field of the viewing screen. This rough setting would be stored after each image is measured and recorded. This mode of operation is particularly effective for a series of parallax plates or spectra of similar objects. The operator would make the final setting but the manual process of locating each image or spectral line would be eliminated.

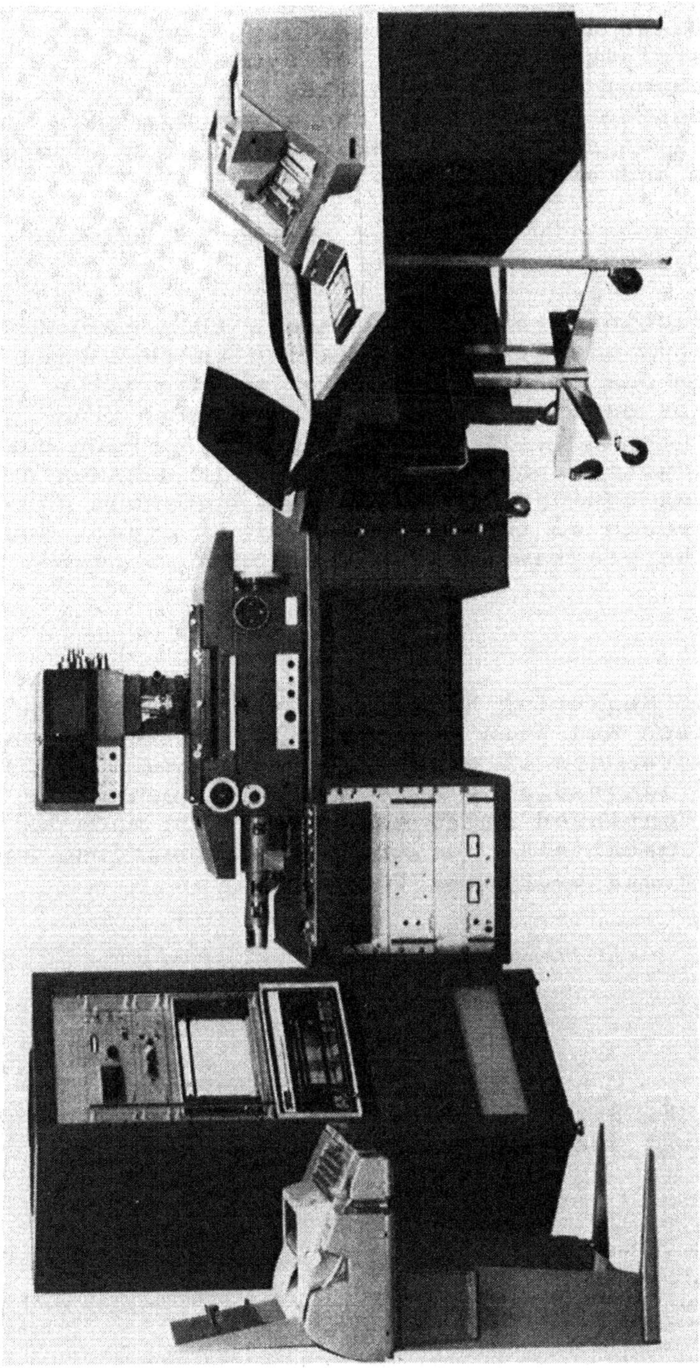

Fig. 3. Single coordinate measuring engine in spectrographic comparator configuration with computer controller and data system.

2. Automatic image measurement can be made by a program which scans the measuring slit across an image and locates the image center by determining the center of symmetry or the center of a fitted profile. This type of measuring process has the advantage that it can give additional information, e.g. image size and asymmetry.

3. Automatically generated raster or boustrophedonic image scans.

Data reduction programs can range from provisional spectral line wavelength determination to full data reduction from one or more series of measurements. If the computer is equipped with disk memories, line printers and plotters, it is possible to do very complete data reduction programs. The Grant engines are currently being equipped with Digital Equipment Corporation computers of the PDP 8 and PDP 11 type. Future systems will be programmed with the Forth software system.

SUMMARY

The Grant Measuring Engine is a high precision instrument which has been successfully applied to many types of measurements of photographic images. It is versatile and adaptable to many different measuring operations. Continued improvements will be made as the technology of materials, construction, detectors, and digital components continues to evolve.

REFERENCE

Tomkins, F. S., and Fred, M: 1951, Journal of the Optical Society of America, 41, 9.

THE U. S. NAVAL OBSERVATORY'S AUTOMATIC MEASURING MACHINES*

K. Aa. Strand

U. S. Naval Observatory
Washington, D. C.

ABSTRACT

The SAMM automatic measuring machine acquired in 1966 was built for the specific purpose of measuring parallax plates obtained with the 61-inch astrometric reflector located at the Observatory's Flagstaff Station in Flagstaff, Arizona. More than 50,000 plates have been measured with this machine.

The STARSCAN, acquired in 1974, is similar in principle to the first machine but reflects certain improvements in solid state circuitry, minicomputer technology, and data handling.

*References for the text are:
Publ. USNO, 2nd Ser., XX Part I, p.19, 1971.
Effective Systems Integration and Optical Design, Seminar Proceedings, Society of Photo-Optical Instrumentation Engineers (SPIE), 54, p.17, 1975.

de Jager/Nieuwenhuijzen (eds.), Image Processing Techniques in Astronomy. 209.

THE TWO-COORDINATE GRANT MACHINE IN LONG-FOCUS ASTROMETRY

John L. Hershey

Sproul Observatory, Swarthmore College
Swarthmore, Pennsylvania

ABSTRACT. A two-coordinate Grant machine at the Sproul Observatory has been in successful operation for the past three and one-half years. Approximately 25,000 plates taken with the 61 cm long focus refractor have now been measured, including plate series ranging up to several thousand plates. The speed of measuring has been greatly increased over manual methods and the error of measurement appreciably reduced. The machine is well suited to the particular needs of differential astrometry.

1. INTRODUCTION

The Sproul Observatory, at Swarthmore College near Philadelphia, Pa., has been continuously engaged in long-focus photographic astrometry since 1911 using the 61 cm. refractor. The focal length of the objective is 10.93 meters, which forms measurable stellar images with a minimum diameter of 40 microns and more typically 100 to 150 microns in diameter. Photographs of stellar fields are taken on 12.7 x 17.8 cm. plates covering about one-third of a square degree. The research is directed towards the determination of stellar trigonometric parallaxes, double star orbits, masses, mass ratios and long term surveillance of nearby stars for evidence of orbital motion caused by possible unseen companions.

The primary desideratum for a plate measuring machine in this type of astrometry is the most accurate determination of changes from plate to plate of the relative positions of star images in two coordinates. Formal probable errors of parallaxes from long plate series have been reaching below 0.1 micron or 1000 Å which

de Jager/Nieuwenhuijzen (eds.), Image Processing Techniques in Astronomy. 211-216.

is well under one-thousandth of the diameter of a typical image.

Over 120,000 plates (two as a rule on one piece of glass) have been taken and the ever increasing collection is kept in an underground vault under controlled temperature and humidity. During the past thirty years the speed of the photographic plates had been gradually increasing and exposure times were reduced, so that the rate of plate taking began to exceed by several fold the rate at which plates could be measured by cross wire bisection and manual recording.

The urgent need for faster measuring techniques in the mid-1960's prompted planning with other observatories and the Grant Instrument Company of Berkeley, California for the design and construction of a two-screw measuring machine satisfying the Sproul specifications. Four machines were made simultaneously with a number of options to satisfy the needs of each prospective customer. At that time the one-coordinate Grant spectrum comparator had proven successful in spectroscopy at numerous institutions. At present two of the large Grant machines are in use for long-focus astrometry; the Sproul machine has been in daily use since its installation in 1971 and approximately 25,000 plates have been measured.

2. DESCRIPTION OF THE GRANT MACHINE AT THE SPROUL OBSERVATORY

The two-coordinate machine is much larger in weight and volume than the single coordinate machine as a result of the capacity for two dimensional motion over 25.5 x 25.5 cm. plates. The size and external design of the machine may be seen in the photograph, Fig. 1. Fast motion is controlled simultaneously in both coordinates by a single variable speed control lever or "joy stick." Handwheels are engaged automatically by magnetic clutches when the slewing motors are stopped. The rotation of the measuring screws is detected by rotary pulse generators giving 2000 counts per revolution, corresponding to one-half micron linear units.

The digital hardware was designed and built by the Zehntel Corporation of Berkeley, California. Excellent documentation of the digital system was provided. The machine as a whole, including the digital system has performed for three and one-half years with time out of service for repairs amounting at most to two or three percent of the total. Most of this time loss was due to problems and repairs on the Trump-Ross rotary pulse generators during the first year. The digital system is interfaced with an 029 IBM card punch. Thumbwheel switches are read out to the first ten columns of each card and two thumbwheel switches are read onto the card along with each of five available 12-digit fields for the X and Y positions.

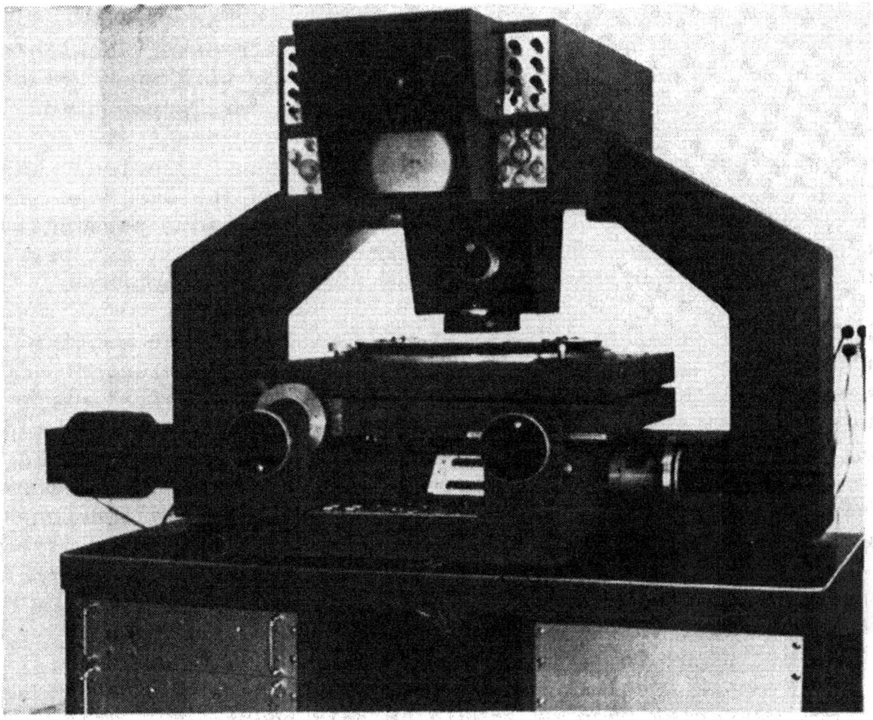

Fig. 1. The two-coordinate Grant machine at the Sproul Observatory.

The measurement of a star image position is accomplished, as in the one-coordinate Grant machine, by superimposing an image density profile upon its mirror image on an oscilloscope screen. The profiles are superimposed on separate screens for each coordinate, under manual control of the screws by the handwheels. By judicious settings of the scanning slit heights, width, and the oscilloscope electronics, well peaked profiles of the circular images can be formed for precise settings.

3. PERFORMANCE

The 25,000 plates measured on the Grant machine during the past few years yield many comparisons with measurements of plate series by visual bisection with cross wires on the old "classical" one screw machines. Usually a decrease of 30% in the error of reduced star positions is achieved with the Grant machine, corresponding to a doubling of plate weight. This accuracy is achieved with one setting per reference star image of the Grant profiles as compared with four settings of visual bisection with cross-wires. The high accuracy of the Grant machine could conceivably be approached by a large number of visual cross-wire bisections, by an expert measurer with intense concentration. However, the Grant machine's accuracy can be achieved by an unskilled operator with much greater ease of operation and at much higher speed, and for longer time intervals of measuring. Moreover the Grant measurements are far less personal than measurements obtained by visual bisection. The Grant machine makes the greatest gain in accuracy on the large images from brighter stars or poor seeing. Although some personal judgment is required in overlapping asymmetric profiles, any personal differences are independent of the direction of approach.

Most of the plates have been measured with only one setting on reference star images and two settings on the central star which opens and closes the measuring sequence. The typical error of repeatability of the position of one image after plate reduction in a stellar reference frame is approximately one micron which is less than the intrinsic error of the photographic process. Experiments with several plate series have been made with two independent measurements of each plate. By comparing the reduced positions from two measurements on each of several hundred plates the error of one measurement of the Grant machine has been found to be approximately one-half the error of the photographic positional information of images in a stellar reference frame. Thus the maximum improvement possible in the present photographic technique by a perfect measuring machine should be only about 10%. This limit would predict a five percent decrease in probable error when the mean of two independent measurements is taken. Reductions in error of 5% or less have been found experimentally where parallax solutions have been made on several plate series from single measurements and from the mean of two measurements. It is a matter of personal opinion whether it is wiser to double plate processing time or increase the number of plates by 10% or less to achieve an equivalent level of accuracy.

Although the setting on images is done manually on the Grant machine, the operators have found that the process of superimposing the profiles in both coordinates, in an iterative fashion, soon becomes a matter of reflex action and is accomplished in a matter of seconds with no physiological strain. The manual mode of positioning allows the operator to avoid errors from obvious image defects, such as adjacent dust specks, which an automated bisection device would not recognize. An automatic centering system would center on an image in a few seconds less; however an appreciable fraction of the measuring time is spent in moving from star to star, inserting and removing plates, and recording plate data and date of measuring. Hence an automatic centering system would not speed the entire measuring process by a large factor when only 20-50 images are measured per plate.

At present the Grant machine is operated daily at somewhat under 8 hours per day, and less in the summer, but plates are being measured at approximately the same rate that photographs are taken at the telescope. This rate represents a factor of five to ten in the speed of the complete process of extracting positions from plates, even though the average number of reference stars measured has increased somewhat. The increase in speed is due in part to the use of the computer for plate reductions from the punched measurement cards.

The measuring process in long-focus astrometry is differential in nature, requiring only the measurement of changes in relative positions of star images from plate to plate. As a result, an exhaustive program of testing absolute positional performances of the Grant machine has not been carried out at the Sproul Observatory. There is evidence of a "hysteresis" error on the level of one micron which shows as a difference in position depending upon the preceding direction of slewing motion. Since all plates in a series are always measured in the same orientation, within a degree or two, and in the same sequence of reference stars, the hysteresis affect should not have harmful effects in differential astrometry. In spite of this problem, the agreement of mean relative position of the central star in several plate series between the Grant and the old machines, has been very close, typically at the level of one-half micron.

Tests have been made for table rotation with an autocollimator, which revealed rotation of a few seconds of arc across the full travel of the table. Again the measuring in the same orientation and star sequence and the use of only one fourth to one half of the full travel of the tables, eliminates this as a problem in differential astrometry.

The long-term stability of the machine is being monitored by measuring a test plate, two or more times each month for the past three years. Some evidence of drift up to one micron exists on a time scale of one year, but no cause in the machine or test plate has been identified. However, plate series are measured, within a few months, in random order to randomize any effects of this nature. It is hoped that the machine is sufficiently stable and impersonal that future additions to plate series can be measured and included without a measuring equation. In any case a comparison with past measurements can be made in a short time. Thermal effects in the machine might be a problem where a long interval of time is required to measure a plate. The Sproul plates are usually measured in less than ten minutes which is short compared with any expected changes due to mechanical or electronic thermal effects. Furthermore a plate reduction for scale and orientation reduce any effects which occur in a continuous manner during plate measurement if the pattern of measuring is circular or at least non-random, and, as mentioned before, the central star opens and closes the sequence of measurements.

4. CONCLUSION

The two-coordinate Grant machine has been in continuous successful operation at the Sproul Observatory for the past three and one-half years and operates at a speed and accuracy which has made it worthwhile to measure several long plate series of which a substantial fraction had been measured on the old machines, such as Barnard's star, VV Cephei, Ross 614, 61 Cygni, M1b 377 and VW Cephei. Numerous shorter series have also been measured and the results of the longer and shorter series have now been published. Some representative examples with comments on the machine performance are referenced below. A number of series are in press or in preparation and will soon appear in publication. The rate of plate measuring by the Grant machine should satisfy the anticipated needs of the Observatory. For the particular needs of the Sproul Observatory the Grant machine has provided a good compromise between low cost and complete automation.

The two-coordinate Grant machine was provided by a grant to Swarthmore College from the National Science Foundation.

REFERENCES

Heintz, W. D., 1974. A.J. 79, 819, "Astrometric study of four visual binaries."

Hershey, J. L., 1973. A.J. 78, 935, "Astrometric analysis of the triple star BD + 66°34."

Lippincott, S. L., and Hershey, J. L., 1972. A.J. 77, 679."Orbit, Mass Ratio, and Parallax of the Visual Binary Ross 614."

Lippincott, S. L., 1973. A.J. 78, 303. "Orbital motion and parallax of the two astrometric binaries BD + 6°398 and BD + 67°552."

van de Kamp, P., 1973. A.J. 78, 1009. "Parallax and mass ratio of the visual binary 61 Cygni."

THE COSMOS FACILITY AT THE ROYAL OBSERVATORY EDINBURGH

N. M. Pratt and R. Martin
Royal Observatory, Edinburgh, Scotland.

L. W. G. Alexander, G. S. Walker and P. R. Williams
Computer Application Services, Heriot-Watt University, Edinburgh, Scotland.

INTRODUCTION. COSMOS, which was conceived in February 1970 and is now in the final stages of development at the Royal Observatory, Edinburgh, finds and measures the Co-Ordinates, Sizes, Magnitudes, Orientations and Shapes of images of stars and galaxies at rates of up to 4,000 images per second. The measurements are recorded on magnetic tape, which may then be analysed on any suitable computer.

REQUIREMENT. The advent of large telescopes in the southern hemisphere, such as the 150" Anglo-Australian telescope at Siding Spring, led the UK Science Research Council to provide a 48" aperture wide angle Schmidt camera, also sited at Siding Spring, to survey the southern skies. Improvements in telescope technology and photographic materials have made possible the detection of much fainter and more distant objects, with better resolution. The performance of the new 48" Schmidt camera is such that the survey can be used directly to study the structure of the Universe in addition to providing data for more detailed studies by larger narrow angle telescopes. These circumstances clearly required the provision of a facility for the rapid and accurate detection of the images of the many faint galaxies present on the plates and the recording of this information, at high speed, in a form suitable for computer analysis.

A study was undertaken to provide a detailed specification to meet a number of requirements. To detect and measure all the images on a 355 mm. square plate in a time not exceeding one day. This implies a means of measuring the position, size, photographic density and shape

de Jager/Nieuwenhuijzen (eds.), Image Processing Techniques in Astronomy. 217-222.

of each image at rates of approximately one million/hour, with an accuracy suitable for statistical studies and also to allow computer analysis to separate the images of stars and galaxies. In addition, it was to be able to make similar, but very precise, measurements of selected stars and galaxies necessarily at lower speed, and to measure the positions and densities of lines in the spectra of the light of all stars and galaxies on photographs obtained using an objective prism with the telescope.

The study led to the formulation of a three mode method of operation, the modes being capable of independent use but arranged to provide complementary output, matched to the various astronomical requirements. Speed of operation being a prime consideration, a method of scanning was necessary which provided output at a rate matched to the data handling speeds of magnetic tape. Overall control by a dedicated mini computer ensured flexibility. In addition, it was foreseen that some aspects of plate measurement might require the digitisation of transmission values on a fine grid of points on all of a plate or, more probably, on selected areas. In order to provide precise information of the photographic magnitude of stars and of the photographic magnitude, axial ratio and orientation of galaxies, a second scanning system of high precision was envisaged which would examine selected images located by a previous high speed scan. It was accepted that the high resolution demanded by these considerations would limit the measurement rates.

Fortunately, the requirements permitted the use of the bed of the machine then in use with the Galaxy system of plate measurement: the mechanics of this instrument were designed and manufactured by Sogenique.

DESIGN IMPLEMENTATION. The instrument, which weighs about 5000 Kgm, consists of a Meehanite base casting carrying a compound horizontal X-Y carriage driven by hydraulic rams. Measurement of carriage position is by moiré fringe gratings providing a measurement interval of $\frac{1}{2}\mu m$. The linearity and orthogonality of the carriage motion are better than 2 arc seconds. The light sources for the two scanning densitometer systems are Ferranti micro-focus CRTs: these, together with the optics necessary to focus an image of the spot in the plane in the emulsion, are mounted above the plane of the carriage. (See Fig. 1)

Each densitometer is of dual path type allowing the continuous monitoring of the spot brightness. The light transmitted by the plate is collected and measured by photomultipliers, housed in a central well. The equipment is housed in a controlled environment with the temperature held to $\pm\frac{1}{2}^{\circ}C$.

Fig. 1. The COSMOS Instrument and Control Cabinet.

1. COARSE MEASUREMENT (Fig. 2)

The operation of this mode makes use of a linear scan CRT together with continuous controlled motion of the Y-Axis. The position of the spot in the scan is monitored, permitting this to be directly related to the main carriage measurement system. The optical components image the scan in the plane of the emulsion with a magnification of $\frac{1}{4}$. The scan operates on the basis of a fixed number of increments - 128 - whose size may be chosen to be 8μm, 16μm, or 32μm, the spot size being adjusted to match the chosen interval. The plate is scanned in lanes of width 1 mm, 2 mm or 4 mm respectively.

Hardware logic and storage analyse the images and produce 8 parameters/image with position information at the end of each scan in which one or more images terminate. These are transferred to magnetic tape. The limiting transfer rate is 4,000 images/sec which gives a handsome margin on the actual average image density of less than 20/square mm, giving 40 images/sec in practice with an 8μm increment.

(a)

(b)

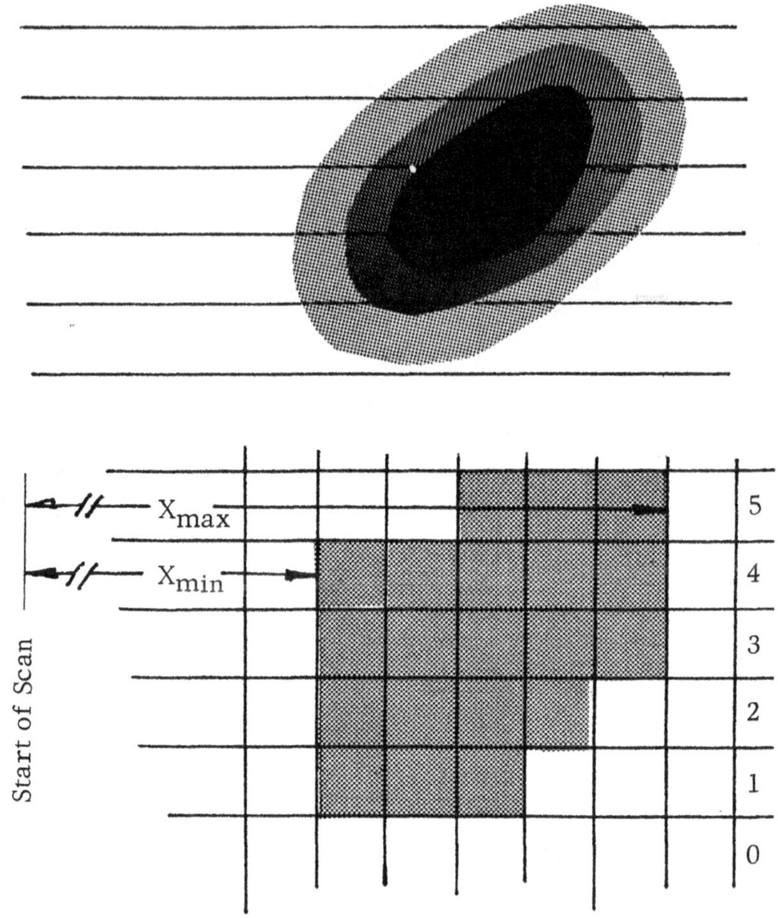

(c)

$\Sigma\,(X_{min} + X_{max})$
Σ Y change
} for calculation of centroid

A = area above threshold
X_{max}
X_{min}
N = number of scans in image
} for calculation of shape factor

Q = 0 or 1 for 1st or 2nd quadrant

T_{min} = minimum transmission in image

Fig. 2. The parameters of the images in Coarse Measurement which are transferred to magnetic tape. (a) Dissected image. (b) Image over threshold as picture elements. (c) List of parameters used.

The hardware also measures the maximum transmission encountered in each scan line and passes this to the computer. This measure permits the introduction of computer controlled background following. An analysis of this background is performed over a moving mm^2 to provide the hardware with a digital threshold level with which the digitised transmission level is compared, thus allowing the separation of images from the background.

The equipment allows the introduction of synthetic computer generated information to the CM system or, indeed, the output of a previous mapping mode output for the purpose of evaluating the validity of thresholding and image detecting techniques. This permits the repeated analysis of such images as double stars, re-entrant images, etc., permitting a rapid method of checking and refining the techniques.

2. MAPPING MODE

The output of the photomultipliers is integrated over the increment distance using the same scanning methods as for coarse measurement. The transmission value is digitised to an accuracy of 1% and transferred to magnetic tape.

This mode allows the detailed examination by off line computer analysis of areas of doubt or complexity such as bridges between galaxies or galaxies with complex structure.

3. FINE MODE

Having determined the positions and other parameters of images either directly using the CM or indirectly via a computer analysis of a mapping mode output, a punched paper tape containing the plate co-ordinates of interesting images is prepared at the off line computer. This tape is used as input to the fine mode and will contain the co-ordinates, ellipticity and an approximate orientation for non-circular images. The fine mode utilises the alternative densitometer system which consists of a generalised elliptical scan generator driving a micro focus CRT together with the optics necessary to focus an image in the plane of the emulsion. The generator produces a concentric family of 1024 ellipses sequentially. In this mode a large magnification factor may be employed 1/240, which entails the use of a 4 mm flat field objective, the depth of focus of such an objective is small, very much smaller indeed than the deviations from the plane of the plate itself. In order to overcome this, the objective rides on the surface of the plate, the separation being provided by an annular collar, with the force on the plate being reduced to 3 gm by a pneumatic balancing system.

The sequence of operations in fine mode consists of positioning the image close to the axis of the elliptical scan generator. The final precise centring is controlled by the output of the densitometer with the generator providing a circular scan. Using the approximate measures of ellipticity and orientation provided, there then follows a series of iterations of the scan generator under computer control to match ellipticity, orientation and the radial transmission profile. Two levels of iteration are envisaged to minimise cross talk between the parameters. On completion of these steps, the data is output to magnetic tape and will contain position co-ordinates of the image to $\pm\frac{1}{2}\mu$m r. m. s., together with ellipticity and orientation to an accuracy of 2%. Objective prism spectra will be examined using an elongated ellipse which is orientated parallel to the lines and constrained to search for emission or absorption events providing a measure of this position and intensity.

PERFORMANCE. All measurement systems are ultimately evaluated by results, however elegant the implementation. COSMOS is no exception and thus it is with no small satisfaction that we are able to state that the equipment has so far greatly exceeded our expectations. The initial plates from the 48" Schmidt camera at Siding Spring are presently being measured and are also being used to evaluate the ability of COSMOS to separate the images of stars and galaxies. This depends on the ability of the system to provide accurate determinations of the area above threshold and the value of minimum transmission in an image, the parameters which are crucial if this separation is to be effective.

Work carried out to date by Mr. Harvey MacGillivray has indicated that a separation efficiency of 95% has been achieved. This makes possible the rapid evaluation of the distribution of galaxies in the southern skies which facts, in conjunction with the previously stated facts relating to the penetration of the camera system, will undoubtedly have considerable repercussions in the field of cosmology. The analyses to date indicate that the number/density counts of galaxies found by Dodd et al. (in press)[1] have been confirmed independently by computer analysis of COSMOS produced data.

REFERENCE

1. Dodd et al., A study of 3000 faint Galaxies, M.N. R.A.S. (in press).

THE ESO S-3000 MEASURING MACHINE

R. M. WEST

ESO Sky Atlas Laboratory c/o CERN
1211 GENEVA 23, Switzerland

ABSTRACT. The ESO S-3000 measuring machine which was installed in 1974 in the Sky Atlas Laboratory is capable of measuring photographic plates up to 14 x 14 inch^2 with a formal accuracy of ± 1μm and ± .02 D in position and density. It is controlled by a 8 k Alpha 16 minicomputer. Software has been written for the ESO-TP HP 2100 computer system, that permits recognition and accurate measurement of stellar and extended images and filtering of plate faults.

1. INTRODUCTION

The ESO 1 m Schmidt telescope started taking plates from La Silla in 1973. The principal observational programme is at present the socalled ESO Quick Blue Survey of the Southern Sky (cf. West (1)) that is now (March 1975) about 1/3 completed. Other, smaller programmes have been carried out during the past two years and about 1000 plates in different colours and with different exposure times have now been taken.

For those involved in the ESO Schmidt project, the need arose early for some way of evaluating the images on these plates. Discussions were undertaken among astronomers inside and outside ESO and as a result it was proposed to acquire a measuring machine that would be capable of measuring large Schmidt plates at a reasonably fast rate. Main restraints for the choice of machine were of budgetary nature. The following tentative specifications were set : 1) measuring area 35 x 35 cm, 2) positions to ± 1 μm, 3) data rate at least 2 kHz, preferably 5 kHz, 4) density to ± 0.02 D in the interval 0-4 D.

de Jager/Nieuwenhuijzen (eds.), Image Processing Techniques in Astronomy. 223-228.

An investigation of the market for raster-scan machines in 1973 finally lead to the decision to acquire an Optronics S-3000, 14 x 14 inch measuring machine from OPTRONICS International, Chelmsford, Mass., USA. After initial debugging of various (in particular electronic) machine faults, the machine is now in regular operation at the ESO Sky Atlas Laboratory, Geneva.

In chosing this machine rather than a special-purpose machine, emphasis is placed on the ability to measure any object whatsoever that is seen on a Schmidt plate. It is believed that it is very important to learn from the beginning the process of image recognition and precise measurement, whether the image is stellar (i.e. round), or extended (galaxies and nebulae). In particular, plate faults may be numerous on a large Schmidt plate and an understanding of how they can most efficiently be eliminated at an early stage in the measuring procedure is of obvious importance. A raster-scan machine also offers the advantage of fast and economic measurement of objective prism spectra.

With this versatility, the speed of measurement and subsequent reduction procedures are inferior to those of specialized machines. So far, however, the machine has been fast enough for most reasonable measuring tasks that have arisen in the connection with the ESO Schmidt plates.

With the rapid development in the field of astronomical automation it is foreseen that a more advanced machine will be desirable at a future date (possibly a diode-array machine). The experience with the present machine may then serve as an excellent base for the future machine.

In what follows, some features of the ESO S-3000 machine (hardware and software) are discussed, that may be of particular interest to the users of this and other similar measuring machines in astronomy.

2. HARDWARE

The ESO S-3000 system consists of a 14 x 14 inch flat-bed scanning microdensitometer which is interfaced with an Alpha-16 computer with peripherals.

The photographic plate rests on an aluminium plate-holder that is driven in the X-direction by a precision lead screw. This stage is placed on top of a granite Y-stage, also moved by a precision lead screw. Both stages move on air bearings and the (X, Y) positions are read to $\pm$ 1 μm by two 14 inch linear Heidenhain encoders mounted on the central axes. The travel rate depends on the sampling interval and varies from 10 μm/sec to 5 cm/sec. The microphotometer is equipped with endswitches and air pressure control.

The density is recorded through a telecentric optical system (magnification 25 X - 400 X). Eight fixed square and rectangular apertures are available. An EMI photomultiplier records the transmitted light and the measured density (or transmission) is read into the computer. With the photographical plate mounted in horizontal position, a certain plate sag cannot be avoided and the S-3000 therefore has a focussing motor which is controlled by the computer.

The control system consists of a 8 K Alpha-16 minicomputer (1.6 μs cycling time) and the following peripherals : Two Kennedy 8108 digital magtape units with formatter, a Remex papertape reader, and an ASR-33 teletype. The computer is connected to the microdensitometer through an interface made by Optronics.

The control software occupies about 3K and is very efficient from the users' point of view. It has commands for positioning of the microphotometer table, for data transfers from microphotometer to magtape (i.e. plate scanning), from magtape to keyboard, etc... Data transfers take place in blocs of 500 density-values through two buffer arrays in the computer.

The raster-scan parameters are chosen by the operator through the teletype keyboard before the actual transfer (scan) takes place. Scans may be made from point to point, or along one axis with steps along the other axis, either uni-directionally or alternatingly direct-reverse (bi-directional mode).

The control programme which was delivered by Optronics, has been expanded with a routine for consecutive raster-scans, the individual scan parameters being input through the paper-tape reader.

The S-3000 was designed for a 5 kHz data transfer rate, but tests show that the present logarithmic amplifier is not quite as fast, and we have therefore prefered to use a 2 kHz rate for the moment. A new and faster circuit is being developed by the ESO-TP Division.

3. SOFTWARE

Good software organization and efficiency is of a paramount importance for a raster-scan machine. Great care must be taken in order to achieve a reasonable speed in image restoration and computation of the image parameters, and the astronomical goal - to measure physical parameters of celestial objects- should never be forgotten in the jungle of modern image handling techniques.

The output from the Alpha-16 computer consists of a series of records of density values on magtape. From this material the images in the plate field are reconstructed and the values of characterizing parameters are computed. For stars these are the position (geometric center or barycenter), the radius, the integrated density and the density profile, and for galaxies, the position, the size, an elongation measure and a rough morphological classification. At the moment only direct plates are reduced, but programmes for the automatic interpretation of objective prism spectra are being developed.

This analysis is done with a series of programmes (mostly in Fortran) which have been written for the ESO-TP Division HP 2100 computer system (32 K). With the operating system, each programme (with data arrays) may take up to 12 K. In what follows some details are given about these programmes.

The main programme first performs the reconstruction of images from the density data on the magtape. The image profiles from each scan line are sorted into individual images and whenever two consecutive scan lines do not any more show a particular image, that image is dumped into a file for detected images. The data in this file consists of the peripheral points of the image (at a certain level above the background), and the integrated density along the profile in each intersecting scan line. At the moment it is possible to construct simultaneously 100 images, which in dense fields corresponds to a maximum possible scan length of about 3 cm on ESO Schmidt plates.

After the detection of the images, another programme subjects each image to tests for squareness, circularity and density distribution. The stellar images are filtered very much in the same way as does the Luyten machine. Images which are not classified as stars normally are either very faint stars, galaxies or plate faults. Galaxies are recognized as reasonably large images that may deviate significantly from circularity, and for which the density profile is not the same as that of a star of the same size. The distinction between galaxies and plate faults is very difficult, and until now we can only filter the plate faults with certainty when they have a very steep density profile or a peculiar shape.

The programme collects the positions, the radii and the total integrated density over the image surface in a disk file. This file is then be used for further reduction, as for instance, when astrometric and photometric standard stars are available in the field, to compute equatorial coordinates and stellar magnitudes.

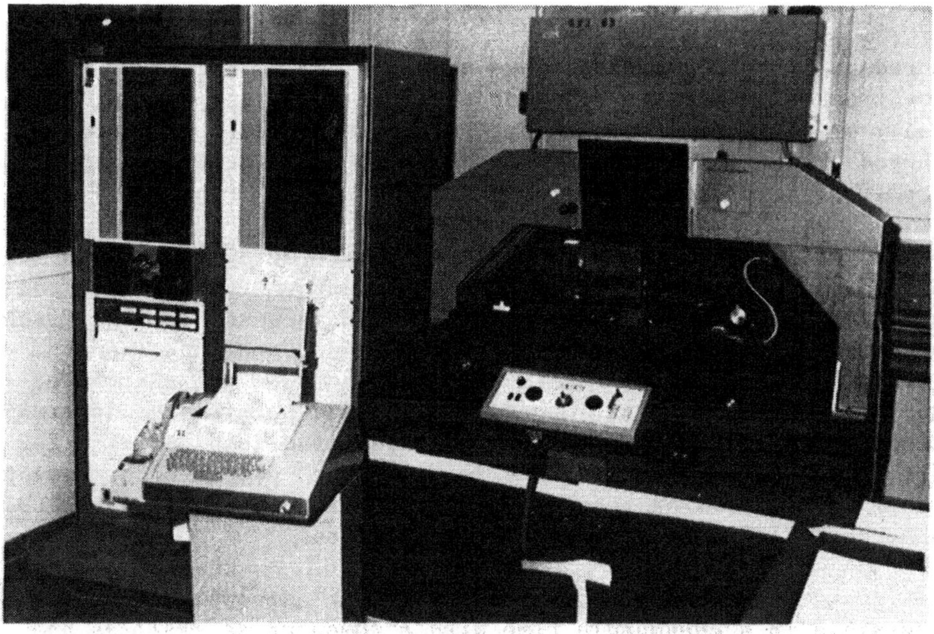

The ESO S-3000 Measuring Machine

Some figures may be appropriate at this place. A complete scan, covering 3 cm in X and 0.5 cm in Y (with 2 μm step in X and 10 μm in Y) took a little more than 1 hour with the S-3000. In a dense field, close to the Milky Way, 3700 images were found in this area, of which about 3000 were classified as stars with radius $\geq$ 10 μm. The rest were mainly very faint stars. The reduction time with the HP-2100 was about 3 hours, corresponding to a total time of 5 sec per star, roughly comparable to that of Galaxy in the measuring phase.

The positional r.m.s. accuracy, which has been estimated from repeated scans, is about $\pm$ 0.8 μm in X and $\pm$ 1.1 μm in Y for the geometric center. The r.m.s. of stellar radii is $\pm$ 0.5 μm for images with radii between 20 μm and 80 μm, the medium size images being the most accurate. These figures correspond to 2 μm steps in X and 10 μm steps in Y. When repeating a scan after 4-6 hours, small zero-point offsets are noted (about $\pm$ 1 μm both in X and Y).

If so desired, a paper tape is punched that can be used with the Alpha-16 computer to position the stage at the centers of the detected images and thereby verify visually the correct assignment of parameters to the images. This feature is of great importance for the determination of threshold values for the image classification parameters.

The reduction procedure described above is now in the process of being improved. There is no doubt that it can be speeded up considerably by software streamlining (perhaps by a factor 2), and it is hoped that the positional and the photometric accuracy may also be somewhat improved.

Various astronomical measuring programmes are being undertaken. The machine is evidently most efficient in dense fields where the image/clear-plate surface ratio is as large as possible. The first programme is therefore the measurement of some stellar clusters, for which ESO Schmidt plates in different colours are available. Comparison of the image files from each plate permits the determination of magnitudes and colours and thus membership decision for faint stars. Another possibility is to "blink" plates of different epochs. As a preliminary test we intend to intercompare on ten LMC plates a small field with many variable stars.

It is gratifying to notice that this kind of astronomical research is possible in a reasonable time with a computer as small as the HP-2100. To avoid the storage of the density data on magtape, the possibility of a direct connection between the S-3000 and the HP-2100 system is now under study.

4. SUMMARY

The experience with the ESO S-3000 measuring machine has shown that it is capable of performing useful astronomical measurements of Schmidt plates at a reasonably fast rate by means of two-dimensional raster scanning. The application of a diode array would shorten the measuring time with a factor equal to the number of diodes. The installation of a diode array in the present machine (substituting the slit) would therefore greatly enhance its capacity. Studies are now under way at ESO to investigate the feasability of such a system with on-line reduction.

ACKNOWLEDGEMENTS

It is a pleasure to acknowledge the help of Dr. D. Wiskott of CERN with the initial testing of the S-3000. Thanks are also due to the ESO-TP Control Section for support in the fields of electronics and programming.

REFERENCE

1. R.M. West, ESO Bull., 10, p. 25 (1974)

IRIS - A TWO AXIS COMPARATOR AND MICRODENSITOMETER USING TWO DIFFERENT SCANNING MODES

N. Åslund, N. von Gersdorff, R. Norberg and J.A. Nordin
Physics IV, Royal Institute of Technology,
100 44 STOCKHOLM 70

ABSTRACT. IRIS (Image Reading Instrument System), built by SAAB-SCANIA in cooperation with Physics IV, KTH, Stockholm, has been conceived with the intent of allowing interaction between three parties: an operator, a measuring machine and a computer.

Seen from the computer the photographic plate in IRIS is a read-only memory, similar to other mass storages. As such it has the additional advantage (e.g. compared with magnetic tape) of allowing simultaneous, random access in two dimensions. At the same time it is also available for visual observations, by means of optical and electronic displays.

To provide for this twofold role of the plate IRIS is equipped with two different scanning systems. A small, rectangular region around the measuring point is covered by means of fast scanning, with a repetition frequency fitted to the speed of human apperception. In addition scanning can be performed by moving the tables, as is common in most measuring machines.

The fast scanning allows for visual displays on CRT screens of the intensity profiles along two perpendicular directions through the central measuring point. Moreover it supplies the measuring machine with the information necessary to perform automatic settings on objects on the plate, e.g. on stars. Further, it makes it possible for the computer to calculate the gradient of the density distribution, e.g. in order to track equidensity curves. It also offers the capability of fast raster scanning.

To make certain a high photometric accuracy IRIS has a calibration system which relates all measurements to a scale defined by two

de Jager/Nieuwenhuijzen (eds.), Image Processing Techniques in Astronomy. 229-236.

separate reference levels. This calibration is synchronized with the fast scanning and it compensates for changes in illumination, slit width, PM-tube and amplifier gain.

Accounts are given on the application of IRIS to different measuring tasks.

INTRODUCTION

An early example of the use of two different scanning systems in a measuring machine is given by the so called Spiral Reader. This machine was originally built by prof. L. Alvarez at Berkeley to measure bubble chamber plates. Here a fast rotating system provides for an efficient use of the measuring time by allowing a desired concentration on the objects of interest. The movement from one such object to another is accomplished by moving the measuring tables.

The same basic philosophy has been adopted in IRIS. However, IRIS is a general purpose machine, that can be used to measure any kind of patterns recorded on photographic plates. Hence the fast scanning has been designed to supply very basic information about the density distribution around the measuring point, e.g. the size and direction of the density gradient. Like the Spiral Reader this new machine has been designed to work with a computer. Further, the idea of making use of interactive procedures has been extended and has led to a design, that allows efficient communication between three parties: the operator, the measuring machine and the computer.

These objectives sometimes necessitate an unorthodox design, since they may lead to conflicting demands. As an example we may take the problem of illuminating the plate. To display the plate to the operator, a large part of it has to be illuminated. On the other hand the photometric readings communicated to the computer should not be distorted due to straylight. This raises a demand for a very small illuminated area. The method chosen in IRIS is to use light of different spectral composition for the different purposes.

Another example comes from the demand that the operator and the machine be close to each other, physically, which is in conflict with the demand that mechanical or thermal disturbances should be avoided.

In addition there have also been some more general principles employed in the design work:

1) To acquire the high accuracy (both in positioning and photometry) rather by automatic calibrations than by large-scale

use of high precision components.

2) To make use of the on-line computer as a tool in the process of refining the design.

The design of the machine reflects the cooperative effort of a research group and an industry. Thus a basic idea has been to create a multi-purpose platform that can be suited to different measuring tasks by modular expansions. This platform comprises both hardware and software.

The present exposition serves to exemplify how these different objectives and principles materialize in the machine and in the use of it.

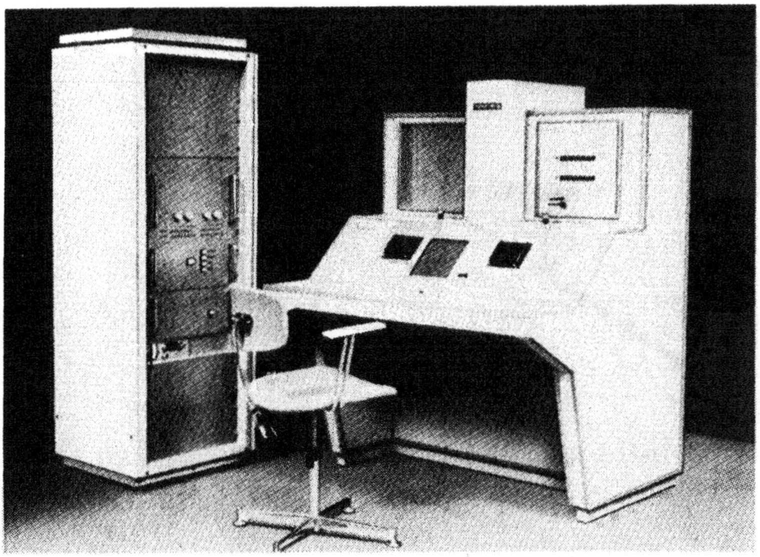

Fig.1. The operator controls the measurements from a console that includes both optical and electronic displays. The plate is mounted into the machine from behind. The outer cover is mechanically isolated from the inner parts of the machine.

THE TWO SCANNING SYSTEMS

The measuring tables, which constitute the slow scanning system, are vertical to avoid bending of the plate due to gravitation. Actual positions of the tables are measured by separate linear digital encoders.

The measurement area is 300 mm x 300 mm and the resolution of the encoders is 0.5 micrometer.

The fast scanning is performed simultaneously in the X- and Y directions. It is a back-and forth scanning which covers about 2 mm on the plate in both directions. The frequency is about 65 Hz, which allows non-flickering displays of the intensity profiles on CRT screens. The scanning is accomplished by means of vibrating prisms, employing an optical principle that is free of distorsions, Åslund (1965). The vibrating prism is carried by a dynamically balanced swing, that moves completely free of mechanical frictions, due to the use of flexure pivots.

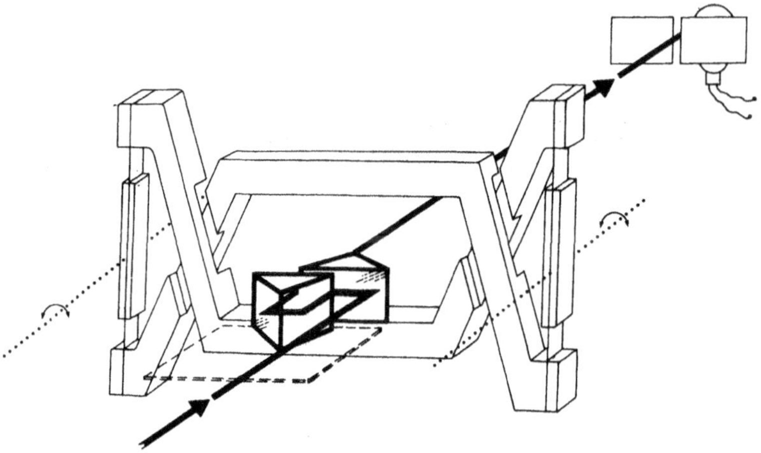

Fig.2. Fast periodic scanning is performed by means of a vibrating prism.

THE TWO REFERENCE CHANNELS

The illuminated area on the plate has a diameter of 20 mm. The image of this area is projected on a viewing screen (magnification 8x), and on two other planes, which each contain a detector slit (Fig.3). This separation is accomplished by means of beam-splitting Abbe cubes. These cubes also unite the light from the plate with the light that comes from the two reference channels.

The one reference channel is for positioning. It includes the index mark. The image of this mark is superimposed on the image of the plate. The two images are produced by light of different spectral composition and are detected separately, behind the slit of each of the two detector assemblies, by detectors of different spectral sensitivity. The beams forming the two superimposed images pass either of the two fast scanning devices, which effectuate X- and Y-scanning respectively.

Refocussing can be performed during the course of the measurements, by moving the detector assemblies. An unintentional, sidewise displacement of the slit when doing this will introduce no parallaxes, since it will not change the position of the plate image in relation to the image of the index mark. Further the imaging is telecentric so that the magnification will not be affected by this refocusing.

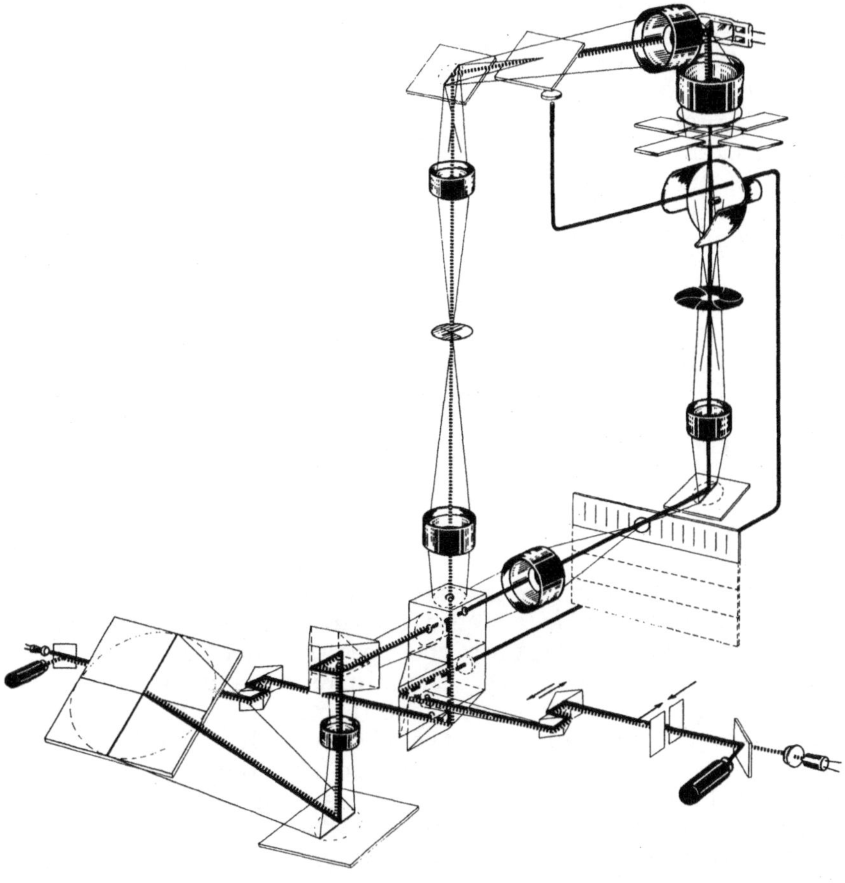

Fig.3. Reference channels for positioning (in the front) and for intensity (fiber optics) allow continous comparisons with fixed references. The plate is indicated in the figure but not the measuring tables.

An example of the expansion capability of the design is provided by the index mark. It can be replaced by a grid of black squares, which define length scales for the fast scanning motions in the X- and Y directions. This makes possible very fast raster scanning with a fixed step length. The first grid produced for this purpose has a step length of 25 micrometers.

The second reference channel is for intensity. Once every cycle of the fast scanning the illumination of the plate is turned off by means of a chopper which is synchronized with the movement of the vibrating prisms. This defines a reference level for darkness. Further, while the illumination of the plate is off, a light pulse that bypasses the plate is supplied to the detectors. The magnitude of the resulting electronic pulse relative to the darkness level is compared with a fixed reference voltage. The electronic amplification is regulated to keep this magnitude constant. In this way drifts, not only in the lamp but also in the PM tube and in the amplifier are compensated for. Consequently the PM tube may be driven with a higher current than normally, since slow changes of this tube are of no consequences. This benefits the dynamic range of the photometric measurements (presently the dynamic range is just below 4 decades).

Since the slit is also included in the chain of elements that is encompassed by this regulation, the slit width may be changed without changing the intensity scale. The optimal slit width can thus be set very conveniently by operator interaction, since a change of the slit will only change the form and not the size of the profile on the CRT screen.

PROCEDURES TO TEST AND TO REFINE THE MACHINE

Interactive procedures have been established to verify and refine the merits of the machine as a densitometer and as a comparator. To test the photometric qualities the operator decides what objects shall be investigated, e.g. the fields of a gray scale, and directs the machine manually to these objects. The position coordinates are stored, together with the photometric data from the X and Y detectors. These latter values are obtained by integrating the transmitted light over rectangular measuring windows. The one side of such a rectangle is determined by how long a part of each cycle of the fast scanning the detected signal is integrated. (The actual setting is indicated on the CRT screen by a strengthening of the profile along that portion). The other side is determined by the width of a preslit.

The computer will take over and repeat the measurements an arbitrary number of times. This gives a good opportunity for studying the effects of different changes of machine parameters on these measurements.

This is also done when testing the machine as a comparator. A precision grid plate is measured, and the machine makes automatic settings on the crossing points of this grid. This is possible since the machine has the ability of setting on the median in the X- and Y- direction of any intensity distribution on the plate, Åslund (1965). The (rectangular) areas of integration employed at these settings can be choosen at will. In the present application the median setting is used both to make the machine follow the mesh lines and to make it perform repeated settings on the crossing points.

During these measurements, which allow a separation of the random and the systematic errors, the computer also collects data from different sensors placed at arbitrary points of the machine, measuring voltages or temperatures. A data analysis is performed to reveal any correlation between the systematic errors observed at the measurements on the grid and the changes of these parameters. In this way it has been possible to trace the main error sources in the machine and to take appropriate counter measures. Presently the environment of the machine is not temperature controlled but at night test, with temperature changes within ± 0.5 ^{o}C, the machine has remained stable for several hours within ± 0.5 micrometers.

APPLICATIONS OF THE MACHINE

One major application project concerns the measuring and evaluation of optical spectra. An interactive procedure is employed, where the operator decides whether to employ the median setting procedure or to scan the line. The latter can be done step by step or by continuous sweeping. From the identified reference lines the computer calculates a coarse calibration curve, which can be improved by including more reference lines. Further, from this calibration curve the wavelength (or wave number) of a line under investigation can be displayed immediately, on a data screen. A main feature of the procedure is that every determination can be supplied with a standard deviation, to accompany it during the subsequent calculations.

Another application concerns the measuring of spectral reflectances of trees on aerial photographs of forests. The individual trees are approximated by polygons, defined manually by the operator. The computer refinds the same tree on other plates, representing recordings in other spectral regions, e.g. to determine the relative frequencies of different density values within these polygons. The results are presented as histograms on a data screen, and the aim of the study is to find out what makes it possible for a human interpreter to identify different types of trees on an aerial photograph. The method employed is representative for a basic idea of this machine, to make possible conversations between operator, computer and machine.

REFERENCES

Åslund, N: 1965, Dissert., Dept. of Physics, University of Stockholm.

MACHINE A MESURER DE L'OBSERVATOIRE DE HAUTE PROVENCE

Ch. FEHRENBACH et A. VIN

Observatoire de Haute Provence, Saint-Michel l'Observatoire, France

The measuring machine - Automatix 250 - has been designed at the Haute Provence Observatory, to reduce the plates of the coudé spectrographs of the Observatory. The automation of this machine has made it well adapted to repetitive line position measures of series of spectra of same dispersion. Its first application is the measure of radial velocities.

We have built the actual machine from an "one-axis" measuring machine of the Marseille Observatory.

The plate is placed on a carriage which can be moved of some 260 mm.

The observer sees the image of the spectrum on a screen, and shifts the carriage by means of a micrometer screw (Fig. 1). When the spectral line to be measured is in coïncidence with the reticle, the measure is recorded.

The automation of this machine has been done as follows :

The measure of a good number of plates of IAU radial

de Jager/Nieuwenhuijzen (eds.), Image Processing Techniques in Astronomy. 237-243.

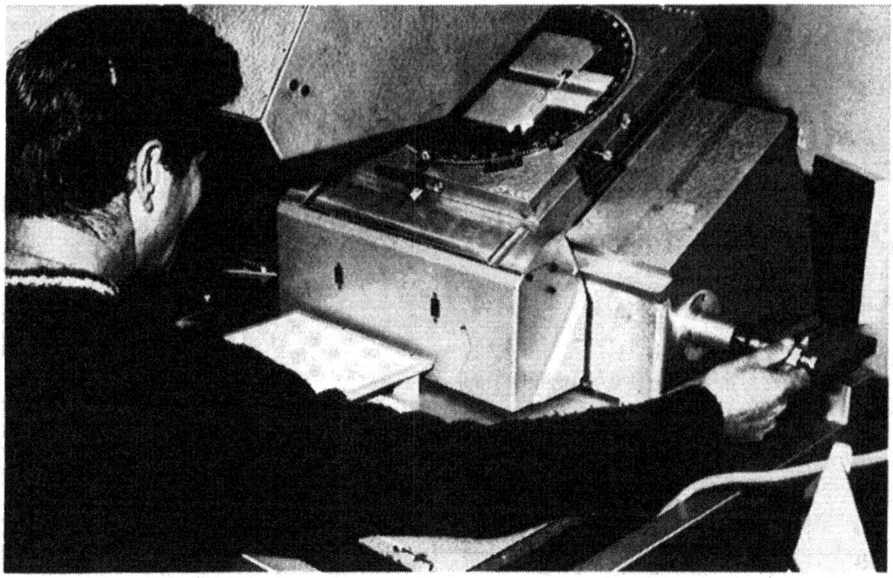

Figure 1.

velocities standards has enabled us to select a number of lines whose wavelength are not altered in a certain range of spectral types.

For instance, for stars of spectral type F5 to M, we have selected 21 lines, of which 9 have been already used by Petric (1) and 6 others have been published in the IAU transactions (2).

For wavelength calibration, we have selected 23 lines of the iron arc (3).

The theoretical abscissae of the selected lines are computed from the dispersion formulae of the spectrograph, and are stored in the computer.

Before measuring, the observer gives the necessary indications to the computer, such as spectral type and dispersion, which selects the adequate table.

Then, the computer drives the carriage such as the

selected lines come in sight of the observer. For each line, it indicates if the line to be measured is a stellar or iron arc one. The observer measures each line by mean of the micrometer screw. For each line, four measures are made, and the abscissa is automatically given to the computer.

At the end of the measure, the computer calculates the difference ΔXi between the measured and theoretical abscissae of comparison lines, and calculates also the parameters of the least square straight line.

$$\Delta X = AX + B$$

and $$X'i = Xi - \Delta X$$

It communicates to the observer A, B and ΔX, which enables him to eliminate a bad measure, for which ΔX is too big. (Fig. 2)

If a line is eliminated, the computer does all these calculations over again, and communicates to the observer the new results. If it appears that the line has been wrongly eliminated, the observer can reintroduce it.

When A and B have been computed for the comparison lines, the correction ΔX which should be applied to the measured abscissae of stellar lines to make them comparable to the table X_{*T} is computed for each line.

Let X_{*M} such an abscissa, the corrected abscissa is $X_{*M} - (AX_{*M} + B)$ and the radial velocity is

$$K\,(X_{*M} - (AX_{*M} + B) - X_{*T})$$

* 1767-
V

CLICHE : GB 2319
OBJET : UAI 005
DATE : 22/3/73
OPERATEUR : LR

SPECTRE DE COMPARAISON
A = +.9501407E-04 B = -.1046769E+02

RAIE	ABSCISSE THEORIQUE	DIFFERENCE	RESIDU	RA
05	+063699,6	-06,4	-02,0	(
06	+067024,8	-04,3	-00,3	(
07	+072476,2	-04,4	-00,9	(
08	+076267,8	-02,4	+00,8	(
09	+079208,7	-04,3	-01,4	(
10	+082521,2	-02,3	+00,3	
11	+088452,4	-01,6	+00,4	
12	+094386,7	+01,3	+02,7	
13	+099999,4	+00,1	+01,0	
14	+101780,4	-00,7	+00,0	
15	+106215,4	+00,2	+00,5	
16	+113531,9	+01,6	+01,2	
17	+118976,3	+00,7	-00,2	
18	+121730,5	+01,3	+00,2	
19	+127737,5	-00,1	-01,8	
20	+136159,8	+04,3	+01,8	
21	+138539,1	-00,5	-03,2	

CONTROLE OPERATEUR
-21,

SPECTRE DE COMPARAISON
A = +.1125105E-03 B = -.1196313E+02

RAIE	ABSCISSE THEORIQUE	DIFFERENCE	RESIDU	R
05	+063699,6	-06,4	-01,7	
06	+067024,8	-04,3	+00,1	
07	+072476,2	-04,4	-00,6	
08	+076267,8	-02,4	+00,9	
09	+079208,7	-04,3	-01,3	
10	+082521,2	-02,3	+00,3	
11	+088452,4	-01,6	+00,4	
12	+094386,7	+01,3	+02,6	
13	+099999,4	+00,1	+00,8	
14	+101780,4	-00,7	-00,2	
15	+106215,4	+00,2	+00,2	
16	+113531,9	+01,6	+00,7	
17	+118976,3	+00,7	-00,8	
18	+121730,5	+01,3	-00,5	
19	+127737,5	-00,1	-02,6	
20	+136159,8	+04,3	+00,9	

CONTROLE OPERATEUR

Figure 2.

The computer, besides giving the mean value of the radial velocity, gives the difference ΔV between the mean value and each singular value, to enable the observer to eliminate a line whose result seems to be bad (Fig. 3).

The computer calculates the internal mean error

$$\sigma = \sqrt{\frac{\Sigma(\Delta V)^2}{n(n-1)}}$$

which materializes the overall quality of the measure.

The whole system is driven by a Multi 8 computer built by Intertechnique, its core size is 16K x 8 bits (Fig. 4).

The computer controls directly the DC motor which moves the slide, and knows its position by counting the pulses of the incremental encoder fixed on it. The resolution of this encoder is 1 micron. The accuracy of the primary approach of the spectral lines to be measured is within 15 microns.

The observer gives his orders to the computer by means of a keyboard which displays the actual abscissa, the number and the nature (star or comparison) of line, as well as some function codes. The nature of the line to be measured is also repeated near the screen.

The accuracy of the radial velocity is in average $\pm$ 0.5 km/sec for the 20 Å/mm dispersion spectra, and $\pm$ 0.25 km/sec for the 12 Å/mm dispersion spectra, obviously for F5 to M spectral types. If we put it in terms of microns, this represents a mean error of $\pm$ 0.3 micron on the plate.

Almost a thousand spectra have already been measured. The average measuring time of one spectrum is of some 30 minutes.

SPECTRE ETOILE

VR = +020,5 KM/S

RAIE	ABSCISSE THEORIQUE	VITESSE	RESIDU	RAIE
01	+072478,0	+019,2	+01,3	01
02	+076960,1	+021,9	-01,4	02
03	+081596,3	+020,5	+00,0	03
04	+082521,0	+018,0	+02,5	04
05	+083969,2	+020,4	+00,1	05
06	+084635,8	+021,1	-00,6	06
07	+087082,5	+022,1	-01,6	07
08	+089552,4	+011,7	+08,8	08
09	+094391,4	+020,1	+00,3	09
10	+096891,6	+023,1	-02,6	10
11	+099157,8	+020,1	+00,4	11
12	+101161,0	+026,8	-06,3	12
13	+105301,7	+022,7	-02,2	13
14	+106493,7	+023,2	-02,7	14
15	+111707,3	+019,5	+01,0	15
16	+116866,4	+021,8	-01,4	16
17	+117478,9	+018,2	+02,3	17
18	+118973,2	+018,9	+01,6	18
19	+121732,7	+020,1	+00,4	19
20	+123427,7	+020,5	+00,0	20
21	+123786,7	+021,2	-00,7	21

SIGMA : +0,61

CONTROLE OPERATEUR
-08,

SPECTRE ETOILE

VR = +021,0 KM/S

RAIE	ABSCISSE THEORIQUE	VITESSE	RESIDU	RAIE
01	+072478,0	+019,2	+01,7	01
02	+076960,1	+021,9	-01,0	02
03	+081596,3	+020,5	+00,4	03
04	+082521,0	+018,0	+02,9	04
05	+083969,2	+020,4	+00,5	05
06	+084635,8	+021,1	-00,2	06
07	+087082,5	+022,1	-01,2	07
09	+094391,4	+020,1	+00,8	09
10	+096891,6	+023,1	-02,1	10
11	+099157,8	+020,1	+00,8	11
12	+101161,0	+026,8	-05,9	12
13	+105301,7	+022,7	-01,8	13
14	+106493,7	+023,2	-02,3	14
15	+111707,3	+019,5	+01,4	15
16	+116866,4	+021,8	-00,9	16
17	+117478,9	+018,2	+02,7	17
18	+118973,2	+018,9	+02,0	18
19	+121732,7	+020,1	+00,8	19
20	+123427,7	+020,5	+00,4	20
21	+123786,7	+021,2	-00,2	21

SIGMA : +0,45

CONTROLE OPERATEUR

Figure 3.

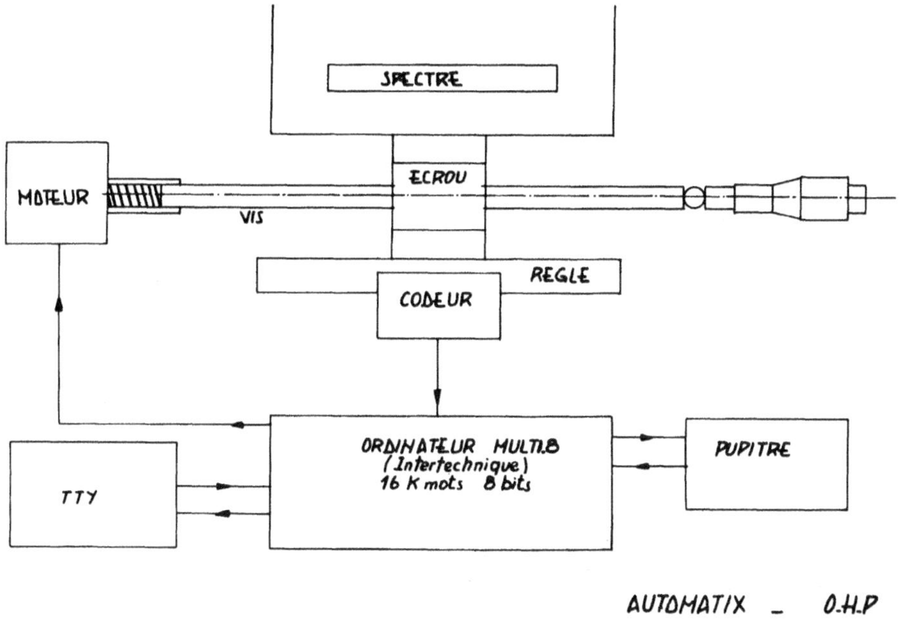

Figure 4. -

This machine will be connected to a larger computer, this will allow us to build up several different measuring programs. We shall then put in operation a rotating mirror self measuring device.

REFERENCES

1. R.M. Petrie, J. R. astr. Soc. 1946, XL, 325.
2. Trans. int. astr. Un. 1935, V, 196.
3. Ch. Fehrenbach, Astron. & Astrophys. 1972, 19, 427.

THE DESIGN OF THE NEW S.R.C. AUTOMATED PHOTOGRAPHIC MEASURING SYSTEM

E.J. Kibblewhite, M.T. Bridgeland, T. Hooley, D. Horne

Institute of Astronomy, Madingley Road, Cambridge, England
CB3 OHA

ABSTRACT. A general purpose automated photographic measuring system is being developed at Cambridge consisting of a novel laser beam scanning microphotometer controlled by an on-line computer. The system can be programmed to carry out any type of photographic analysis and will not only be able to measure parameters of images down to low isophote levels but will collate and compare those data taken on successive plates of the same sky area with a minimum of off-line computing.

Analysis of non-stellar images often requires us to use information very near the background level of the plate. Machines such as COSMOS and the CDC-Luyten machine delineate images from the background using a threshold set some distance above the background. Spurious images due to grain noise appear above the threshold as small spikes which can easily be eliminated by the computer. As the threshold is reduced the noise images become very much more numerous (at a threshold of one standard deviation $> 10^5$ cm^{-2} using a 10-micron diameter scanning spot) and much larger in area. At these low thresholds it becomes essential to accurately measure the background and use a number of different image shape parameters to decide whether or not an image is due to noise. These two factors require extensive computing facilities. The volume of data is so large ($\sim 10^{10}$ bits or 300 reels of tape/plate) that on-line data reduction is the only economic solution to this problem, the plate itself acting as the mass-storage device. On-line facilities become even more important for other work (e.g. objective prism spectra) near the plate limit.

de Jager/Nieuwenhuijzen (eds.), Image Processing Techniques in Astronomy. 245-246.

We have been designing an automatic photographic measuring system at Cambridge to overcome some of these problems. The plate is mounted on a precision x-y table and scanned in discrete areas of 1 cm^2 by a Laser Beam Scanner. Data is digitised from the plate into a 14 bit transmission estimate at a rate of 300,000 samples/sec. This data is converted into intensity estimates using a calibration look-up table and fed into a special purpose image processing computer. The computer determines the background level of the plate very accurately and measures the position, integrated intensity, eccentricity and position angle of all images. In this mode the plate will be scanned in 6 hours.

Using a software simulation of the hardware we achieve 100 per cent detection of all images two of us decide are real on a IIa O 1.2-metre Schmidt plate at a threshold level two standard deviations above background. We believe we pick up more data on image shape than the eye and thus more reliable distinction between faint galaxies and stars. This is possible because the eye tends to delineate images from background on the basis of gradient (galaxies seem to have sharp edges), whereas our software can effectively integrate noise out over a larger area than the eye. The radial profile and deviation of shape from a circle or ellipse (expressed as a Fourier series) can also be computed on individual images at a rate of 30,000/hour.

Data from the Image processing computer is fed into a NOVA computer for further processing collation with previously scanned plate and/or off-line storage. Except for stellar images, which are entirely specified by position and size, large scale storage of data on magnetic tape is not cost-effective since different features of images are needed for different astronomical programs. Use of the on-line computer allows us to use the system as a filter to detect and measure images of specific interest to the astronomer. The speed of the system is such that plates can be scanned again for different programs.

The microphotometer is being built by the Cranfield Unit for Precision Engineering and the whole project supported by the Science Research Council. The measuring system will be available as a national facility in two years.

THE COMPUTER CONTROLLED COMPARATOR ASTROSCAN

P. W. H. Blansjaar and J. van Kuilenburg
NIWARS-dataprocessing, The Netherlands[+]

ABSTRACT. A description is given of the computer controlled comparator densitometer "Astroscan". This instrument consists of a monocomparator and a digital computer which controls the comparator stage. The density measurements of a plate or film are performed by using a multi-element diode array, which is also under computer control. The "Astroscan" is fully programmable in a high level programming language.

1. INTRODUCTION

The construction of the automatic comparator Astroscan was started in 1970 in order to fulfil the needs of a digital processing facility for advanced aerial imagery. The Leyden University Observatory joined the project in an early stage, with the aim of using the instrument for automatization of astronomic measurements. The combined specifications for the instrument concerning speed, accuracy and flexibility, resulted in the design of the final Astroscan version which was completed last year.
The use of instruments like the Astroscan concerns many diciplines of which important ones are nuclear physics as reviewed by Royston (1972), astronomy as discussed by Strand (1971), and remote sensing as discussed by Morain (1966) and Beers (1974). All these applications have in common the need for some kind of random access image input to a computer, with the aim of extracting numerical information, removal of distortions, or enhancing features. Recently a considerable range of equipment became available to solve the image processing input problems. For reviews and details in this respect the reader

+ Netherlands Interdepartmental community for the Application of Remote Sensing techniques, Kanaalweg 3, Delft, Netherlands.

de Jager/Nieuwenhuijzen (eds.), Image Processing Techniques in Astronomy. 247-253.

is referred to the other contributions announced for this conference

2. TECHNICAL DESCRIPTION OF THE ASTROSCAN COMPARATOR DENSITOMETER

The Astroscan instrument was designed using a monocomparator (D.W.Mann) and a minicomputer (PDP11/40). The instrument is shown in figure 1 while the basic component are depicted in figure 2. The controller connects the computer to the comparator, which was modified by adding stepping motors to control x and y motion, and by adding a 128 element linear diode array (a Reticon product as described by Hög (1973))for the optical read-out. The design of the Astroscan resulted in a hybrid machine which consists of hardware and hard software for the basic operation, while the application software can be fully coded in a high level language (Fortran). The main specifications of the instrument are summarized in Table I and the basic Fortran control instructions collected in table II.

Table I Astroscan specifications

Spotsize	: min. 10 μm
Accuracy	: 1 μm
Stepsize	: 5.0 μm
Dynamic range	: 1 : 100
Max. density	: 3D
Density read-out	: 10 bits
Diode array	: 128 elements
Opt. read-out time	: 3msec (128 el.)
Max. speed	: 4 mm/sec
One-step speed	: 20 msec/step
Mono Comparator	: David W. Mann
Computer	: PDP 11/40
Memory size	: 32 Kbytes
Control language	: Fortran IV

3. THE ASTROSCAN PROGRAMMING

The essential instructions are "call DENS", which provides for the density measurement, and "CALL MOVE", which provides for the mechanical motion of the stage. By the simple Fortran statement "CALL DENS" the read-out of the 128 densities is performed using a hard software subroutine. This subroutine stores the densities as integers in a COMMON area after checking for hardware-errors. The main part deals directly with the controller hardware and controls the optical read-out from the self scanning linear diode array, which delivers a video signal with an amplitude proportional to the light level. After integration, amplication and sampling, this signal is converted to digital values and stored sequentially in the computer core.

By the simple Fortran statement "CALL MOVE (DELTAX, DELTAY, VELOCITY)", the mechanical movement of the table is initiated along the axes by means of two steppingmotors. The distance to be travelled by the table relative to the current position is given by DELTAX in

TABLE II BASIC ASTROSCAN INSTRUCTIONS

PROGRAM INSTRUCTION	EXPLANATION
CALL PROL	Open hardware functions
CALL EPIL	Close hardware functions
CALL DENS	Optical read-out into computer core
CALL MOVE (DX, DY, V)	Move stage DX and DY steps with maximaal speed V.
CALL WAITM	Wait for the completion of the current MOVE instruction
DX=RESIDU (1)	Variable DX contains the number of steps in X-direction still to be travelled during current MOVE instruction
DY=RESIDU (2)	Variable DY contains the number of steps to be travelled in the Y-direction
B=RESIDU	Boolean B will be "true" if the current MOVE instruction is completed, else B will be "false"
CALL FILABS (X, Y)	Initialisation prior to MOVEABS instructions
CALL MOVABS (X, Y, V)	Move stage like MOVE instruction to absolute position (X, Y), with the origin as defined by FILABS
CALL DENSAV (ADR, N)	Take the average of N read-out instructions CALL DENS

Figure 1 : Picture of the computer controlled comparator "Astroscan". The comparator (right) is positioned by two stepping motors according to the computer program, while the density readout by a 128 element diode array can be performed indepently by the computer. The control computer is a standard PDP 11/40 (left) which communicates with the opto-mechanical device through the special purpose interface computer behind the comparator.

the x-direction, and DELTAY in the y-direction. The integer VELOCITY can have three values, corresponding with three rotation speeds of the motors. The use of some instructions making the Astroscan perform a task, is illustrated by the annotated listing in Figure 3.

Stepping motors can run very accurately with high rotation speeds, when moving heavy loads, only if they are properly accelerated and decelerated before starting and stopping.

A certain maximum change in speed must not be exceeded in order to prevent the loss of steps. Therefore the software driver computes for all commands an accelerationcurve. It also computes the "MACRO-instructions" for the controller, giving the hardware information about the number of steps to be carried out by the controller, the velocity, the rotation direction, and the information about acceleration and deceleration. A request for a certain table motion is handled by the driver by issuing movement commands of at most 31 steps at a time. The first "MACRO-instruction" starts the movement. As a result of a following hardware request (interruption) the software driver will compute the next "MACRO-instruction". This will continue until all the steps, specified in the FORTRAN subroutine call, are carried out.

When the computer goes down, or signals an error, the comparator screws can be damaged, thus they are protected by means of an "emergency" routine, started by the hardware.

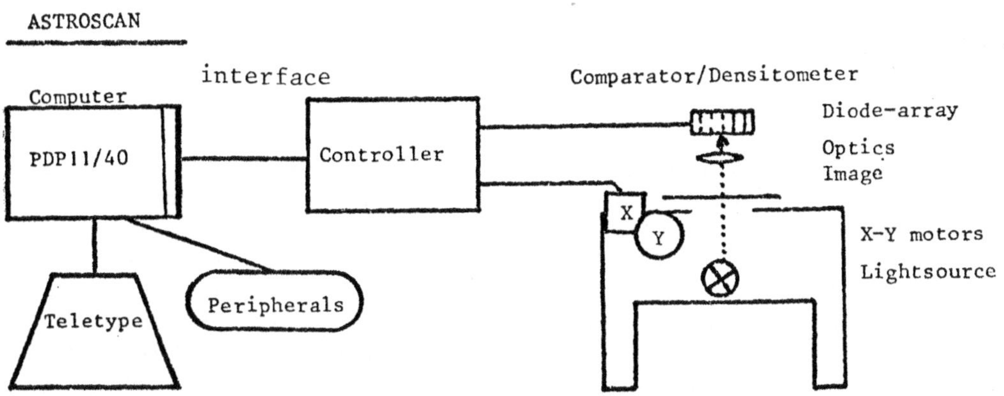

Figure 2 : Diagram depicting the basic components of the Astroscan.

PROGRAM TEXT	COMMENT
`COMMON/DENSZ/ID (128)`	core space for optical read-out
`CALL PROL`	open hardware operation
`WRITE (5,1)`	print header
`1FORMAT (' ASTROSCAN SAMPLE PROGRAM')`	
`CALL MOVE (-10,-10,1)`	remove backlash, run backwards
`CALL MOVE (0,10,1)`	run forwards in Y for backlash
`DO 2 I=1,10`	step in Y direction
`CALL MOVE (10,0,1)`	run forwards in X for backlash
`DO 3 J=1,200`	step in X direction
`CALL DENS`	optical read-out
`WRITE (5,4)(ID(I).I = 1,128`	print result
`4 FORMAT (' ',20I4)`	
`CALL MOVE (1,0,1)`	move in X direction
`3 CONTINUE`	
`CALL MOVE (-210,256,2)`	move in Y direction and backwards
`2CONTINUE`	
`CALL EPIL`	
`STOP`	
`END`	end of program

Fig. 3. A Fortran program ilustrating the use of some Astroscan instructions to perform a simple task. The program shown measures a grid with a sample distance of 1 step (5 um). The grid contains 200 gridpoints in the X direction and 1280 gridpoints (10 diode array widths) in the Y direction. The backlash of the instrument is removed by measuring densities after moving the stage in the positive direction along the X- and Y-axis.

4. THE ASTROSCAN APPLICATION SOFTWARE

The on-line and off-line programming is still at the development stage. A number of utilities have been designed to calibrate the diode array, the density scale and the comparator srews and ways. A general program to digitize a rectangular grid relative to two points is run for service tasks. A special routine has been developed to locate holes and dots in an predefined area, in order to locate pricked control points which have been premeasured with low precision. The same procedure locates stellar images. A more elaborate program digitizes grids on films which are regularly spac in a given coördinate system (Fig.4). About 20000 gidpoints per hour are measured and processed by this program. Separately a special procedure for measuring astronomical plates is designed by the Leyden Observatory using predefined rough stellar data on a pencilfollower. About 100 stars per hour are processed by this pro A search-mode is not planned to be implemented on this machine, as other devices are thought more appropriate for that purpose.

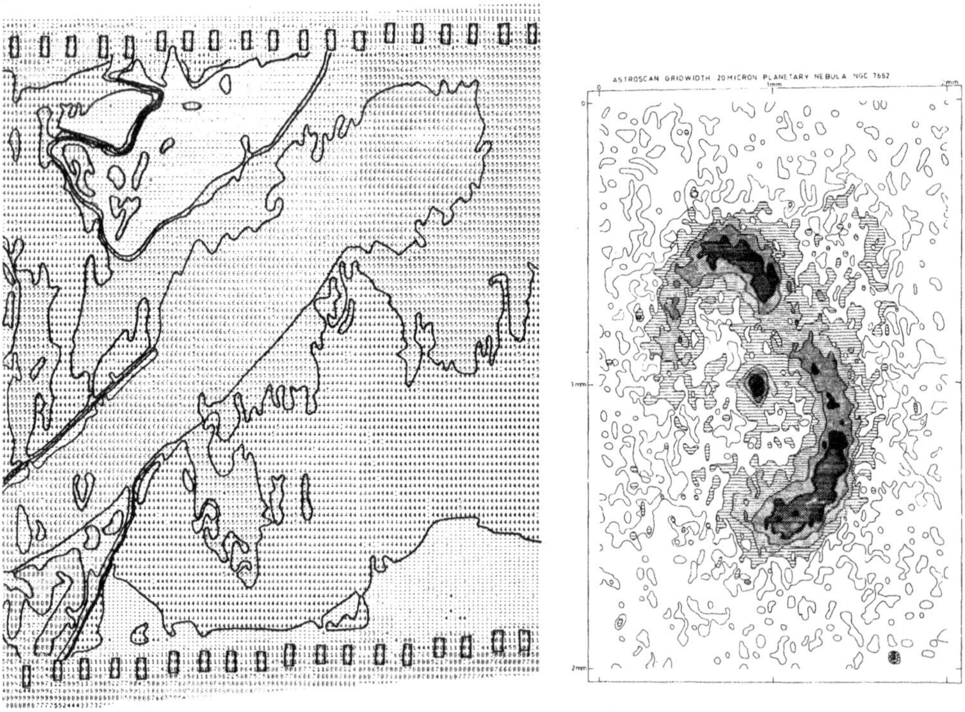

Fig.4. A temperature map derived from digitized thermal scanner imagery (IRLS) imagery is shown on the left. The map pictured covers the Rotterdam coastal region.
A digital representation op a picture planetary nebulae NGC 7662 is shown on the right.

REFERENCES

Beers, J.N.P., Kuilenburg, J. (1974): "Automatic processing methods applied to multispectral photography", Proc.Intern. Symp. on Remote Sensing III, Ed. Sharokhi, U. of Tennessee.

Hög, E., Wiskott, D. (1973): "Automatic measurement of photographic plates with a photodiode array", ESRO note 73-4.

Morain, S.A.. Simonett, D.S. (1966): "Vegetation analysis with radar imagery"Proc. 4th symposium on remote sensing of environment (Michigan), pp 605.

Royston, R.J. (1972): "Applications in nuclear physics" ch. 4 Conf. Series 13 on Machine Processing of Patterns and Pictures, Inst. of Physics (London).

Stand, R.C. (1971): "Computer controlled digitizers of optical image data extraction" Prof. Conf. on Photographic Astrometric Technique, NASA Cr-1825, pp 15.

A SIMULTANEOUS SIX-CHANNEL MICROPHOTOMETER WITH COMPUTERIZED DATA ACQUISITION

D. H. Andrews and K. O. Wright

National Research Council of Canada,
Dominion Astrophysical Observatory,
Victoria, B.C., Canada

ABSTRACT: The instrument described is a single purpose device for the microphotometry of stellar spectrograms. Optical transmissions of up to six strips along the plate may be recorded simultaneously at intervals from one to 64 microns selected independently for each channel.

The instrument described in this paper is intended for the spectrophotometry of normal stellar spectrograms and eventually, in some cases, for the determination of absolute wavelengths and radial velocities. With this instrument, the transmission of up to six strips on the plate, near and parallel to the stellar spectrum can be determined simultaneously. Thus measurements can be made more quickly, and, since comparison and stellar spectra can be scanned together, it does not matter if the plate cannot be returned to exactly its original position. The microphotometer is essentially a flying-spot scanner. Its disadvantage is that much less light is available than in other systems, and it is affected by statistical fluctuations of the number of photons, but, fortunately, these increase the normal plate-to-plate scatter of a few percent by only a fraction of one percent. The accuracy can be improved by reducing the operating speed below the usual 0.8 mm/sec. The transmission and accuracy of the instrument should also be considerably increased if the multi-element lenses now used were replaced by simple lenses.

The optical layout is depicted in Fig. 1. The spot source is an inexpensive cathode-ray tube having a very low persistence P16 phosphor which emits in the near UV and blue region. Immediately below the CRT is a translucent plastic cylinder which diffuses the light for the monitor photomultiplier, PM2.

de Jager/Nieuwenhuijzen (eds.), Image Processing Techniques in Astronomy. 255-260.

Lens L1 images the CRT spot on the first slit. Immediately below is a diagonal mirror which may be inserted to allow the operator to inspect the plate when it is illuminated from below by the viewing lamp. The mirror, M2, directs the viewing light vertically and seals the sensitive lower section. Lenses L4, L3, and L2 then complete the viewing train. In scanning operation, L2 reimages the spot and first slit on the plate emulsion at a CRT-to-plate reduction of 13:1. The spot image has a typical half-intensity width of six microns and 90 percent of the light falls within 11 microns. The spot may be further narrowed by the adjustable first slit.

Flexible magnetic strips clamp the plate firmly along both its edges, emulsion side down, to one of the plateholders carefully machined for each plate size used at this Observatory. The holder forms part of the plate stage of a screw-driven measuring engine and keeps the emulsion at a fixed distance from the projection lens, L4. This arrangement eliminates one focus adjustment. The effects of variations in the plate thickness can be removed by measuring a control strip of "clear" glass.

Lens L4 focusses the plate and spot image on the analyzing slit at a magnification of 25X. Slits of effective widths 2, 4 (normal) or 8 microns can be interchanged; their effective length on the plate is 5 mm. Focus, slit rotation, and similar adjustments are usually extremely stable for several months; nevertheless devices such as Hartmann tests are built in to simplify any necessary changes. A Fabry lens, L5, mirror, M3, and final lens, L6, produce an image of L4 on the photocathode of the primary photomultiplier, PM1, an EMI 6094.

The instrument is supported by an equilaterial triangular base with the optical axis running vertically through its centre. The most important feature of the design is a steel disc, of diameter one metre, rotatable about the optical axis, which carries a screw-type measuring engine, its motor, a rotary encoder, and ancillary equipment. Spectrograms with lines not perpendicular to the dispersion can be scanned without any realignment of the slits by turning this disc. The rotating disc is most important if one wants to reduce computation and storage in the computer by averaging both sides of the comparison spectrum. Since our calibration spectra run parallel to the stellar spectra (Fig. 2A), rotating the disc, through 90° brings the calibration into position for scanning, without any other adjustment to the plate being necessary.

Fig. 2A depicts the plate and the fixed trajectory, MN, of the flying spot. The speed of the plate-carriage motor is highly regulated to better than 0.1 percent. An integral generator whose output together with the encoder pulses, P1, are used to produce

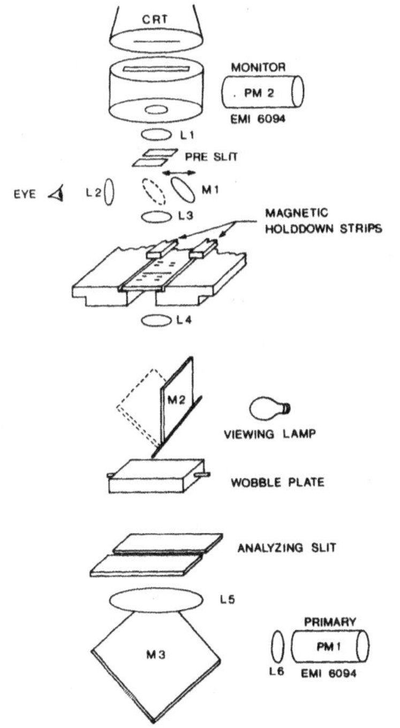

FIGURE 1 OPTICAL LAYOUT

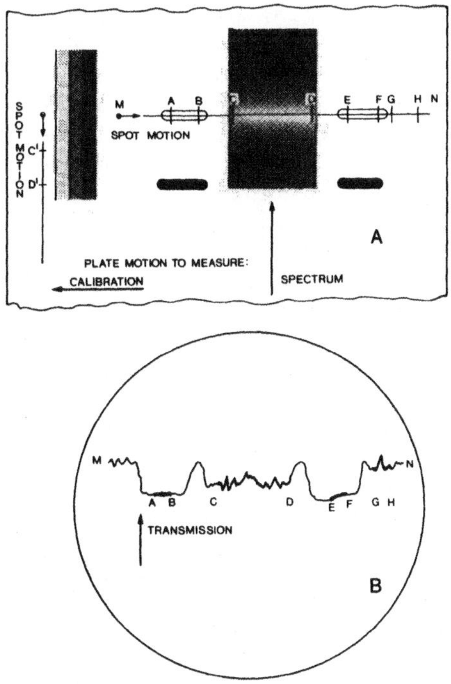

FIGURE 2 PLATE & DISPLAY

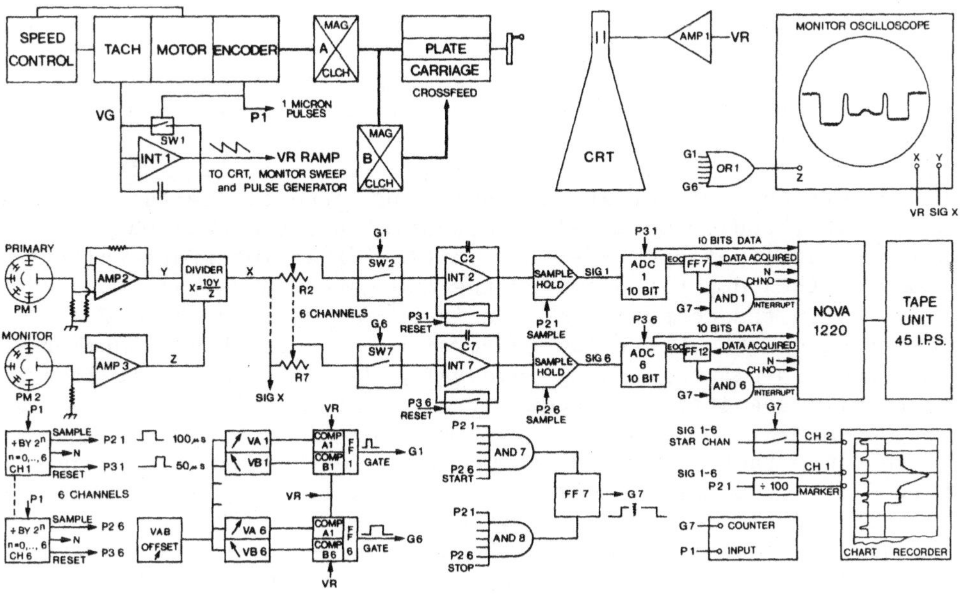

FIGURE 3 6 CHANNEL MICROPHOTOMETER CIRCUIT BLOCK DIAGRAM

a ramp voltage, VR. This voltage returns to zero at the end of every micron of plate travel, and its amplitude is insensitive to motor speed. This ramp is used as the sweep for the flying-spot CRT, the monitor oscilloscope, and to produce the gating signals, G1-G6, which steer the photomultiplier output to the appropriate channel. The monitor oscilloscope generally displays the photomultiplier output as depicted in Fig. 2B. The heavy sections indicate a brightening of the trace by the gating signals which allow the operator to choose the position and lengths of what normally would be a microphotometer's operating aperture. These gates may also be steered as a unit by the operator to correct for tracking errors or to follow the "S" distortion of some electronographic spectra. Such a system was far simpler to implement than a computer-controlled two-coordinate plate stage.

Two voltages, VAn and VBn (n is the channel number), offset by the steering bias, VAB (Fig. 3) for each channel and the ramp, VR, are compared by COMPAn and COMPBn which in turn generate gates G1-G6. Thus the gate may be adjusted to any length and positioned anywhere during the sweep. Each channel's gate operates an electronic switch, SW2-SW7, to allow the primary output, divided by the monitor's output, to be integrated during the gate period by an analogue integrator, INTn. This integration continues for a preset number of sweeps which may or may not vary from channel to channel over a range of 1 to 64, in powers of two.

Pairs of pulses, the Sample, P2n, and Reset, P3n, are produced by binary counters from the encoder pulses, P1. The Sample pulse immediately precedes the Reset pulse and is used by a sample/hold module to acquire its channel's integrator voltage after the present interval, for example, after every 8 microns of plate travel if 8 microns is that channel's digitizing interval. The Reset pulse, P3n, sets the integrator voltage to zero in readiness for the next digitizing interval. P3n also initiates the conversion cycle of its channel's analogue to digital converter (ADC). The "End of Conversion" output of the ADC sets the appropriate flip-flop, FF7-FF12, initiating an interrupt which is passed to the computer provided the AND gates, ANDn, are enabled by G7. When the computer has acquired the 16-bit data word, consisting of the 10-bit ADC conversion, a 3-bit channel number code, and a 3-bit digitizing interval code, it resets flip-flop FFn to kill the interrupt request.

An important feature of the circuit is the data gathering gate, G7. The Sample pulses, P2n, from each channel are relatively synchronized; for example, every fourth 4-micron digitizing interval pulse will coincide with another channel's 16-micron digitizing interval pulses. When the operator presses the 'START' button for data gathering the signal, G7 does not commence until all P2n Sample pulses for channels in use occur simultaneously;

unused channels are turned off and do not gather extraneous data. When the operator presses the 'STOP' button, G7 continues to allow data to be gathered until all the sample pulses are again coincident. This enables the computer to check that the number of data points gathered in each channel, multiplied by the channel's digitizing interval is equal to the number of microns of plate travel during the data-gathering interval when G7 is on. By using G7 as a counter gate and counting the encoder pulses, P1, the computer can check each channel to determine if there is any extraneous or lost data. The operator may then rerun the plate if he deems it necessary.

The two-channel high-speed oscillosgraphic chart recorder also uses the data gate G7. The analogue output of any channel or combination of channels may be recorded. One oscillograph channel usually records the average of the two sides of the comparison spectrum continuously; the other is gated by G7 and records the stellar spectrum only during the data-gathering interval. Channel one's Sample pulses, P21, are counted down by a factor of 100 so marker lines of known separation, for example 400 microns, may be impressed across the chart recording. This chart forms a temporary record which is useful for the subsequent processing of the magnetic tape data by a larger computer. For example, it is often useful to know the approximate data-point number for a specific wavelength and this may be readily determined from the recorded comparison spectrum, the start of the stellar record, and the known spacing of the marker lines.

A Data General NOVA 1220 computer with 8K of memory is used exclusively with the microphotometer. It drives a tape recorder at 45 inches per second and the tapes are processed by a larger computer at a CRT terminal. At present, the data-reduction software is a slightly modified version, restricted to a single-channel reduction, of a program used to process the output of a single-channel microphotometer. The NOVA prompts the operator so that a data wordlength code for the plate (spectrum) number, recorded with each type of run, may later be used to identify run type on the magnetic tape records. An entire multi-spectrum plate is recorded between file marks and may contain only one set of records of calibration spectra, identified by a pre-fixed wavelength, whereas each spectrum may have several types of run such as a short record of clear glass and zero percent transmission, a record with a reference filter, and, of course, the main spectrum. Each of these runs is prefaced by a plate or spectrum number, padded with blanks to give a different word length. The reference filter run is made with a grain-free filter of appropriate density to establish relative zero points and gains between channels so that the corrections derived by the background-level channel may be applied to the stellar spectrum.

Test results from the microphotometer are still preliminary. A comparison of processed intensity traces from this and another single-channel microphotometer were in very good agreement for a spectrum which had deep lines and whose continuum was heavily exposed.

In general, the fractional error, En, due to system noise is given by

$$En = AT^{-\frac{1}{2}}$$

where En is the ratio of the standard deviation of a single point to its transmission level, T, as determined by measuring a series of filters. The constant, A, ranges from 0.005 to 0.01 depending upon such factors as the carriage speed. This accuracy may not compare favourably with other devices since it implies a density error of 0.013 to 0.043 for a density of 2. However, a similar expression for the error introduced by grain is:

$$Eg = 0.074(1-T)$$

where again Eg is the ratio of standard deviation to transmission for a 0.5 mm wide strip of an average IIaO plate. We have calculated that the error caused by grain in estimating the area of a Gaussian profile of half-width 100 microns, whose continuum is at 0.01 transmission and whose centre is at 0.31 transmission, is 0.9 percent. Such a continuum would be considered over-exposed at Victoria. At normal running speeds, this error is increased to 1.3 percent and at one quarter normal speed the error is increased 16 percent due to grain and system noise. The latter calculation for the depth of this line, 0.75 on a rectified scale, shows that its error is increased from 1.45 percent to 1.54 percent in the worst case. For a line with the same continuum but whose depth is only 0.25, the error is increased from 12 percent due to grain to 16 percent due to grain and system noise. The latter calculations are based on the depth being determined by a single 8-micron point and a linear relationship was always assumed between intensity and density. The ten-bit ADC imposes a fundamental lower limit of approximately half a bit to the standard deviation of any average quantity measured. Thus for a transmission of 1.0 percent the error of measurement cannot be less than 5 percent. We feel that the system noise will rarely, if ever, significantly affect the accuracy of measurement of our normal spectrograms.

The authors wish to acknowledge the considerable help provided by Mr. J.M. Fletcher during many discussions and particularly for assistance in programming the computer. Mr. Eburne has done much of the testing as well as electronic construction and our thanks are due to him and the rest of the technical support staff.

A VERY SIMPLE DIGITAL MICRODENSITOMETER-COMPARATOR

Robert J. Rutten and H. Jacques van Amerongen

Sterrewacht "Sonnenborgh", The Astronomical Institute, University of Utrecht, Utrecht, the Netherlands.

In contrast to the preceding contributions on sophisticated and expensive measuring machines this paper is meant to be a poor man's guide, intentioned for those people who use an old-fashioned microdensitometer to analyse stellar spectra, who have only a modest electronics workshop and who would like to take part in the digital era without being able to spend much money or effort.

At Utrecht, we found ourselves in that position five years ago when we had to digitise a variety of solar and stellar spectrograms and could not wait for the completion of the VAMP machine that was then on order. We therefore modernized our old Moll-type microdensitometer in we think about the simplest way possible and obtained a quite useful instrument. We never published this since similar instrumentation had been described already, for instance by the Frascati group (Gratton et al., 1971) and, most extensively, by Hoekstra (1969). However, we received quite a few requests for detailed information on our instrument, so it seems worthwhile to describe it as a do-it-yourself recipe of how to modernize in a practical way old-type instruments used for tracing stellar spectrograms.
The microdensitometer may be of any type or construction provided it has a detector that monitors the transmission or density of a spectrogram while this is moved continuously through the measuring beam. The first step is to have simultaneous digital recording beside the paperchart recorder. The budget choice is between papertape and magnetic cassette or cartridge tape, depending on what the off-line computer to be used can read. Papertape still has many advantages, especially that the punched information can be checked by eye when things go wrong. But punching is slow and large reels of papertape are difficult to handle. One should use the most

de Jager/Nieuwenhuijzen (eds.), Image Processing Techniques in Astronomy. 261-264.

compact output format possible: one 8-bit character for each density measurement. The simplest way is to punch the output of an 8-bit binary analog-to-digital converter ("ADC") that samples the detector signal. The maximum resolution so obtained is 1:256 or 0.4%. Although this is rather bad, it is sufficient since the noise due to plate grain is usually worse. However, since cheap 10-bit ADC's are now abundant (in contrast to five years ago when we had to build our own 8-bit ADC), we would now sample in 9 or 10 bits but still use papertape and punch only the lower 8 bits. The operator would then have to determine from the simultaneous chart recording in which half or quarter of the full range a scan started and the software that reads the tape would have to keep track of cross-overs into other zones. This way of increasing resolution without sacrificing compactness is possible because stellar spectrogram traces are continuous.
The compact output format also permits scanning speeds that are much higher than obtainable with a chart recorder or a curve follower. We use a 120 character/s punch and work with a scanning speed of 100 samples per second. For tracing stellar spectrograms this proved fast enough: usually setting-up takes more time than the actual scanning.

We now turn to positional accuracy. Often expensive high-accuracy screws and stepping motors are installed to drive the plateholder in high-precision steps.This is not necessary since to obtain high positional accuracy in the samples there is no need to position the plateholder itself exactly but only to know precisely where it is. Therefore it is simpler to attach an accurate linear position encoder to the plateholder and to use its position read-out to trigger the ADC at the desired sampling intervals. In this manner one samples "on-the-fly": while the spectrogram moves continuously through the measuring beam, samples of the detector signal are picked at the moments the proper positions pass the detector.
We selected an encoder that consists of an etched nickel-in-glass grating with an optical monitor[0] . We use the change-over of the last digit of the position display to trigger the ADC and we automatically suppress punching at the beginning of a scan until the zero position comes along. In this manner a simple but accurate comparator is obtained: all parallel scans through the stellar spectrum and the comparison arc spectra are started somewhere

---

[0] We used a 2 μm resolution Heidenhain grating and monitor head (Joh. Heidenhain, 8228 Traunreut, W-Germany). These optical encoders do not need special precautions and are reliable and easy to use. However, the accompanying Heidenhain position display is expensive and not easily interfaced; we can supply diagrams to build one instead.
A more modern alternative is the steel rule magnetic encoder supplied by Sony Magnescale Inc. (Tokyo 141, Japan).

before the desired position, at which the position display has previously been reset to zero. All traces are then exactly aligned, sample by sample, and there is no need for accurate positioning by hand or punching positional information.
The minimum sampling interval in on-the-fly measurement is set by the vibrations between plateholder and encoder rule on the one hand, and measuring spot and encoder monitor head on the other hand. To ensure that a given position results in triggering only one sample even when this position is seen three or more times because of vibrations, there must be a dead time after its first pass in which no extra sample commands are accepted. The duration of this dead time must correspond to a distance larger than the largest vibration amplitude and smaller than half the sampling interval. Also, the time that the ADC takes to complete a sample and the RC-time of the detector's amplifier must be short compared with the time interval between successive samples. If this is not the case systematic shifts will be caused by the motion of the spectrogram during the actual measurement, resulting in spurious asymmetry in spectral line profiles.
These constraints can easily be determined and checked empirically from trial scans of high-resolution transmission gratings since these show signals that change appreciably on the micron scale. With our instrument we found sampling at 2 μm intervals to be possible, and we chose 4 μm as smallest interval.
Repeatibility and absolute precision depend on vibrations, temperature drifts and intrinsic errors of the position encoder. Our instrument shows a sample-by-sample repeatibility of better than 2 μm , and makes a precision of 1 μm in dispersion-curve determination quite feasible: vibrations and small-scale rule errors are averaged out in the determination of the position of a spectral line from many samples, and differential temperature drifts are small during the short scanning times. The large-scale rule errors are small for the high-resolution encoder we selected; when they are too large they may be determined and corrected for, and even this is not necessary when many spectral lines are used to determine the dispersion curve.
The nice thing in triggering by encoder is that the linearity of the plateholder motion is not important; with our instrument one can even disengage the screw, push the plateholder by hand and still obtain 1 μm precision!
Also, proper orthogonality between the scanning direction and the perpendicular plateholder adjustment is not needed. Together with any error in the alignment of dispersion direction and scanning direction, the deviation from orthogonality is easily determined and corrected for from symmetric traces through both comparison arc spectra.

This system was completed with an extensive reduction software package that included Fourier-filtering and -restoration, following Brault and White (1971). An extra advantage to the reduction

procedure is that the simultaneous records from the still present chart recorder are condensed by the high scanning speed into a form that in many cases is very suitable to draw the continuum level on, even if the recorder's response is too slow to follow line cores correctly. Using these records the spectrum can be scaled to a "true" astronomer-drawn continuum already in the first pass through the reduction software.
The instrument has been used daily for two years, until it was superseded by the much more sophisticated and automated VAMP machine; in the meantime it proved to be a quite useful, reliable and astronomer-proof instrument.

ACKNOWLEDGEMENT

We thank Drs. T. de Groot, J. van Nieuwkoop, H. van de Stadt and C. Zwaan for comments on this paper.

REFERENCES

Brault, J.W. and White, O.R.: 1971, Astron. Astrophys. 13, 16.
Gratton, L. et al.: 1971, Pub. Roy. Obs. Edinburgh 8 nr. 11, p. 142.
Hoekstra, R.: 1969, Thesis, Amsterdam.

PHOTOELECTRIC SPECTROSCOPIC PLATE MEASURING MACHINE BUILT AT THE MEUDON OBSERVATORY

Françoise LAUNAY

Département d'Astrophysique Fondamentale
Observatoire de Paris-Meudon, F-92190 MEUDON, France

ABSTRACT. A measuring machine has been built in order to measure both laboratory spectra recorded on 18 inch long plates and astronomical spectra. Three spectrograms are simultaneously measurable by means of three adjustable field diaphragms optically associated to three photomultipliers. The position of the lines is set by the Tomkins and Fred photoelectric method and the abscissae are read by a Heidenhain Moiré fringe device with a precision of 1 µm on the whole of the 18 inch length of the plates.

INTRODUCTION

The first aim of this program is to provide a facility able to measure in one run, and with a good accuracy, the laboratory spectra recorded on 18 inch long plates with the 10 meter vacuum UV spectrograph set up at the Meudon Observatory. In this case, we have two spectra for one experiment : the spectrum of interest and a reference spectrum.

But we are aware that the astronomers also need a good comparator which would allow an easy simultaneous measurement of three spectrograms, since two reference spectra are generally taken on each side of the spectrum of interest.

In view of these requirements, the main features of the machine are the following :

- measurement of the line abscissae with a precision of 1 µm,

de Jager/Nieuwenhuijzen (eds.), Image Processing Techniques in Astronomy. 265-268.

- maximum size of the plates : 90 × 500 mm,
- X measurement range : 450 mm,
- scanned field : 5.5 × 1 mm (1 mm in the X direction),
- 3 simultaneously measurable spectrograms,
- 2 spectrograms are simultaneously displayed on the oscilloscope screen,
- diameter of the optically monitored field : 15 mm,
- magnification (for both measured and monitored field) : 10,
- rough measurement of the line amplitudes : 5 to 10% accuracy,
- recording : digital readout on punched IBM cards.

Let us see now how these specifications are obtained.

DESCRIPTION

The displacement measuring system is a Heidenhain Moiré fringe device in which the longer grating attached to the plate carriage is 500 mm long. The position of the line is set by the well known Tomkins and Fred (1951) photoelectric method. In our design, the rotating prism is a plane parallel plate of 30 mm thickness driven by a 1300 rpm synchronous motor. The time basis is provided by an alternator which is attached to the same axis as the motor.

The figure shows the layout of the optical system. The plate is horizontal, however the plate-holder has been carefully designed to avoid any bending of the plate under measurement. Of course the plate-holder can be rotated in order to have the spectrogram normal to the travel.

To avoid scattered light coming from the monitored field, its illumination has been made independant from that of the measuring field by the use of two separate quartz iodine lamps and dichroïc beam splitters. Köhler type condenser systems are used to obtain uniform illuminated areas on the plate. This design makes it possible to introduce into the measuring beam three adjustable field diaphragms sharp images of which can be formed in the plane of the plate.

The adjustable diaphragms are obtained by the combinations due to the relative translation of two thin metallic foils allowing about ten arrangements of one, two or three rectangular apertures each.

The defects of the dichroïc plates, due to their use at 45° are compensated in the measuring channel by means of a third dichroïc plate cutting off the residual transmitted green light.

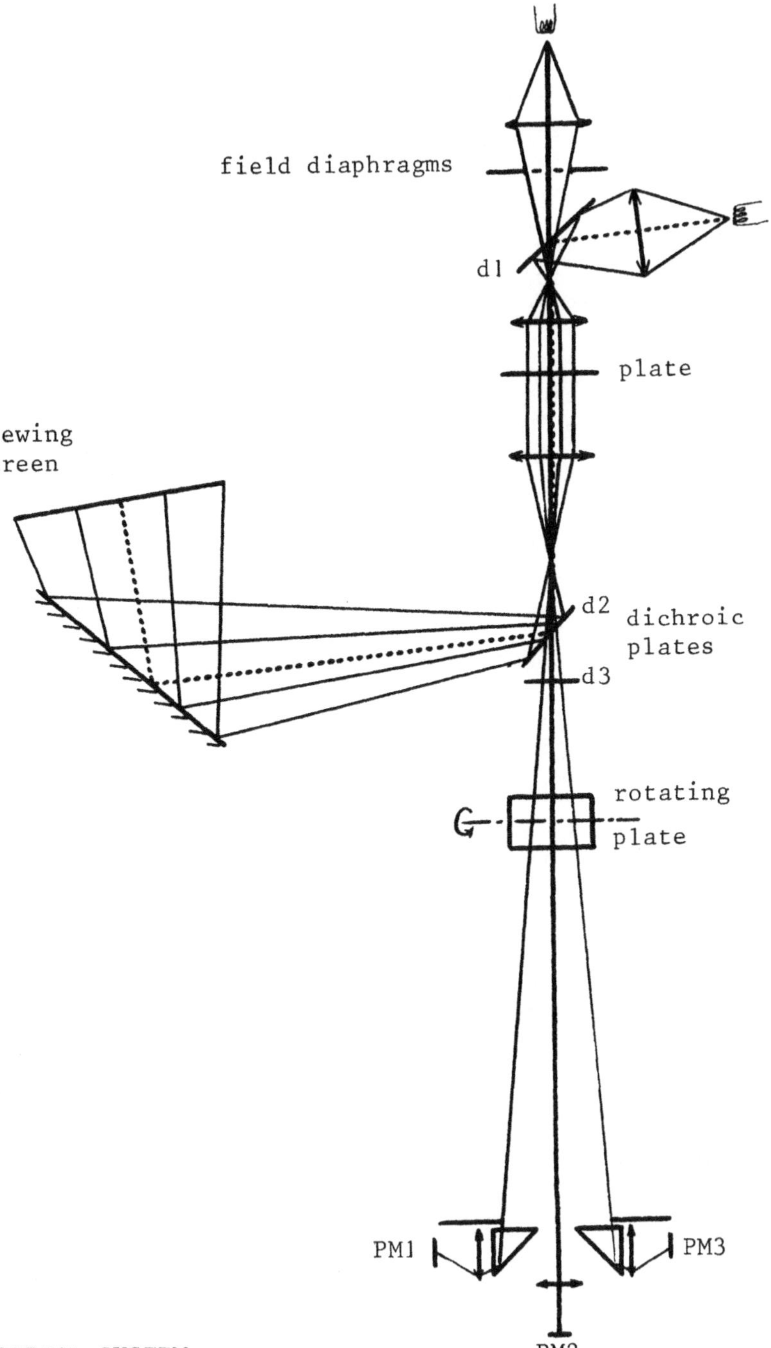

OPTICAL SYSTEM

This same light is used to display the measured field on the 15 cm diameter control screen where it appears in green on a yellow background.

The blue measuring beam goes through the rotating plate which sweeps the three measured images over the slit. The slit width may be adjusted between 0 and 800 μm while its length is fixed to 55 mm. The three spectra selected by the field diaphragms are fed to three separate photomultipliers tubes by two 90° reflecting prisms set with a fixed 12 mm spacing behind the slit. Thus a central photomultiplier receives directly the spectrogram of interest while the two side ones measure the reference spectograms.

The housing containing the rotating plate, the slit and the photomultiplier tubes may be rotated round the vertical optical axis in order to make the slit parallel to the image of the spectrum lines.

A stepping switch makes it possible to choose the two spectra which are to be displayed on the oscilloscope screen.

The measurement of the line amplitude occurs at the coincidence time (i.e. when the rotating plate is horizontal). A field effect gate is open for 5 μs during which time the photomultiplier tube output is stored in a capacitor. During the rest of the scanning cycle, the charging voltage of the capacitor is read by means of a voltage to frequency converter.

After the setting of the line, the measurement is transfered to the punching machine. The automatically recorded values are :

- the reference of the photomultiplier tube in use (1 figure),
- the measure index (3 figures),
- the line amplitude value (2 figures),
- the line abscissa in μm (6 figures).

REFERENCES

Tomkins, F.S. and Fred, M. : 1951, J. Opt. Soc. Am. 41, 641.

PART 4

SOFTWARE TECHNIQUES

SOFTWARE TECHNIQUES FOR IMAGE PROCESSING

Daniel A. Klinglesmith

Goddard Space Flight Center

1. Introduction

The astronomical community has been involved in image processing for centuries. We have gone through a series of levels of sophostication in our image processing using the unaided eye, telescopes, photographic plates, image intensifiers, television type sensors and computers. All of these devices have been developed and used in order to obtain a "better" image. For each of us the word "better" has had many meanings. For example; higher spatial resolution, fainter limiting magnitude, higher signal to noise ratio, greater dynamic range, more uniform photometric response, greater spectral response, etc. It is stating the obvious to say that no single device and/or image processing system can satisfy all of these desires for a better image simutaneously.

This review will describe several software techniques that can be used to get the "most", again a relative term, out of an image, without destroying or adding any information. Also, a simple overview of the components of an image processing system will be given and a list of the image processing systems with which I am familiar will be included.

2. Image Processing Systems

In order to understand what can be done with software techniques for image processing, a general image processing system will be described. This system is the type that can be best

de Jager/Nieuwenhuijzen (eds.), Image Processing Techniques in Astronomy. 271-283.

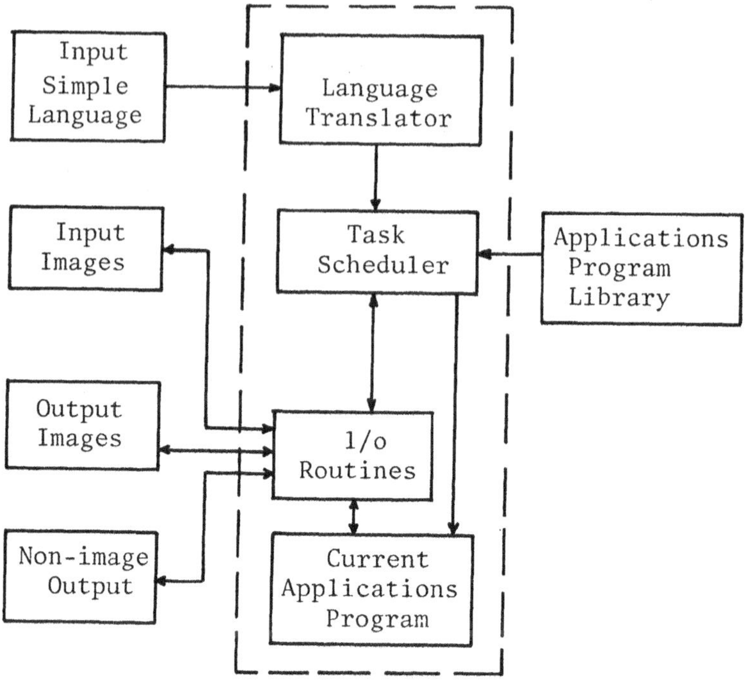

Figure 1. A General Image Processing System.

used on large computers in a batch mode of operation. However, interactive mini-computer systems would have the same basic structure plus an interactive communications handler.

The block diagram is shown in Figure 1. All of the components inside the dashed box are resident in core at one time. The language translator operates on a very small, freely structured set of verbs and modifiers outputing the necessary control language, program requests and parameter data sets. This frees the user from needing to know the internal details of the computer and its control language. The task scheduler receives this information and proceeds to set up the necessary input and output data sets, checking for completeness in the process and then brings into core the particular applications program that has been requested from the library of image processing programs. The applications program once in core, can begin immediately to perform image processing without worrying where the input images are or where the output images are going to be. The I/O routines and task scheduler have already taken care of those housekeeping type problems. Thus the applications programs can be easily modified or replaced without having to be concerned with I/O and housekeeping tasks. This results in simpler, more efficient coding with fewer mistakes. With this type of open ended system where applications programs can be freely added or replaced, the image processing tasks can proceed unimpaired as the nature of the task evolve.

A number of systems of this type already exist. Table 2 lists the systems that I am aware of. The list is most probably not complete. The table contains a name, the computer and the responsible person. Table 3 compares the various systems for content and type of applications programs.

These systems have already been developed and are in daily use on major computers. Each one represents several man years of software design and testing. Therefore, I urge the newcomer to think carefully before redesigning an image processing system.

Table 1

Image Processing Systems	
DIMES CDC 6600	- Digital Image Manipulation and Enhancement System Victor E. Shely U. S. Army Engineer Topographic Laboratories Fort Belvoir, Virginia
ESPR CDC 3200	- Electrographic Stellar Photometry Reductions Barry Newell KPNO/950 Cherry Ave./P.O. Box 26732 Tucson, Arizona 85726
IDAMS CDC 3200	- Image Display and Manipulation System William Afford, Code 933 GSFC/Greenbelt, Maryland 20771
IDECS IDECS CPU and PDP 1520	- Image Discrimination, Enhancement and Combination System P. L. Currier Remote Sensing Laboratory University of Kansas Lawrence, Kansas 66044

MSFCIP IBM 7094/I and Interdata 70	- Marshall Space Flight Center Image Processing W. K. Polstorff, Chief Computer Systems and Simulation Division Computation Laboratory MSFC/ Huntsville, Alabama
PAX II UNIVAC 1108	Azriel Rosenfeld Department of Computer Sciences University of Maryland Greenbelt, Maryland
PECOS IBM 360 and IBM 370	- Picture Enhancement Computer Operating System James E. Burke Electrographic Systems Laboratories 495 Java Drive Sunnyvale, California
PECOS II IDIM HP 3000	James E. Burke Electrographis Systems Laboratories 495 Java Drive Sunnyvale, California
SMIPS IBM 360/91 and IBM 360/75	- Small Interactive Image Processing System Johannes G. Mock Computer Sciences Corporation 8728 Colesville Road Silver Spring, Maryland 20910
VICAR IBM 360/44 and IBM 360/75	-Video Image Communication and Retrieval System Fred C. Billingsley JPL/4800 Oak Grove Drive Pasadena, California 91103
WLCIP PDP 11/20	- White Light Coronagraph Image Processing Russ Howard Naval Research Laboratory 4555 Overlook Ave. S.W. Washington, D.C.

Table 2

Image Processing Software Summary

	DIMES	ESPR	IDAMS	IDECS	MSFCIP	PAX II	PECOS	PECOS II	SMIPS	VICAR	WLCIP
<u>Image Combination</u>											
Mosiacing	X		X		X		X	X	X	X	
Differencing	X						X	X	X	X	
Averaging	X	X					X		X	X	X
Linear Combination, General	X						X	X		X	
Ratio	X				X		X	X	X	X	
Histogram Equalization	X		X				X	X			
Correlation	X		X				X		X	X	
Modification			X						X	X	X
<u>Image Enhancement</u>											
Contrast Stretch	X		X		X		X	X	X	X	X
Image Shading	X	X	X		X				X	X	
Convolution Filtering	X		X				X		X	X	
Recursive Filtering					X						
Non-Linear Smoothing		X	X						X	X	
<u>Geomtric Manipulation</u>											
Change of scale Linear	X		X		X			X	X	X	X
Change of scale Non-Linear	X		X		X				X	X	
Rotation, General					X				X	X	
Rotation, 90° only					X		X	X	X	X	
Transposition					X		X	X		X	
<u>Frequency Domain Analysis</u>											
FFT	X	X	X		X		X	X	X	X	X
Filtering	X	X	X		X		X	X	X	X	X
FHT, (Hadamard)					X		X				
Filter Generation	X		X		X		X	X	X		
<u>Utilities</u>											
Image Generation	X		X		X		X		X	X	
Grid Overlay	X		X		X		X	X	X	X	
Printer Listing	X	X	X		X	X	X	X	X	X	X
Image Copy	X	X	X	X	X	X	X	X	X	X	X
Histogram	X	X	X	X	X	X	X	X	X	X	X
Text Generation	X		X		X		X	X	X		X

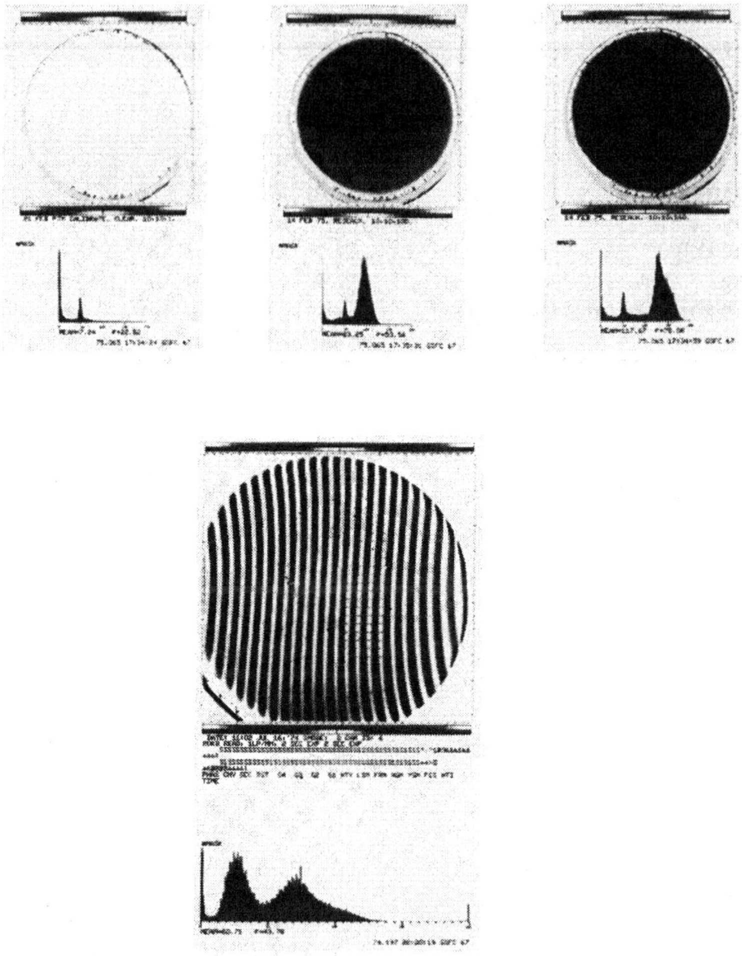

Figure 2. Histograms of SEC Vidicon imagery.

3. Image Processing Techniques

In the process of preparing an image for analysis many tasks are performed. These tasks can be divided into three types: image diagnostics, subjective enhancement and quantitative restoration. Image diagnostics are those tasks that give both qualitative and quantitative information concerning the image; for example, Histograms, Differences and Fourier Analysis. Subjective enhancement tasks give only qualitative information concerning the image; for example, Contrast Stretching, Edge

Enhancement and Mosiacing. These techniques are very much orientated toward each individual image. There is no best way for all images, because each image was created for a different purpose. Quantitative restoration tasks give quantitative information concerning an image; for example, Photometric Correction, Contouring and Geometric Correction. These techniques are used singlely and collectively to obtain the "best" image for a particular problem. Examples of these techniques will be given in the following sections.

3.1 image diagnostics

One of the first problems that has to be understood about an image is its quality. Figure 2 has several images obtained with digital SEC Vidicon camera systems. All have "Histograms" below the image. The histogram is a plot of the number of times each grey level occurred in the image as a function of the grey level. In Figure 2a, 2b and 3c; which are images of uniform fields, the histograms indicate two basic facts about the images. First, the level of exposure can be estimated from the average value of the large peak, which is seen to shift to higher grey levels as the exposure level is increased. Second, the lack of uniformity of response is seen in the width of the peak. Incidently, the small peak at about 64 DN is due to the target ring of the SEC Vidicon and is the same for all three exposures. The histogram in figure 2d also gives two types of information about the bar pattern image. First, the two peakes in the histogram are the average value of the minimum and maximum amplitudes in the bar pattern. Therefore, an estimate of the overall average of the MTF could be computed quickly without ever making a direct measurement on the image. Second, this particular TV system was having trouble with the D/A converter in the video line, as can be seen by the 10 - 20% variation between adjacent even and odd values. This was finally traced to the fact that the least significant bit conversion was not working properly and has since been corrected.

The histogram which can be obtained quickly contains a large amount of qualitative information about an image.

The "difference" of two images is another useful diagnostic tool. Astronomers have been using differences for years to obtain colo-color plots for point sources. Recently, Worden has used the differences of digitized images to obtain the color structure in M51 (1974) and the degree of polarization within the Serpens nebula (Worden and Grasdalen, 1974).

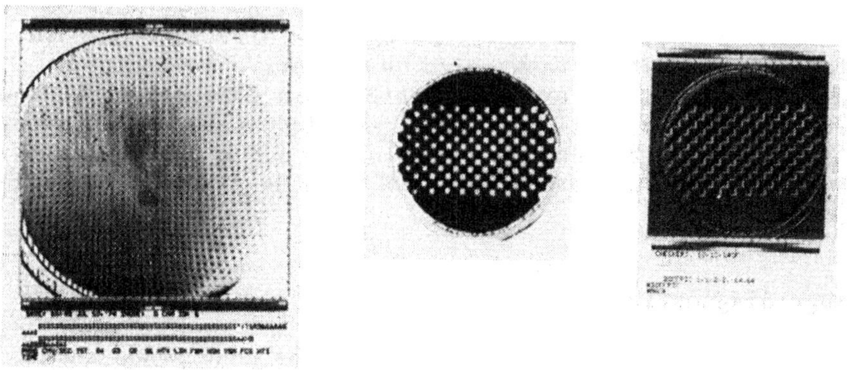

Figure 3. Difference Images. a) Difference of two pin hole pattern images with a change in the distortion. b) Checkerboard pattern obtained with an SEC Vidicon system. c) Line spread function obtained by differencing (see text).

Two examples of differences that have been used in analysing SEC Vidicon systems are shown in figure 3. Figure 3a is the difference of two images of a pin hole pattern that was used for geometric distortion analysis. The black pin holes are from one image and the white pin holes are from another. We were studying the method of readout and the difference image clearly shows that the geometrical distortion was different for the two methods used.

Another use for the difference technique is to obtain the line spread function for an imaging system. Figure 3b is an image of a checkerboard pattern seen by an SEC Vidicon system. The system is seen to have poorer resolution around the edges. In order to obtain a quantitative measure of the loss of resolution the image was shifted one sample to the right and one line down and a difference taken. This is effectively the derivative alone a line at 45° to the scan direction. In figure 3c, this difference is shown with a zero derivative being grey (128 DN), a positive derivative being black (255 DN) and a negative derivative being white (0 DN). The width of the vertical and horizontal edges of the checkerboard are a direct measure of the line spread function in that direction.

The last example of diagnostic techniques is one of Fourier analysis. Figure 4 has an image and its Fourier Transform (4a and 4b). The Fourier Transform has its origin in the middle of 4b, ω_x and ω_y range from -0.5 to +0.5 cycles/pixel. The pattern of reseaux seen in the image is clearly retained in the Fourier image, since the Fourier Transform of a bed of nails is a bed of

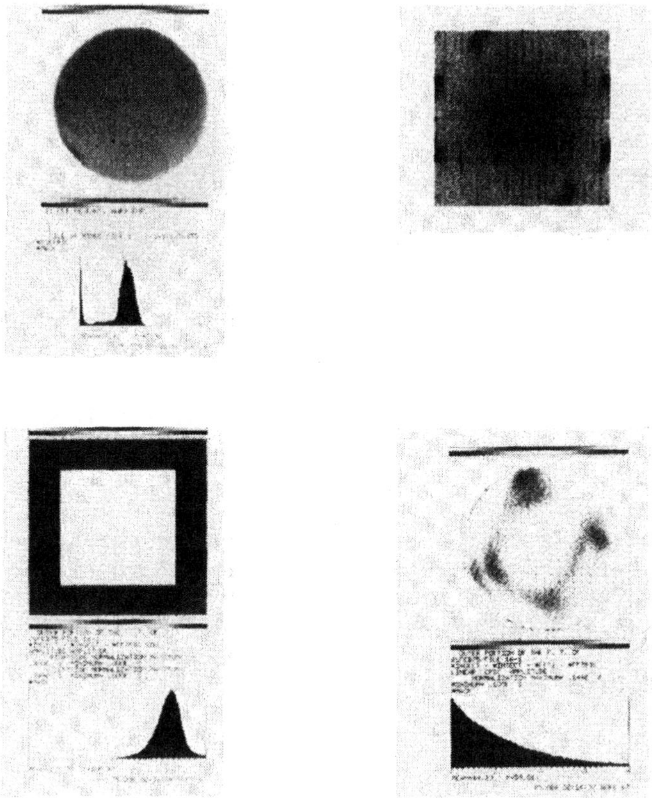

Figure 4. Fourier Transform as a diagnostic tool, see text for detailed description.

nails. However, several fuzzy patches are also seen in the Fourier image at high frequencies. Figure 4c is a high pass filtered transform of 4b that was created by zeroing out the low frequency values. Figure 4d is the inverse Fourier Transform and shows quite clearly the noise pattern effects that are present in the original image. Notice, also, that the circularity of image and the sharpest reseaux are also returned.

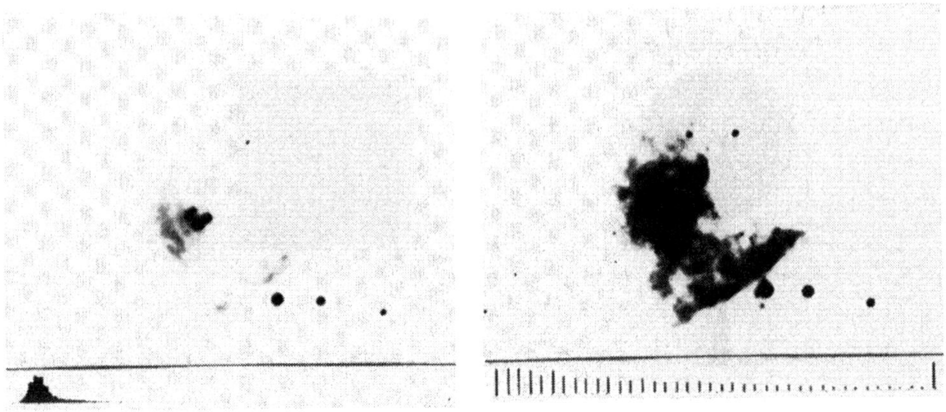

Figure 5. Contrast Enhancement. a) Original low level exposure, b) Linear contrast enhanced image.

Fourier analysis of this type must be done very carefully with a rather detailed understanding of the nature of the image under study or else in the process of removing annoying high frequency noise, the signal is also modified. This has been emphasized by Brault and White in a comprehensive paper on astronomical Fourier analysis (1971).

3.2 subjective enhancement

There is no such thing as best image. The information content of any image is very large, usually larger than the mind can register in a single glance or the computer can analyze in a single pass. As a result of this, each image can be enhanced to show only some of the information that it contains. Two of the techniques for doing this type of task are contrast stretching and edge enhancement.

Some images have an inherently low contrast or dynamic range. One of the best examples of this type are the images obtained by JPL with their Mariner 9 Television Images of Mars (Green et al., 1975). Another example of contrast stretching or enhancement is shown in figure 5; 5a is an original digitized image of the central portion of the Orion nebula. The histogram shows this to be a low exposure level image. However, by stretching out the digitized levels between 24 and 63 to cover the range 0 to 255, we see in Figure 5b that there really is a lot of information in the original image. This has been done solely to bring out low contrast features.

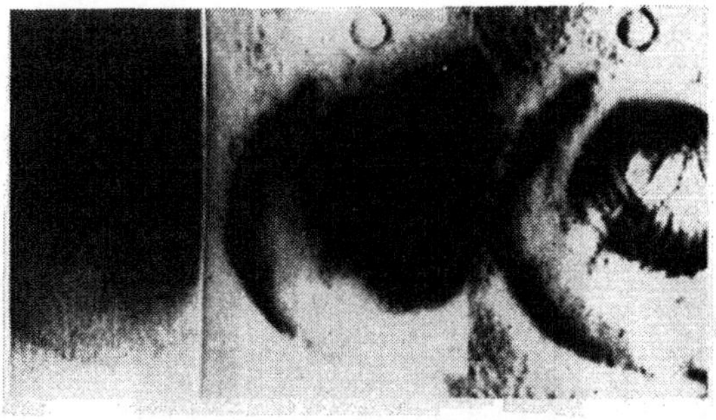

Figure 6. Enge Enhancement results from the Mariner 9 Mars imagery.

Another type of enhancement that has been used quite effectively is that of edge enhancement. Figure 6 shows an example of this, again from the Mariner 9 Mars imagery (Green et al., 1975). Here the first image is the original image. The middle one has been contrast enhanced and the last one has had an edge enhancement algorithm applied to it. The algorithm is similar to the high pass filtering discussed in the section or image diagnostics except that it was done in image space by convolution. Obviously, a convolution in image space and a filter in frequency space are equivalent (Bracewell, 1965).

Psuedo-Color displays of an image lend themselves very nicely to showing small changes in image intensity provided that the scene is smooth enough. The color image displays of lunar spectral reflectance ratios (Billingsley, 1972) are a good example of this type of enhancement. Wray (1972) has also used color techniques to map the B-V color of the surface brightness of Galaxies.

3.3 quantitative restoration

Ultimately, the quantity needed from an astronomical image is the number of photons coming from an image element, be it a star brightness, wavelength interval or surface area. All systems that are used to collect these photons degrade the image in some manner or other. The task of returning from instrumental system of units to useful scientific units is one of quantitative restoration. The calibration process that many of us are familiar with when dealing with photographic film are fine examples of the types of problems that can be encountered. However in these modern times one has to add to all of those problems, ones that deal with non-uniform photometric response and variable geometric distortions when dealing with TV systems.

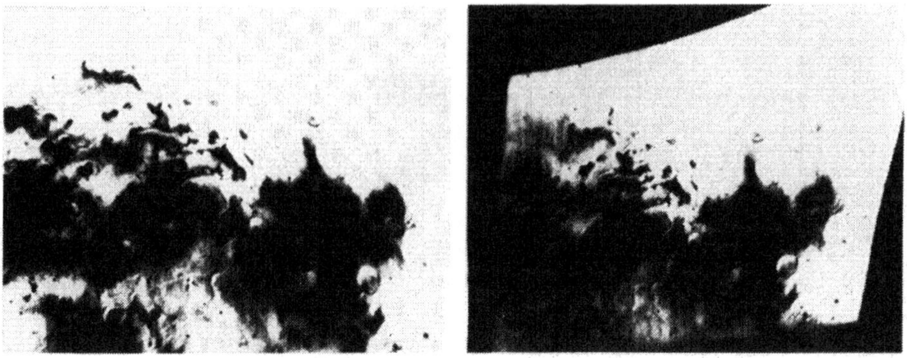

Figure 7. Geometrical Distortion. a) Mars image distorted by large view factor. b) Corrected image.

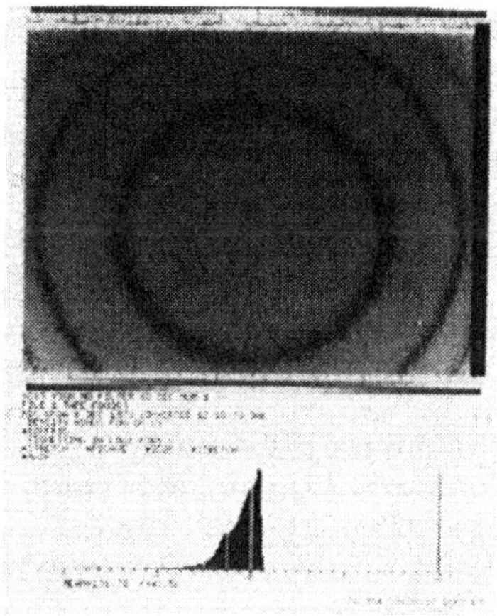

Figure 8. Contouring of a smoothly varying scene by setting every eighth level to saturation.

An example of both non-uniform photometric response and geometric distortion will be given at this conference by Dunford and Klinglesmith (paper 5.3) in which they show the magnitude of the effects and how they can be corrected. The IUE system is an extreme example because the entire tube faceplate is used where most other TV systems stay away from the edges.

An example of geometric distortion that is not an instrumental effect but rather a view angle effect is shown in figure 7 which shows, in 7a, an original image that has been completely calibrated except for the view angle effect and in 7b, which has had the orthographic projection correction made. In comparing the two images the corrected image shows round rather than ellipsodial craters. This work was done by staff at the image processing laboratory at JPL and kindly supplied to me.

Contouring is also used as a means of obtaining quantitative information from smoothly varying imagery. Figure 8 is a contoured image of a uniform flat field exposure taken by the Apollo 15 low light level camera. This camera system had rather severe vignetting as will be shown by Alvord et al. (paper 5.9). Figure 8 was produced by setting every eighth grey level equal to 255. It is a very fast operation which does not need complicated contouring algorithms.

4. Conclusions

There are many corrections or enhancement techniques that can be applied to an astronomical image. It was the purpose of this review to show some of these techniques. Also, it was suggested, that since these techniques already exist on many different computer systems, it would be foolish to repeat the extensive amount of effort to develop these routines again.

I would like to acknowledge the very detailed and generous help given to me by the staff of the Image Processing Laboratory at JPL particularly Mr. Fred Billingsley and Mr. William Green. Without their help I would not have any image processing systems or knowledge to talk about.

References

Billingsley, F. C., 1972, Photo. Sci. Engng., 16, 51.

Bracewell, R., 1965, The Fourier Transform and its Applications, McGraw-Hill, New York.

Brault, J. W., and White, O. R., 1971, A. & A. 13, 169.

Green, B. W., Jepsen, P. L., Kreznar, J. E., Ruiz, R. M., Schwartz, A. A. and Seidman, J. B., 1975, App. Opt., 14, 105.

Worden, S. P., 1974, PASP. 86, 92.

Worden, S. P. and Grasdealen, G. L., 1974, A. & A., 34, 37.

Wray J. D., The Role of Schmidt Telescopes in Astronomy, 1972, ed. Ulrich Haug, Hamburg.

PROBLEMS OF MACHINE INDEPENDENCY AND INTERACTIVE MATTERS

H. Nieuwenhuijzen

Astronomical Institute, University of Utrecht, Utrecht, the Netherlands

ABSTRACT. A review is given of software developed for image processing in astronomy by a number of astronomical institutes. The presented programs are the result of a questionnaire survey and the list is to be considered as preliminary and incomplete.

1. OBJECTIVE OF THE QUESTIONNAIRE

In planning the conference, it was felt there might be some use in coordinating software programs that are being developed to process data as measured by hardware machines like microdensitometers.

Although the development of basic, machine independent, measurement reduction modules as yet will be an ideal, it is clear that the use of software developed elsewhere might decrease development time for writing programs and might also decrease the time needed for checking for program errors, at least for programs of a general nature.

Before the conference a questionnaire was sent out to all those who had expressed an interest in this conference, inviting descriptions and names of software programs, type of machine(s) for which the software was developed, an indication of core needed for the program, and any other relevant data.

Also names of other scientists not attending the conference were asked, in order to be able to send them the same list of questions.

From the answers received, a list has been compiled, and as a result of the conference it was decided to send all participants a

de Jager/Nieuwenhuijzen (eds.), Image Processing Techniques in Astronomy. 285-299.

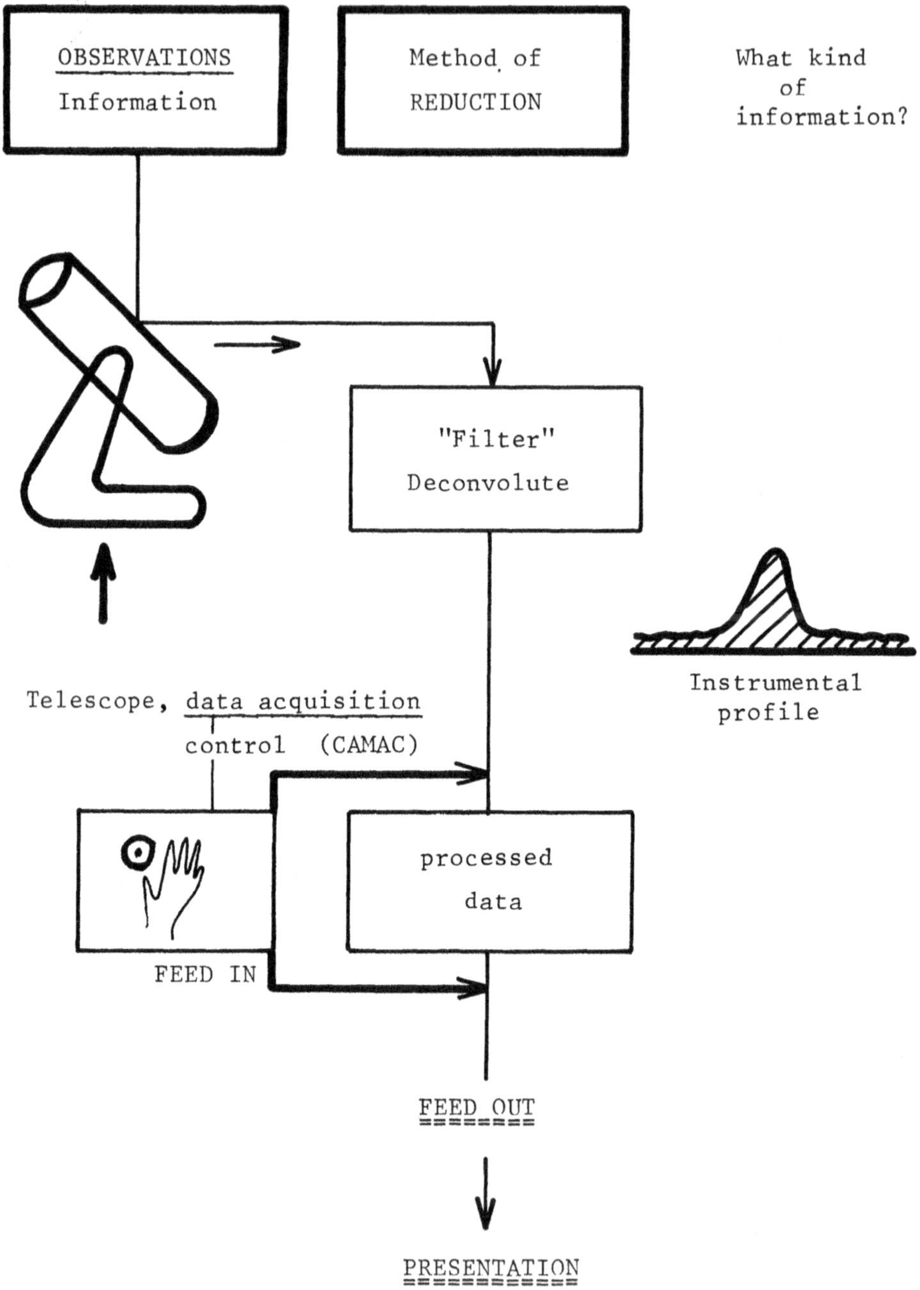

A: TERMS IN WHICH TO DESCRIBE
B: SOFTWARE TO DO THE PROCESS under terms described
C: IN REALITY: 2-DIM INFORMATION

Figure 1

preliminary copy of this list for immediate availability, to give comments and make any changes that seem desirable. The list has been revised before final publication.

2. RESULTS

As a result of the answers received it becomes clear that there is a wish for interactive processing (A). Furthermore I have tried to classify the various software systems (B). In a number of fields some consolidation of programs is taking place (C).

A. Interactive processing

From the answers to the software questionnaire it now appears that the interaction of the astronomer with the processing of his data is considered to be very helpful in order to obtain optimal results. The interactivity is becoming more intensive in a number of leading institutes. There is in short a trend towards greater involvement of the astronomer with his observational data instead of more fully automated processing.

If we combine this with the interaction between the astronomer and his telescope and measurement system, as was stressed by the handsymbol on the cover of the ESO/Cern conference on Auxiliary Instrumentation for Large Telescopes (editors I. Laustsen and A. Reiz, Geneva 1972), we get a system as illustrated in fig. 1 .

Here the hand symbol indicates both data acquisition and data processing. The last taking a number of sequential "passes", with the astronomer deciding after each "pass" what to do next. The data is filtered and deconvoluted for the instrumental profile, and is indicated only for one-dimensional data like spectra. However, the scheme is also applicable to more general two-dimensional processing.

When the processing time is short enough, it is thought to be of advantage to change the system to interactive processing. In practice it turns out that the involvement of the astronomer and his creativity are optimized. However, for very large amounts of data processing, like from the radiointerferometer at Westerbork, the time involved is too long and only multiple pass operation is possible. It may still be possible however, to use interactive methods to obtain the best representation of the finalized data.

B. Classification

In an attempt to classify the various software systems, I have arranged them according to the degree of complexity of data handling into the following three groups:

1A. Two-dimensional picture processing enhance-
ment, distortion, etc.
e.g. Mariner photo's (Klinglesmith)
solar images (Livingstone)

2A. Two-dimensional raster scan or one-dimensional raster scan | to find stars or galaxies

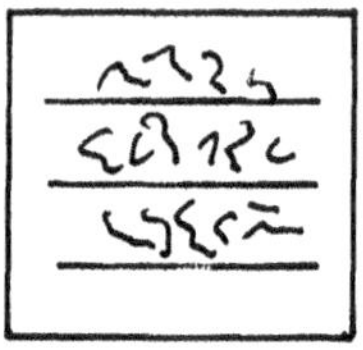

2B. As above, but with <u>on-line</u> computer decision for star or galaxy.

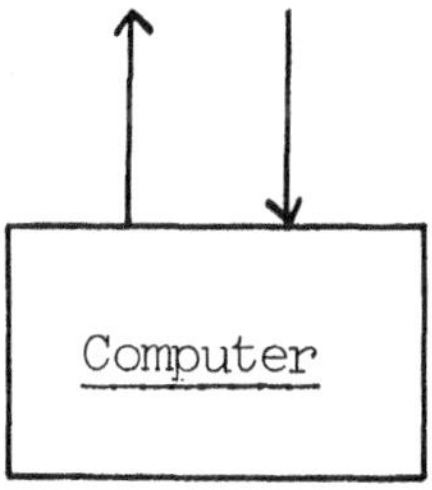

3. One-dimensional line scan (for example: stellar spectra) with (off-line) processing for filtering, deconvolution.
e.g. Reduction program

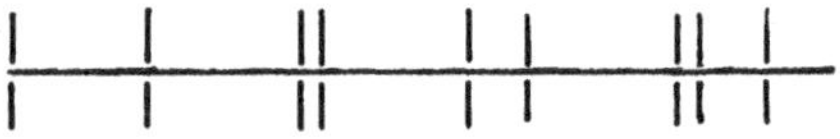

C. Consolidation

From the material presented it seems that there are three main areas in which consolidation of software effort has taken place.

i) The picture processing systems have a large amount of software modules available for general methods in picture processing. Many of the programs have been originally developed by the Jet Propulsion Laboratory for Apollo and Mariner type photo's, and have been used and augmented by various institutes. Those who are interested are requested to contact Drs. Dunford and Klinglesmith.

ii) There is some consolidation in the recognition cycli for stellar and nebular images. It should not be forgotten that for example the Luyten measuring machine (see the paper by J.S. Newcomb in the proceedings of this conference) has buried in it a large amount of software, as has also the Galaxy machine.

iii) Also there is quite a degree of consolidation in the field of one-dimensional image processing.
See also the classical paper by Brault and White (1971).

3. LIST OF PROGRAMS

In the following list I use the order mentioned in paragraph 2.2. After a short indication of general class of programs, the name of the institute and the contact person is given. For persons attending the conference, the full address can be found in the list of participants.

The next columns indicate the sort of program, a typical machine for which the program was developed and an indication of machine core used, followed by a column for remarks.

3.1. a) Image combination
b) Image enhancement
c) Geometric manipulation
d) Frequency domain analysis
e) Utility programs I/O

Institute	Contact person	Software	Computer	Core	Remarks
Laboratory for Atmospheric and Space Physics, University of Colorado, Boulder, Colo 80302, USA.	Bruner, E.	OSO-I image system, interactive picture processing 2-dimensional 1-dimensional			
Science Research Council Appleton Laboratory Slough, United Kingdom	Dunford, E.	Image processing software for IUE satellite, developed from Jet Propulsion Laboratory for processing Apollo and Mariner Pictures See also Klinglesmith	IBM 370/195	modules - 10 k - 50 k -300 k	VICAR program
NASA/Goddard Space Flight Center, Maryland, USA.	Klinglesmith, D.A.		IBM 360/75 IBM 360/91		
Center for Astrophysics Cambridge, Mass. 02183 USA.	Levine, R.H.	Smoothing, filtering scaling, image processing, interactive display	CDC 6400, Comtal 8300 display + PDP 11/10		Harvard experiment on Skylab Mount (20.000 spectroheliograms to be processed) at up to 7 UV wavelengths simultaneous

Institute	Contact person	Software	Computer	Core	Remarks
American Science and Engineering Cambridge, Mass. 02183 USA.	(R.H. Levine)	Analysis of solar X-ray photographs			
Johnson Space Center Houston, Texas, USA.	Lyon, J. in liaison for solar physicists	On-line smoothing, expanding, pattern recognition	IBM 360/75 + on-line display hardware		Earth re-sources Information Processing (ERIPS) system, used by scientists working with Skylab data from the Harvard exp.
MIT Image Processing Program, MIT, USA.	McCord, T.B. Kinnucan, P.F. Fawcett, G. Jn.	Interactive and batch processing for 2-dimensional images and for spectra, many programs. programs developed in interaction with JPL, Stanford U. Research Inst. and US Geolo-gical Survey	IBM 360/65	158 k	batch system 4 years, interactive 1 year, operative
Kitt Peak National Observatory, P.O. Box 26732, Tucson Arizona 85726, USA.	D. Wells	Interactive data processing	CDC 6400, Varian V 74 + Comtal 8300 interactive display		See article by Living-ston

3.1. f) Reduction of radio maps
h) Interactive processing of radio maps

Institute	Contact person	Software	Computer	Core	Remarks
Sterrewacht Leiden Huygens Laboratorium Leiden, the Netherlands	Van Someren Gréve, H.N.		IBM 370/158 IBM 1800	192-256 k	PLI language interactive
Sterrenkundig Laboratorium "Kapteyn", Hoogbouw WSN, Postbus 800, Groningen the Netherlands	Allen, R.J.	interactive processing	PDP-9	24 k + 10^6	Assembler Fortran

3.1. g) Analysis of filtergrams, spectroheliograms

Institute	Contact person	Software	Computer	Core	Remarks
Osservatorio Astronomico Capodimonte P.G. Napoli, Italy and	Falciani, R.	flying spot analysis of solar images	HP 2100	16 k	interactive, Fortran IV
Inst. for Elaboration of Informations, I.E.I., Via S. Marina 46, 56100 Pisa, Italy.	Azzarelli, I.				

3.2. a) Locating star-like images in two-dimensional raster scans

Institute	Contact person	Software	Computer	Core	Remarks
Geneva Observatory Switzerland	Blecha, A.	contours, raster scan, histograms matrix reduction	HP 2100	32 k + 10^6	some programs interactive

Institute	Contact person	Software	Computer	Core	Remarks
Geneva (cont.)		filtering, smoothing calibration, stellar image recognition, stellar profile fitting			
Kitt Peak National Observatory, P.O. Box 26732, Tucson, Arizona 85726, USA.	Furenlid, I.	reduction of area scans	CDC 6400	32 k	multiple operation
Copenhagen University Observatory Denmark	Høg, E. (+ Wiskott, ESO)	evaluation of raster scans of stellar images by means of the "sector method"	HP 2100	16-32 k	Algol
					Compare with Galaxy Software Organization
Instituto di Fisica Via Amendola 173 Bari, Italy	Impedovo, S. and Marangelli, B.	image segmentation and classification in real time	PDP 11/20	3 k	algorithm uses tangential envelopment of contour im.
Laboratorio di Fisica Cosmica e Technologie Relative, Via Celoria 16 20133 Milano, Italy	Mussio, P.	gamma astronomy analysis of spark chamber images (S 133 experiment)	Univac 1106 IBM 360/370	192-288 k	Fortran IV

3.2. a) Locating star-like images in two-dimensional raster scans (continued)

Institute	Contact person	Software	Computer	Core	Remarks
Astronomy Group Imperial College Prince Consort Rd London SW 7 2BZ England	Stephens, C.L.	examines star field, fits curves, displays contours, shows cross section at angle θ to x,y	CDC 6400 CDC 6600	12-24 k	Fortran interactive
Dunsink Observatory Dublin, Ireland	Wayman, P.A.	Galaxy photometry	Data General	2 x 16 k + 10^6	New: BASIC
ESO-TP Division c/o CERN, CH - 1211 Geneva 23 Switzerland	West, R.M.	recognition and measurement of stellar and extended images	HP 2100	12-32 k	image classific-ation (Compare with Luyten Fortran)

3.2. b) Iterative smoothing

Institute	Contact person	Software	Computer	Core	Remarks
Universitätssternwarte Göttingen, Germany	Wittmann, A.	"Van Cittert method" of iterative smoothing instead of filtering by Fourier methods	Univac 1108	48 k	It would be interesting to compare results of iterative smoothing and Fourier processing

3.2. c) On-line with computer to microdensitometer or flying spot scan

Institute	Contact person	Software	Computer	Core	Remarks
NIWARS, Delft the Netherlands (Neth. Interdepartemental Working Group on Application Research on Remote Sensing Techniques)	Van Kuilenburg, J.	star search and calibration, densitometer to on-line computer	PDP 11/40	16 k	Fortran
Lund Observatory Svanegatan 9 S-222 24 Lund, Sweden	Lyngå, G.	analysis of round images, microphotometer and separate table with diode array camera, spectrophotometry, solar spectra	HP 2116 B	16 k	BASIC + CALL routines
Astronomisches Institut der Ruhr Universität D-463 Bochum, Germany	Maitzen Schlosser, W.	control of photometer and rapid spectrum scanner	HP 2114 B	8 k	Assembler, modified BASIC-compiler
Royal Greenwich Observa vatory, Herstmonceux, United Kingdom	Murray, C.A. Nicholson, W.	"Galaxy" star search; sorting and collating data from different plates, reduction	Nova 2/10		changing over to on-line facility

3.3. a) One-dimensional time series

Institute	Contact person	Software	Computer	Core	Remarks
Kitt Peak National Observatory, P.O. Box 26732, Tucson, Arizona 85726, USA.	Brault, J. Slaughter, C. et al.	extremely powerful general reduction program: Reducer	CDC 6400	50 k	multiple pass
Laboratoire d'Astro-physique, Mons, Belgium	Delcroix, A.	reduction of micro-photometric data, calibration	Burroughs B - 6700	13 k	Algol
Kitt Peak National Observatory, P.O. Box 26732, Tucson, Arizona 85726, USA.	Furenlid, I.	reduction of spectra including image tube spectra with special handling of back-ground	CDC 6400	32 k	multiple pass development from Latham, it is import-ant to main-tain operator control, pro-grams will be changed to interactive
Laboratorio di Astro-fisica Spatiale, Frascati, Italy	Gratton, L.	processing of high dispersion spectra, calculation of single line intensities, noise filtering for photo-electric scanner spectra	IBM 370/125 + Univac 1108	16 k	Fortran IV
Center for Astrophysics 60, Garden Street Cambridge, Mass. 02138 USA.	Latham, D.W.	reduces spectrograms and echellograms to intensity versus wavelength, printer	CDC 6400	52 k	

3.3. a) One-dimensional time series (continued)

Institute	Contact person	Software	Computer	Core	Remarks
		plots, equivalent widths. (Program also used and rewritten by Furenlid)			
Sterrewacht "Sonnenborgh" Zonnenburg 2, Utrecht, the Netherlands	Rutten, R.J.	Fourier restoration program	CDC 6500	48 k	ALGOL
Sterrewacht "Sonnenborgh" Zonnenburg 2, Utrecht, the Netherlands	Rosenbaum, J.	processing of stellar spectra for intensity versus wavelength, noise filtering	CDC 6500		
Department of Astronomy University of Michigan Physics-Astronomy Bldg. Ann Arbor, Michigan 48104 USA.	Teske, R.G. Stencel, R.	determines equivalent widths, relative Dopplershifts for spectra lines from microdensitometer output			interactive, in use for two-dimensional scans of solar spectrum plates

3.3. b) Meridian circle

Institute	Contact person	Software	Computer	Core	Remarks
Kopenhagen University Observatory, Denmark	Fogh Olsen, H.J.	Brorfelde meridian circle "Profile search" with on-line computer	HP 2100		HP-Algol

Acknowledgements

I wish to thank the Conference Organizing Committee and especially Dr. Wiskott for a number of helpful suggestions and all those who have so warmly responded and given information about their developed software.

REFERENCES

L. Azarelli, C. Carlesi, R. Falciani, M. Giordano, R. Panicucci, M. Rigutti, C. Roberti, Flying spot analysis of solar images, this conference proceedings.

J.W. Brault and O.R. White, The analysis and restoration of astronomical data via the Fast Fourier Transform, A & A 13, 169, 1971.

M.J. Cullum and C.L. Stephens, Data reduction techniques for direct astronomical electronography, Advances in Electronics and Electron Physics Vol. 33, 757 (1972).

A. Delcroix, Short description of program Reductor, private communication.

E. Dunford, Documentation on image processing, software, VICAR programs, Program Description, Load Module Status Index, Program Timings, private communication.

R.D. Ekers, R.J. Allen and J.R. Luyten, Interactive processing of map data produced by the Westerbork supersynthesis radiotelescope, Astron. & Astrophys. 27, 77, 1973.

H.J. Fogh Olsen, O.H. Einicke, S. Laustsen and H. Schnedler Nielsen, Precision of circle reading and determination of diameter corrections, Astron. & Astrophys. 10, 8, 1971.

I. Furenlid, Spectrum reduction programs for PDS microphotometer data using scans software, Programs for the reduction of PDS area scans, private communication.

L. Gratton, Short description of program, private communication. More programs are available also from: R. Viotti, G. Natali, A. Cassatella and A. Martini of the same institute.

E. Høg and D. Wiskott, Automatic measurement of photographic plates with a photo-diode array, ESO technical Report No. 5, 1974, and proceedings of this conference.

E. Høg, Short description of search and measure of images in a raster scan, private communication.

E. Høg, A theory of a photoelectric multislit micrometer, Astron. & Astrophys. 4, 89, 1970.

D.W. Latham, Computerized microphotometry of stellar spectrograms, Proc. 11th colloquium of the I.A.U., Automation in optical astrophysics, Edinburgh, 1970, Publication of the Royal Observatory, Edinburgh, Volume 8, 1971.
The program has been developed since this publication.

R.H. Levine and J. Flagg, The Harvard Skylab video display system (paper to be presented at the Imaging in Astronomy Conference, Harvard University, June 18-20, 1975), preprint.

W.C. Livingston, The photometry of extended objects: Diode-array detectors and interactive data processing, Proceedings of this conference.
C.A. Murray and W. Nicholson, Proceedings of this conference.
P. Mussio, Computer Phys. Comm. 4, 299, 1972.
J.H.G. Rosenbaum, Documentation of VAMP programs, private communication.
R.J. Rutten, Manuals for spectra 7 and spectra 9, private communication.
J. van Kuilenburg, Proceedings of this conference.
H.W. van Someren Grêve, Program list and software pack reduction group WRST May 1974 list, private communication.
P.A. Wayman, Irish Astr. J. 9, 1, Proceedings of this conference, and see also:
C.J. Butler, Irish Astr. J. 10, 251.
R.M. West, Proceedings of this conference.
A. Wittmann, Description of subroutine image and its application, private communication.
A. Wittmann, On the elimination of seeing effects from solar intensity measurements, Solar Physics 21, 237, 1971.
A. Wittman, Numerical simulation of the Mercury transit black drop phenomenon, Astron. and Astrophys. 31, 239, 1974.

APERTURE SYNTHESIS

W.N. Brouw

Netherlands Foundation for Radio Astronomy,
Dwingeloo, The Netherlands.

1. INTRODUCTION

Radiotelescopes have a diffraction limited resolution, as contrasted with optical telescopes whose resolution is limited by atmospheric irregularities. Hence, from the early days of radio astronomy methods to overcome the mechanical and economical limits to the size of single dish radio telescopes have been devised. Many ingeneous schemes have been and are still being used. The correlation interferometer, consisting of two separate elements connected by cables and receiving equipment to produce the product of the signals received by the individual elements, is one of the most successful schemes. The resolution of such an interferometer is determined by the separation of the elements, rather than by the size of the individual elements. By varying the distance between the elements (the "baseline"), and its orientation with respect to the sky, the autocorrelation function of the electromagnetic field can be measured over an area with a diameter up to the longest baseline used. The sky brightness distribution is the Fourier transform of this autocorrelation function. A map of the sky with a field of view equal to the resolution of the individual elements, but with a resolution depending on the size of the synthesized autocorrelation aperture, can thus be obtained.

One of the earliest aperture synthesis observations was made by Christiansen and Warburton (1955). However, full use of this method had to wait for the general availability of electronic computers. In the late fifties aperture synthesis became a standard tool of radio astronomy, mainly due to the work done by the Cambridge radio astronomy group (e.g. Ryle, 1962). Several large aperture synthesis telescopes are operational today, providing

de Jager/Nieuwenhuijzen (eds.), Image Processing Techniques in Astronomy. 301-307.

maps of the sky at radio wavelengths with resolutions of the order of a few to slightly less than one arcsec. The use of correlation interferometers with intercontinental baselines (VLBI) have attained resolutions of 10^{-4} arcsec.

By measuring the autocorrelation function of the electromagneti field as a function of baseline length, baseline orientation and time, aperture synthesis telescopes can have spectral resolution as well. Most existing aperture synthesis telescopes have incorporat some type of spectral resolution in their list of observational capabilities.

2. SOFTWARE REQUIREMENTS

Digital computers play an important role in aperture synthesis observations. Part of the work of the digital computer could, in principle, be done by analog devices. However, the flexibility and precision of digital computers have, up till now, made them the favoured tool. The increased complexity of the telescopes on the one hand, and the increased capabilities of optical and other analog methods on the other hand, have generated a renewed interest in analog methods. However, digital computing is still the standard method used for aperture synthesis data reduction.

A digital computer is used during many stages of synthesis observing. A rough division can be made into four different areas:

a. Steering telescope and receiving equipment.
b. Calibration and correction of observed data.
c. Fourier transform of data into sky brightness map.
d. Handling of sky brightness maps.

2.1 Hardware steering

The observed autocorrelation function is a complex function. Each value has an amplitude and a phase. Phases are measured with respect to a "fringe-stopping centre" on the sky. Earth rotation, both the regular and irregular components, change the orientation of the interferometer baseline with respect to the sky continuously. Hardware is needed to correct for phase changes due to ongoing changes in the position of the reference phase point in relation to the baseline. A digital computer is necessary to drive this hardware. Many minor corrections, due to imperfections in interferometer elements, imperfect knowledge of the position of the elements with respect to the sky, atmospheric influences etc, have to be made as well.

2.2 Calibration and correction of data

The imperfections mentioned in Chapter 2.1 can be calibrated quite often by observing standard calibration sources, and comparing the receiver output with the expected output for a theoretical behaviour of the instrument. These differences can be used to deduce correction parameters, which can be used to either drive the steering program, or post-correct the observed correlation function. As long as the corrections are relatively small (roughly speaking less than 2π radians), both methods can be used. Although this phase of the data reduction is a very important one, determining the ultimate precision obtained, I will not go into the details here. An overview of the effects occurring can be found in several handbooks (see e.g. Brouw, 1975).

2.3 Fourier transform of data

The observed autocorrelation function is the Fourier transform of the sky brightness distribution. However, since the autocorrelation function is sampled over a limited aperture only, and in this aperture quite often not fully sampled, the sky brightness function, corresponding to an observed autocorrelation function, cannot be determined uniquely. The standard way to obtain an estimate of the sky brightness distribution, is to assume all non observed parts of the autocorrelation function to be zero, and to take the Fourier transform of the complete function thus obtained. The attraction of such a scheme is its relative simplicity on the one hand, and on the other hand the easy way a "synthesized beam" can be defined. There is, of course, additional information available on the map obtained. The sky contains only positive brightnesses, hence negative brightnesses obtained are due to sidelobes, caused by the imperfect knowledge of the autocorrelation function. Furthermore, the structure expected has more or less known characteristics: a pointsource on a map with a perfect ring around it, will be interpreted as a pointsource with a ring shaped sidelobe, *not* as a ring shaped source with a pointsource obtained by the constructive interference of all the sidelobes from the different parts of the ring source. Also, the sky is relatively empty, and disentangling of sources and sidelobe structure is, therefore, possible. Again, much could be said about the actual way the Fourier transform should be done, especially taking computing time, "aliasing", "beamshaping" and other aspects into account. Due to the limited time, I refer for these details to the literature mentioned before, and emphasize the map handling part of the data reduction. Partly because this aspect has a lot of similarity with problems occurring in other astronomical spectral domains.

3. MAP HANDLING

The map obtained from the autocorrelation function sampled on (part of) the synthetic aperture plane is an estimate of the true brightness distribution on the sky. The true brightness map can only be obtained if the autocorrelation function has been fully sampled for baseline lengths from zero to infinity, and all orientations of the baseline. How closely the map obtained represents the true sky depends on the actual sampling of the autocorrelation function, and the way the map has been estimated. The standard way of estimating the brightness distribution, is taking the Fourier transform of the measured autocorrelation, assuming all non-sampled points to be zero. The estimate obtained is than, apart from measuring errors, a convolution of the true brightness distribution and a "synthesized antenna pattern". This antenna pattern will have a finite width, depending on the maximum baseline used, and will have sidelobes. The amplitude of the sidelobes will depend on the weights attached to the different samples, and the regularity with which the autocorrelation has been sampled. Especially in the case of irregularly and undersampled autocorrelation functions, these sidelobes have a large amplitude and are quite extended. But even when the autocorrelation function has been adequately sampled up to a maximum baseline length, sidelobes of a few percent remain. It will be clear that, especially in those cases where one is interested in weak features in the neighbourhood of strong features, these sidelobes hamper an interpretation of the observation considerably. Let us compare it with an optical photograph. One should realize that an average aperture synthesis telescope has maximum baselines equal to about a hundred diameters of the individual elements. Hence, a field of view is presented consisting of about a hundred synthesized beams per linear dimension. This translates in about one square arcmin for optical photographs. The weak fatures one is interested in lie only a few beams (or a few arcsec in the optical case) from the strong source. Differences of up to 7.5 magnitudes have been measured, but an even better "dynamic range" is required. The case of the optical magnitude of Sirius' compagnon should give you an impression of the difficulty of the problem.

What ways are there to improve on the straightforward estimate of the brightness distribution ? There are several ways, all depending on information we know already from other sources about the sky brightness distribution. Some of the methods are:

1. Brightnesses are positive quantities. Biraud (1969) has tried to use this property in the one dimensional case, and could show remarkable improvements. However, the method is clearly limited to observations with a high signal-to-noise ratio; higher than available in most cases.
2. Burns and Yao (1970) have interpolated the observed autocorrelation function to obtain values for non-observed points. The interpolation was based on the limited extend radiosources have on the sky. Results are possible, but for a general method there are some severe limitations.
3. Sources have a finite extent; they are "bandlimited". Hence, the autocorrelation function is an analytic function. By analytic continuation unobserved parts of the autocorrelation function could, theoretically, be obtained (Ville, 1956). Excessively high

signal-to-noise ratios are, however, necessary to make use of this method.

4. Ables (1974) has drawn attention to a completely different way of estimating the sky brightness distribution: the "Maximum Entropy Method" (MEM). From a set of possible maps one chooses the one with the least possible amount of information. The set of possible maps contains all the maps that will represent the observed set of autocorrelation samples. The least possible amount of information is translated into: the maximum amount of entropy. Roughly speaking one selects the map with the minimum number of features able to represent the observed autocorrelation function samples. In the one dimensional case striking results have been obtained. In the two dimensional case no real successes have been obtained as far as I know, partly due to the prohibitive amount of computing time.
5. There are other data adaptive spectral analysis methods, e.g. the Maximum Likelihood Method. None have been used for two dimensional aperture synthesis observations as yet.
6. Högbom (1974) proposed a method which got known as "Clean". This method has proved to be very successful indeed in aperture synthesis observations.

4. CLEAN

Clean is based on two basic assumptions. The first assumption is that the estimated brightness distribution is the true sky distribution convolved with a fully known antenna pattern. The straightforward method of deriving the estimate described above, clearly fulfils this criterium. It produces a completely, apart from measurement errors, known antenna pattern. The second assumption is that the field of view under observations contains a limited number of distinct features only. Furthermore, the number of features should be less than the number of autocorrelation function samples. If the latter is true, a unique solution can be obtained (U.J. Schwarz, private communication). If either of the assumptions is not fulfilled, the method breaks down. If a unique solution is possible, one can be reasonably sure that the solution represents a better estimate of the brightness distribution; however, certainty that this is indeed true does not exist.

The method proceeds as follows. The absolute value of the amplitudes of all, in general discrete, points in the estimate are inspected. The one with the highest value is selected. The antenna pattern, scaled with the amplitude of the point, is shifted to this point, and subtracted from the full map. In general not the full amplitude of the map point is used in scaling the antenna pattern, but only a fraction. This is done because a source will, in general, not be situated at the exact position of a grid point, and may, furthermore, be extended. Hence, leakage from neighbouring points should be taken into account. The value of the fractional depends on the type of observation, but values around a half are commonly used. Theoretically an infinitesimal value should be used. The above process is than repeated, until a certain criterium is met: quite often till the maximum amplitude remaining is below a threshold value.

At this stage the map has been "deconvolved" into a set of delta functions, representing pointsources at map grid positions. The map can now be rebuilt with the aid of an artificial antenna pattern. In general, but not always, an artificial pattern with about the same width as the original, but without the sidelobes,

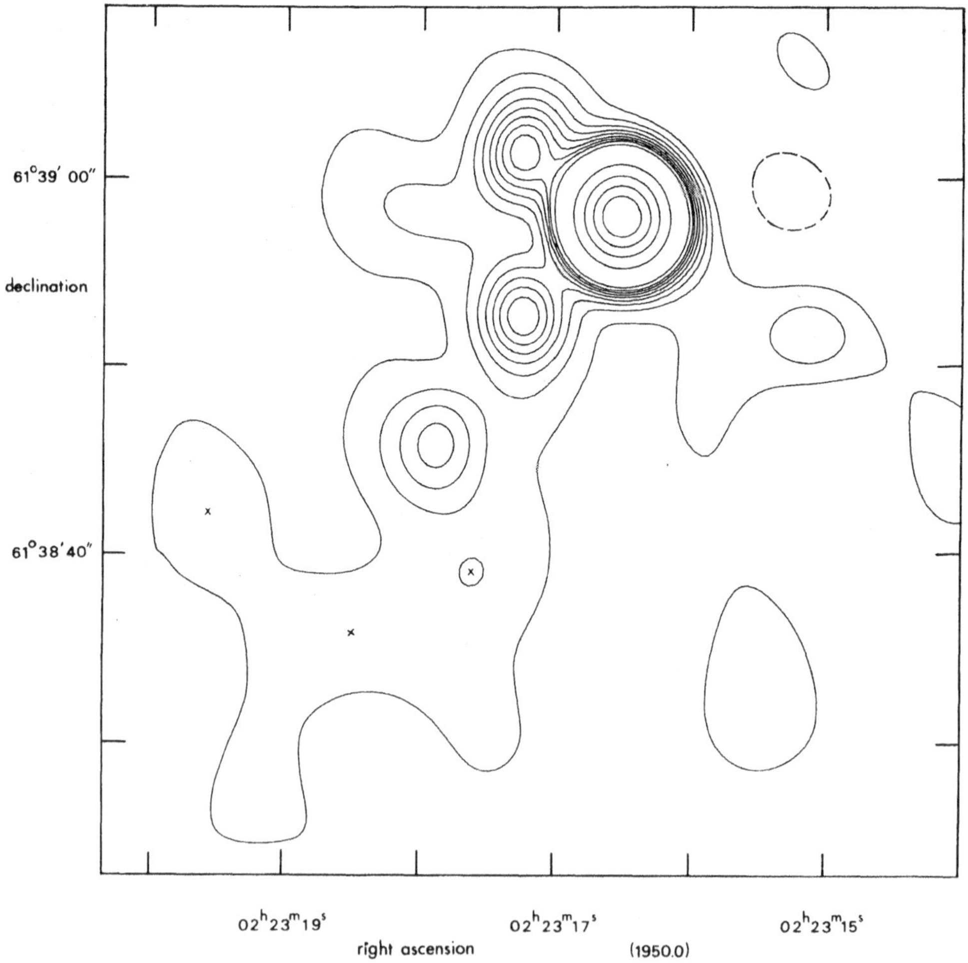

Fig. 1. W3-OH at 6 cm. Observed with a beam of 6.6 x 7.4 arcsec, cleaned, and restored with a beam of 4 x 4.4 arcsec.

is chosen. An example of the effectiveness of the method is given in Figure 1 (courtesy R.H. Harten). It represents an observation with the Westerbork Synthesis Radio Telescope of W3-OH at 6 cm. In this case the original antenna pattern had a width of 6.6 x 7.4 arcsec, the restoring antenna pattern had a width of 4 x 4.4 arcsec. Comparison with observations done with the Cambridge five kilometre array, show, however, that all features are significant.

No literature on the theoretical background of the method exists. Schwarz (1975) is preparing a paper on the subject. Some general comments on the method can be found in Ekers, Miley and Le Poole (1973).

5. DISPLAY

Aperture synthesis telescopes produce of the order of 10^5-10^6 data-points per map, with a dynamic range of about 25 - 30 dB ($6^m.3$ - $7^m.5$). If the spectral information is obtained as well the, three dimensional, brightness distribution has 10^6-10^8 points. A problem occurs in displaying this amount of data in a visual form useful for astronomical interpretation. This problem occurs, of course, in all branches of science where digital computers are used to process large amounts of observed data. Especially the three dimensional maps pose problems. Several attempts have been made up till now: colour displays; layered photographs, where each layer represents a two dimensional cut, on which can be focused easily, etc. No perfect way has been found as yet. Cooperation between different branches of science with the same problem seems very necessary here.

REFERENCES

Ables, J.G., 1974, Astron. & Astrophys. Suppl. **15**, 383.
Biraud, Y., 1969, Astron. & Astrophys. **1**, 124.
Brouw, W.N., 1975, *Methods in Computational Physics Volume 14*, 131 (in press).
Burns, W.R., and Yao, S.S., 1970, Astron. & Astrophys. **16**, 481.
Christiansen, W.N., and Warburton, J.A., 1955, Aust.J.Phys. **8**, 474
Ekers, R.D., Miley, G.K., and Le Poole, R.S., 1973, Neth. Found. Radio Astr. Techn. Note **127**.
Högbom, J.A., 1974, Astron. & Astrophys. Suppl. **15**, 417.
Ryle, M., 1962, Nature **194**, 517.
Schwarz, U.J., 1975, In preparation.
Ville, J.A., 1956, Cables et transmission **1**, 44.

INTERACTIVE COMPUTER REDUCTION AND DISPLAY OF RADIO SUPERSYNTHESIS MAPS

R.D. Ekers and R.J. Allen

Kapteyn Astronomical Institute, University of Groningen

ABSTRACT. This paper discusses first in a general way the problems arising in extracting the interesting information from the large amounts of map data which are presently being produced by modern supersynthesis radio telescopes. Examples are taken from observations made with the Westerbork array. The necessity for visual presentation of the maps and the advantages of certain different types including colour are discussed. A description is given of an interactive data processing system which has been constructed in a medium-sized computer in order to explore and deverlop methods for processing radio map data. Various modular program units can be summoned by the user and the results of major steps in the reduction can be presented on an oscilloscope display for rapid visual examination. Based on his evaluation of these results the astronomer can directly alter his choice of the processing sequences to follow.

The main part of this paper has already been published {1}.

REFERENCES

1. Ekers, R.D., Allen, R.J., and Luyten, J.R. 1973, Astron. Astrophys. 27, 77-83.

de Jager/Nieuwenhuijzen (eds.), Image Processing Techniques in Astronomy. 309.

THE DESIGN OF VAMP SOFTWARE FOR THE MEASUREMENT AND REDUCTION OF STELLAR SPECTROGRAMS

Robert J. Rutten and Gerard P. van Gelder

Sterrewacht "Sonnenborgh", The Astronomical Institute, University of Utrecht, the Netherlands

We present an outline of the software development for the Utrecht VAMP microdensitometer to complement the hardware description given by Heintze et al.

The VAMP software is divided into two quite different categories:

1. - the control programs for the on-line PDP8 that drive the VAMP to trace plates along prescribed paths;
2. - the off-line reduction programs that convert the measured densities on the spectrograms into astrophysical information.

The present control programs, supplied by the VAMP's manufacturers, form an excellent general-purpose operating system that enables the user to program the motion of the measuring spot over the spectrogram incrementally, using smallest addressable steps in the X- and Y-directions of 1 μm.

The pattern of the spot's movement relative to the plate is generated in incremental plotter fashion as a series of line segments between specified co-ordinate pairs.

The specification of these endpoints is obtained either manually with a "mark" button that dumps the location of the spot on a papertape, or automatically by the computer.

Usually they are defined by the operator using a joystick control to move the spot while observing through the viewer. A straight line will be generated between each two co-ordinate pairs in smallest steps of 0.5 μm in X-axis (plateholder) and Y-axis (spot); the operator specifies with buttons and a teletype whether he wants the VAMP to scan or to skip along that line, as well as scanning mode, resolution and density range. A measurement pattern consists of all line segments that together define all necessary scans through a complete spectrogram. The pattern is punched on papertape during its definition; in the subsequent measurement phase the

de Jager/Nieuwenhuijzen (eds.), Image Processing Techniques in Astronomy. 311-314.

control program reads and executes the whole pattern automatically, transferring the measured densities to magnetic tape.
Of the two possible scanning modes the stepping mode is the more precise since in the "on-the-fly" mode the plate moves appreciably during the sampling process. The advantage of scanning on-the-fly is the higher sampling rate of 100 samples/s, which is about four times faster than in the stepping mode.
This operating system works beautifully for any type of plate or problem: however, we have decided to extend it to a system that is designed exclusively for the tracing of large numbers of standard coudé spectrograms. Its purpose is to minimise operator tasks, both at the machine and in the off-line reduction phase, to produce a larger throughput of plates and to achieve complete standardization of scanning patterns as well as output formats. Also, this control program will not only write the measured positions and densities on the magnetic tape but also all the information needed for subsequent calibrations, e.g. the transmissions of step wedges and the wavelengths of comparison lines. The tape can then control the calibration programs on its own. This automation results in high reliability of the whole procedure that converts plates into digital intensity-against-wavelength data.
The way in which this special-purpose operating system will work can best be demonstrated from a simplified example of the setting-up procedure:
After clamping a spectrogram to the plateholder, the operator skips the time-consuming process of aligning the dispersion direction with the plateholder's X-motion direction.
Instead he dumps two positions that together define the dispersion direction. Subsequently he defines with a minimum of dumped points, and in answer to typed requests, the length of the scans, the positions of the comparison arc spectra, of the intensity calibration spectra and of the "clear plate" lanes that must be scanned. The control program then generates parallel and aligned scans through all these spectra, using on-the-fly sampling where possible. In these on-the-fly scans the spot is kept on the non-aligned spectrum by intermittent 0.5 μm steps of the Y-axis. Also, off-line software routines correct the positional shifts caused by this mode. In this manner the position addressing of the VAMP is exploited to generate non-aligned scans while sampling on-the-fly. The comparison arc spectra are scanned in "zip" mode by measuring only selected lines and zipping fast in between. For this purpose the astronomer has to supply a list of the wavelengths of the calibration lines to be used. Their relative positions are determined and checked on the first plate of a batch of similar spectrograms in an interactive process.
Also, the control program tells the operator when and how to change slit orientation and width, for instance at the start of transverse scans through the intensity calibration spectra.
Finally, there are many options such as scanning double Zeeman

spectra, scanning step wedge spectra at specified intervals, scanning inhomogeneously guided spectra in parallel bands, etc. This special operating system will be most useful for large batches of plates that have been obtained with the same camera of the same spectrograph. It will on the average reduce the total time per plate spent at the machine with a factor between 3 and 5, to about half an hour for a complete 20 cm coudé spectrogram.

We now turn to the off-line calibration and reduction programs. Sofar we used for these a HP2100 system; at the moment we are transferring programs to the CDC 6500 that Utrecht University acquired in the meantime. The presently available programs achieve intensity and wavelength calibration in a simple fashion and clean the spectrum by filtering in the Fourier domain. However, since other groups are advanced to a more sophisticated level, notably at Frascati, Victoria and in England, we will give no details of these but rather describe briefly the organizational setup that we envisage in connection with the new VAMP control program.
First we make a clear distinction between the calibration programs that convert the plate densities back to the original intensity-against-wavelength information, and the interpretative programs that do astrophysics. The calibration programs constitute a rather hard part of the system, since they are machine-dependent. They will be operated by the VAMP staff rather than by the astronomer and they will be automated as much as possible. This is relatively easy for wavelength calibration but quite difficult for intensity calibration when that has to include background correction, determination of the calibration curve and the elimination of plate flaws. But however automated, the result will be a new tape ("library tape") containing the original spectrum as intensity-against-wavelength data. These tapes are machine-independent.
The reduction programs that are used on the calibrated data are quite general. They include noise filtering and restoration with the inverse of the instrumental profile, continuum level determination, line finding and identification, radial velocity and equivalent width determination, etc.
These programs should also largely be supplied by the VAMP group since their use is general; the astronomer should operate them, however, with good insight in their capabilities and limitations. We feel that appreciable advances in this field are both possible and valuable; however, the effort needed for this undertaking exceeds the present - reduced - manpower at Utrecht, and a large scale co-operation would be fruitful.
Finally, the remaining programs that do astronomy should be left to the astronomers themselves: e.g. abundance determination, magnetic field strength measurement, binary star orbit determination.

ACKNOWLEDGEMENT

Drs. T. de Groot, Th. Gunsing and H. van de Stadt have taken part in the definition of the new VAMP control program. Mr. J.H.G. Rosenbaum has written all the present reduction programs. We thank Drs. T. de Groot, J. van Nieuwkoop, H. van de Stadt and C. Zwaan for comments on the presentation of this paper.

DEVELOPMENT OF DIGITAL SYSTEM FOR STELLAR SPECTRA REDUCTION AT TORUŃ OBSERVATORY

Z. Turło[°], S. Krawczyk[°°], J. Smoliński[°]

° Polish Academy of Sciences, Institute of Astronomy, Astrophysical Laboratory I

°° Institute of Astronomy of Nicolaus Copernicus University, Toruń, Poland

Reduction of the spectrograms recorded on photographic plates is one of ubiquitous problems of stellar spectroscopy, where application of digital technology appears to be quite straightforward and the results are most likely to repay the initial investments. A number of systems for spectroscopic data processing have been in operation over considerable length of time e.g. Peat and Pemberton (1970), Robinson (1970), Bonsack (1971). Other even more advanced systems are now under development. In this paper we summarize briefly our experiments with digital reduction of stellar spectra since 1972, and describe some recent improvements introduced to the new version of the digital microphotometer.

At the Astronomical Observatory in Toruń we have begun experiments with the digital processing of the spectrophotometric data, using a system based on the already existing instruments which were converted to the digital mode. To describe briefly our first system, up to now in operation with some minor modifications, we could say it was essentially an off-line, minimum hardware, non interactive set-up, perhaps somewhat less powerful than one would desire nowadays. Nevertheless this system enabled us to benefit to considerable extent by the speed, the accuracy and the flexibility inherent to computerized data processing,without a large initial investment and without excessive engineering. Software developed for this system included: scale conversion from density into any other scale defined by the calibration curve, computing the sum or the difference of several spectra reduced with independent input parameters, plate noise filtering by convolving the spectrum with the appropriate weighting profile, normalizing a spectrum with respect to the local continuum level. In order to cope with the

de Jager/Nieuwenhuijzen (eds.), Image Processing Techniques in Astronomy. 315-321.

diversity of input data types and the requirements of the actual reduction problem on hand, software has been constructed in modular form with a number of options enabling to select the range of the reduction routines. To estimate the overall accuracy of our system independently we compared our intensity tracings with those obtained at the Dominion Astrophysical Observatory in Victoria for the same plates. In another test, we compared equivalent widths derived with our system with those published in the Solar Spectrum Atlas, Minnaert et al. (1940). In all cases we have found quite satisfactory agreements. Preliminary results obtained with this system, along with detailed description of the hardware and testing have been published elsewhere, Turło (1974).

Examples of the practical application of the reduction system described here are shown in Figures 1 and 2. Both spectra were converted into intensity scale, convolved with the gaussian weighting profile and normalized to the local continuum level.

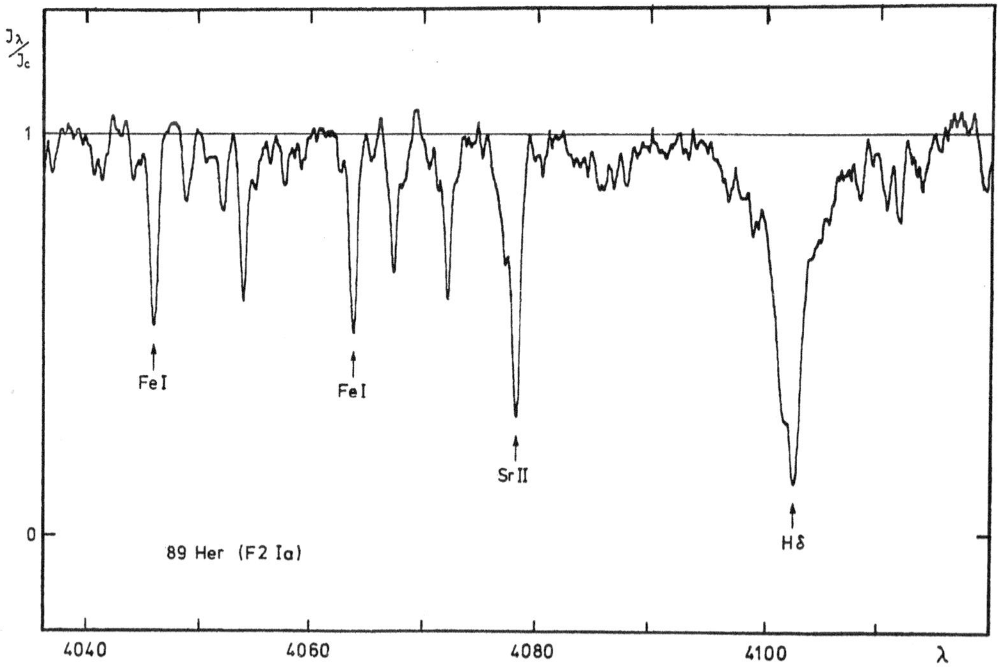

Fig. 1. Intensity tracing of 89 Her (F2 Ia) in the Hδ region. Continuum level is indicated by the thin solid line.

Tracings shown here were reproduced directly from the output tapes obtained from the computer. The spectrum of 89 Her (F2 Ia) shown in Figure 1 is in fact one of the first spectra taken with the Canadian Copernicus spectrograph at the Toruń Observatory. Figure 2 shows in detail the Hα line profile of the ρ Cas (F8 Iap) spectrum with the strong emission features clearly reproduced with our system

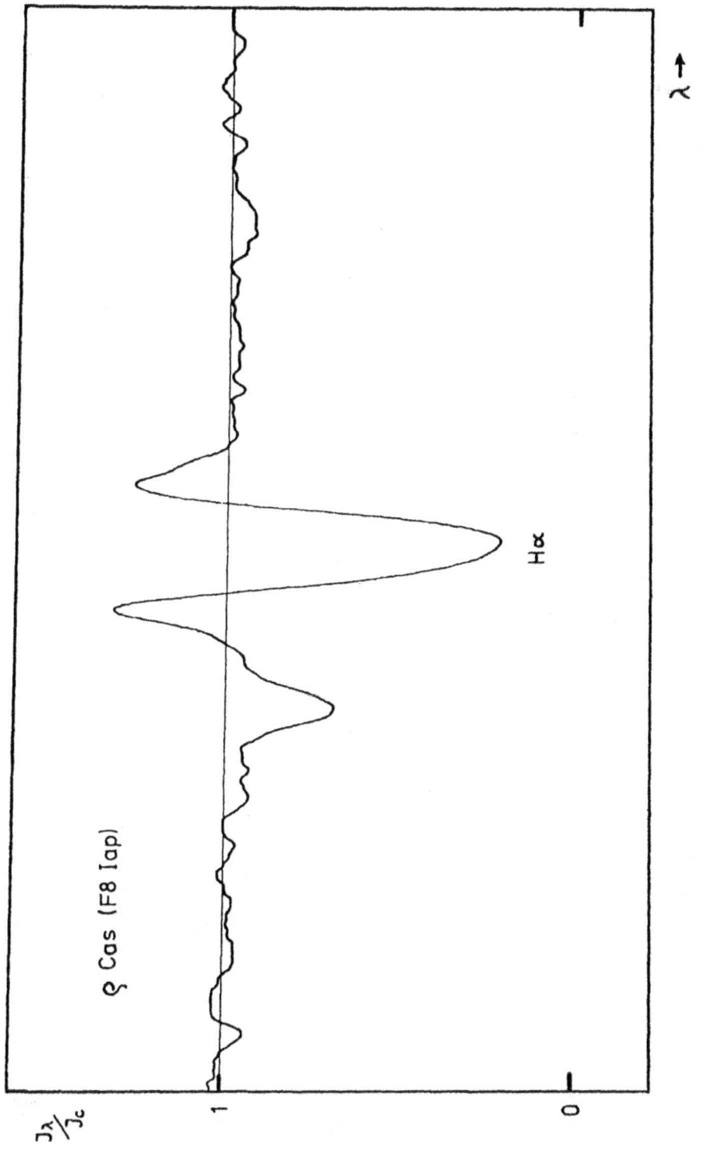

Fig. 2. Hα profile of ρ Cas (F8 Iap), normalized to the continuum.

After two years of operation of the digital microphotometer we became increasingly aware of a number of possible improvements which could be made in order to increase the overall efficiency. In an attempt to reconcile, at the moment, the exact demands of stellar spectroscopy with the technical means at our disposal we have developed a new version of the hardware shown in Figure 3. With this new system operating it is now possible to increase the linear resolution on the plate to 1μ within the 20 cm working range. Density levels can be sampled at the rate of 25 samples per second with up to 200 samples capability in the on-line mode. Amplitude resolution has been improved (up to 1000 density levels) using a differential digital voltmeter. Some new features have been also included in this version which we expect will further extend its application range. For instance, in a number of cases

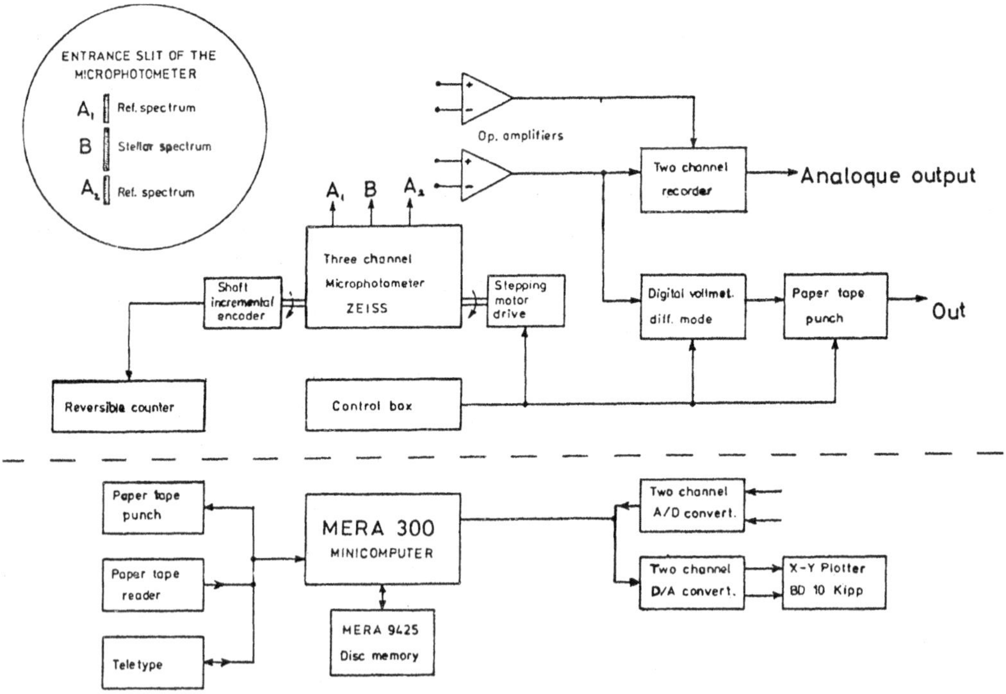

Fig. 3. General layout of the hardware elements of the new version of the digital microphotometer. Extensions for the on-line mode, now under development, are shown on the lower part of the picture.

encountered in spectroscopic data processing it is of paramount importance to restore in the computer the wavelength scale for each spectrum. In the new version we use to this aim a three channe microphotometer, scanning the stellar and the reference spectra simultaneously in the one single sweep through the plate. The position of the three channels on the entrance slit of the microphotometer can be adjusted using a set of appropriate masks (see the inset in Figure 3). This procedure provides in fact for each spectrum the instantaneous calibration of the wavelength scale and is clearly advantageous with respect to the traditional one channel approach in accuracy and speed, especially in the on-line mode of operation.

Another crucial factor bearing on the feasibility of the computerized system is the effective turn-around time. With an off-line system like our first one, it takes typically a day, before results are available to the user in the final form, despite the fact that the actual reduction can be performed in minutes in a large computer. To improve in this respect the turn-around efficiency of our system as well as for a number of other applications, we acquired a minicomputer with basic peripherials, already operational since early 1974 at our Observatory. Mostly for economical reasons we have chosen a minicomputer system MERA 300 of Polish make. This is a fairly inexpensive machine with manufacturer service support available with a number of peripherials including magnetic disc memory and CRT display. The central processing unit operates on the 8-bit words with a 2μ sec cycle, 8k memory. Using a disc-drive unit, memory volume will be expanded in the near future, up to 5×10^6 8-bit words. Some precautions are necessary with this computer in order to assure sufficient amplitude resolution of the recorded spectra within the rather limited memory space. It appears that even with the 8-bit word per sample it is quite possible to obtain a sufficient amplitude resolution taking advantage of the fact that for typical spectra, consecutive density samples change substantially less than the amplitude range of the whole record. As we have found in practice, an improvement by a factor of four can be readily obtained recording the differences between consecutive samples instead of the actual samples themselves. By filtering the noise from the raw density record and computing higher order differences, the amplitude resolution can be further improved at the expense of an insignificant increase of processing time. Figure 4 shows this possibility on a practical example. In the typical case as illustrated in Figure 4 the amplitude resolution can be increased up to 4000 density levels using only a 8-bit word per sample. The system described here is now in the test operation at our Observatory. For the time being it is used in the off-line mode, however it is now possible to shorten turn-around time from days to hours at least for rudimentary reduction. Further work on the development of the software for the minicomputer and implementation of the on-line mode operation is now in progress.

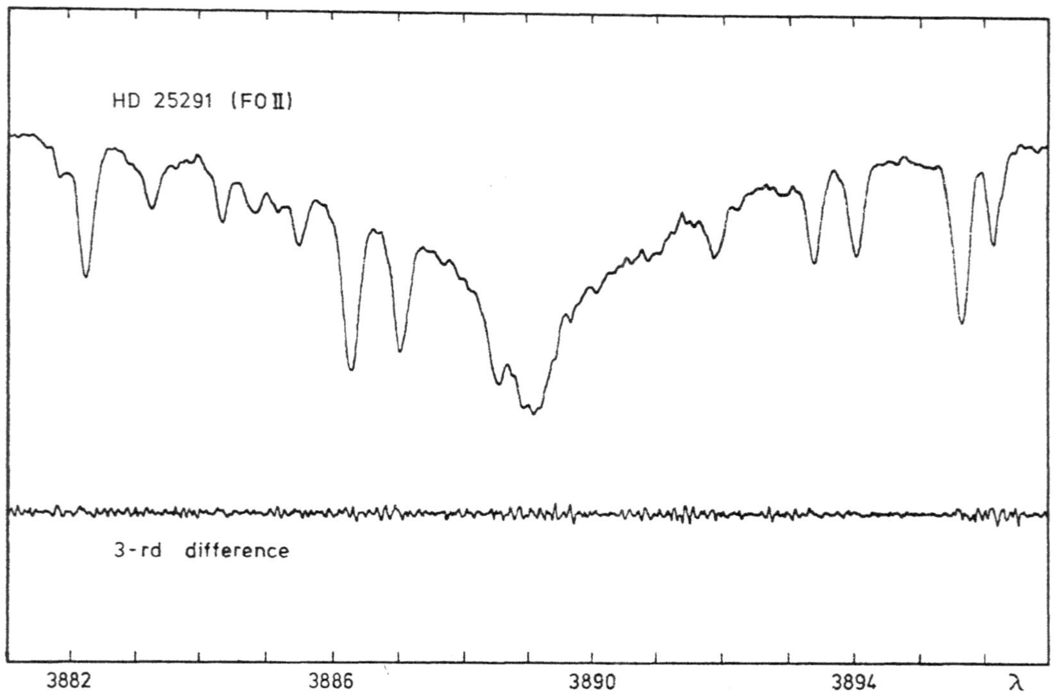

Fig. 4. Tracing of the spectrum of HD 25251 convolved with the gaussian weighting function, with its third order difference plotted in the same scale.

Finally we would like to acknowledge many informative talks with the staff of our Observatory, special thanks are due to Professor W. Iwanowska for kind encouragement during this work.

REFERENCES

1. D.W. Peat and A.C. Pemberton, The Observatory 90, 141, 1970.
2. L.B. Robinson, Proceedings of the Eleventh Colloquium of the IAU Publ. Roy. Obs. Edinburgh 8, 198, 1971.
3. W.D. Bonsack, Astronomy and Astrophysics 15, 374, 1971.
4. M.G.J. Minnaert, G.F.W. Mulders and J. Houtgast, *Photometric Atlas of the Solar Spectrum*, Amsterdam, 1940.
5. Z. Turło, Bull. Astr. Obs. Toruń 53, 9, 1974.

EVALUATION OF RASTER SCANS OF PHOTOGRAPHIC STELLAR IMAGES x)

Erik Høg

Astronomisk Observatorium, Brorfelde,
4340 Tølløse, Denmark

ABSTRACT. Raster scans of photographic stellar images can be obtained with a microphotometer, or still faster with a photo-diode array. A both simple and accurate procedure for evaluating a raster scan is presented. This "sector" method computes the position and size, the non-circular form and the diffuseness of an image. It derives iso-transmission curves for the image, and it can thereby recognize and evaluate partly overlapping images. It is potentially fast enough to evaluate a hundred thousand stellar images per hour with available on-line minicomputers.

Raster scans of a Schmidt plate obtained with a diode array and evaluated with the sector method produced in a first attempt results which were repeatable within $0\overset{m}{.}07$ (m.e.) for photometric measurements (under certain conditions even $0\overset{m}{.}03$ were obtained) and within 0.4μm for astrometric measurements.

INTRODUCTION

The fast devices becoming available for producing raster scans of photographic stellar plates call for fast numerical evaluation procedures to be used in an on-line computer. The larger number of star images evaluated encreases the need for a procedure which is accurate not

x) This work has been supported by The European Southern Observatory.

de Jager/Nieuwenhuijzen (eds.), Image Processing Techniques in Astronomy. 323-328.

only on perfect images, but which also provides data for automatic recognition of disturbed images. The "sector" method presented below is adaptable to different desired levels of accuracy and automatic recognition of pathological images.

In the famous Galaxy machine in Edinburgh the plate is taken out of the machine after the "search" phase and has to be inserted again later on for the "measure" phas As the searching and measuring become faster the more important it is to reduce the time for manual handling and orientation of the plate. Høg and Wiskott (1974, ESO Tecnical Report No.5, hereinafter called Paper I) have described how the search and measure phases can be carried out simultaneously in a proposed photo-diode array machine. Such a machine may be used in two modes of operation: In mode A single stars at approximately known positions on the plate are measured, while in mode B all sta on the plate are measured. It must be noted that the modes A and B both use the simultaneous search and measure technique, but mode A applies it to small areas of the plate with one main star, while mode B treats a large area with many stars.

Paper I discusses these problems, it gives information from the tests of an array of 256 diodes, and from the evaluation of raster scans of a Schmidt plate with the sector method to be described briefly below.

Some astronomers doubt the usefullness of measuring 100 000 stellar images per hour, but some other astronomers want to find very rare objects in this way. Luyten has compared pairs of plates from two widely separated epochs. But treatment in a computer will allow compariso of three or more plates of the same field, e.g. taken in three different colours in order to select stars from special areas of the U-B versus B-V diagram.

THE SECTOR METHOD

The sector method derives the shape parameters for the iso-transmission curves at different levels of transmission. In a polar coordinate system with its origin approximately at the centre of the star image eight points of each iso-transmission curve spaced at 45° are found. Fig. 1 shows the well-ordered partition of the raster elements in eight sectors. Within each sector the elements are given numbers k = 0, 1, ... which are proportional to the distance from the centre, and so that those elements in each sector having about the same distance receive the same number. This ordering is adequat for images which are only slightly elongated and it ensu

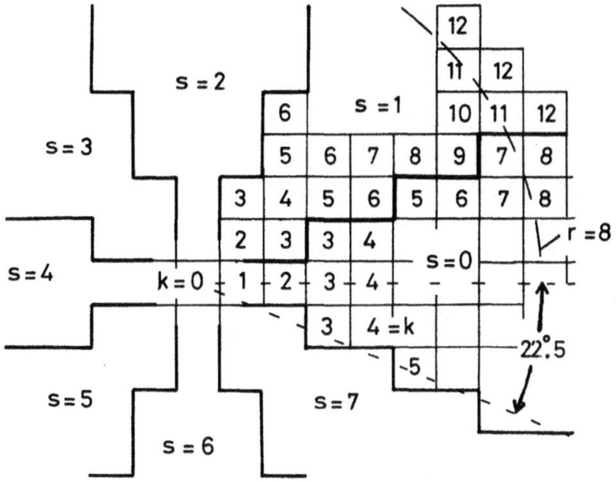

Fig. 1. The partition of raster elements.

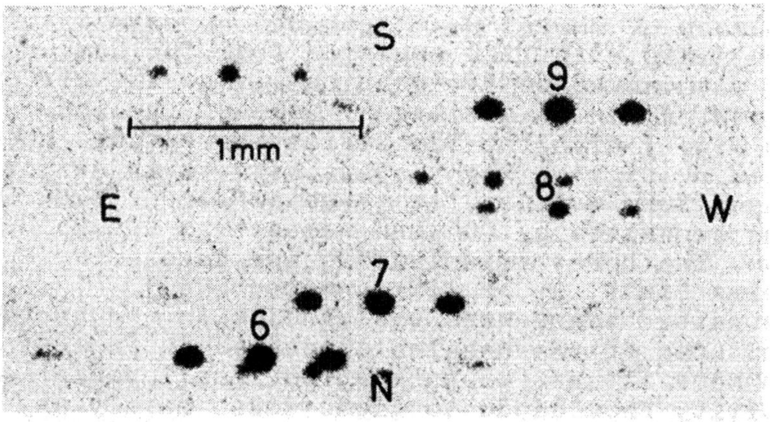

Fig. 2. Star plate obtained with objective grating.

res a fast computation process, since the first step is only to add in each sector all transmission values having the same number k, and to store their sums. The average transmission values as a function af k in each sector form the "sector profiles".

By inverse linear interpolation in the sector profiles the distances from the centre to the four levels of transmission T = 20, 40, 60 and 80 percent are found. A harmonic analysis is performed on these eight radial distances for each iso-transmission curve and the parameters derived have a simple interpretation: the constant gives the diameter of the iso-transmission curve, the first harmonics are the astrometric deviations in x and y, the second harmonics give the ellipticity of the iso-transmission curve, the third and fourth harmonics are indicators for the noise of the measurement. Further details are found in Paper I.

RASTER SCANNING WITH A DIODE ARRAY.

Raster scans were obtained with an experimental device ISIS (Instrument for Stellar Image Scanning) covering in each measurement about one square millimeter of a plate with a self-scanning array RL256, manufactured by RETICON. The plate was moved with constant speed perpendicularly to the row of diodes, the array recording the plate transmission sequentially in 10 x 10 μm^2 areas of the plate at a rate of 16 µs per value. The values were stored in a small HP2100 computer and subsequently punche on paper tape for later evaluation on a more powerful HP2100 computer.

A number of images were measured repeatedly with ISIS on a plate with IIaO emulsion from the Hamburg Schmi obtained with an objective grating giving $1^m.0$ difference between central and side images. Some of the images are shown in Fig. 2 with the identification numbers used in the following figure. Eight stars with their respective side images were measured in three different plate orientations separated by 90° with respect to the direction of motion. The three orientations are denoted R, U and L, North being right, up or left, respectively.

The raster scans were evaluated with the sector method, and some of the results discussed in Paper I, shall be given here. It was shown that the small variations in sensivitivity from diode to diode cause hardly any problems.

The repeatability of one measured position obtained from the 60% transmission curve alone was ±0.4 µm (m.e.) in one coordinate. It was derived from the measured positions of the two side images relative to the central image, measured in the three orientations of the plate.

In case of partly overlapping images, as for star No 6, different centres and different elongations were obtained for the different iso-transmission curves. The elongation and the higher harmonics of a certain iso-transmission curve indicated the degree of disturbance of the

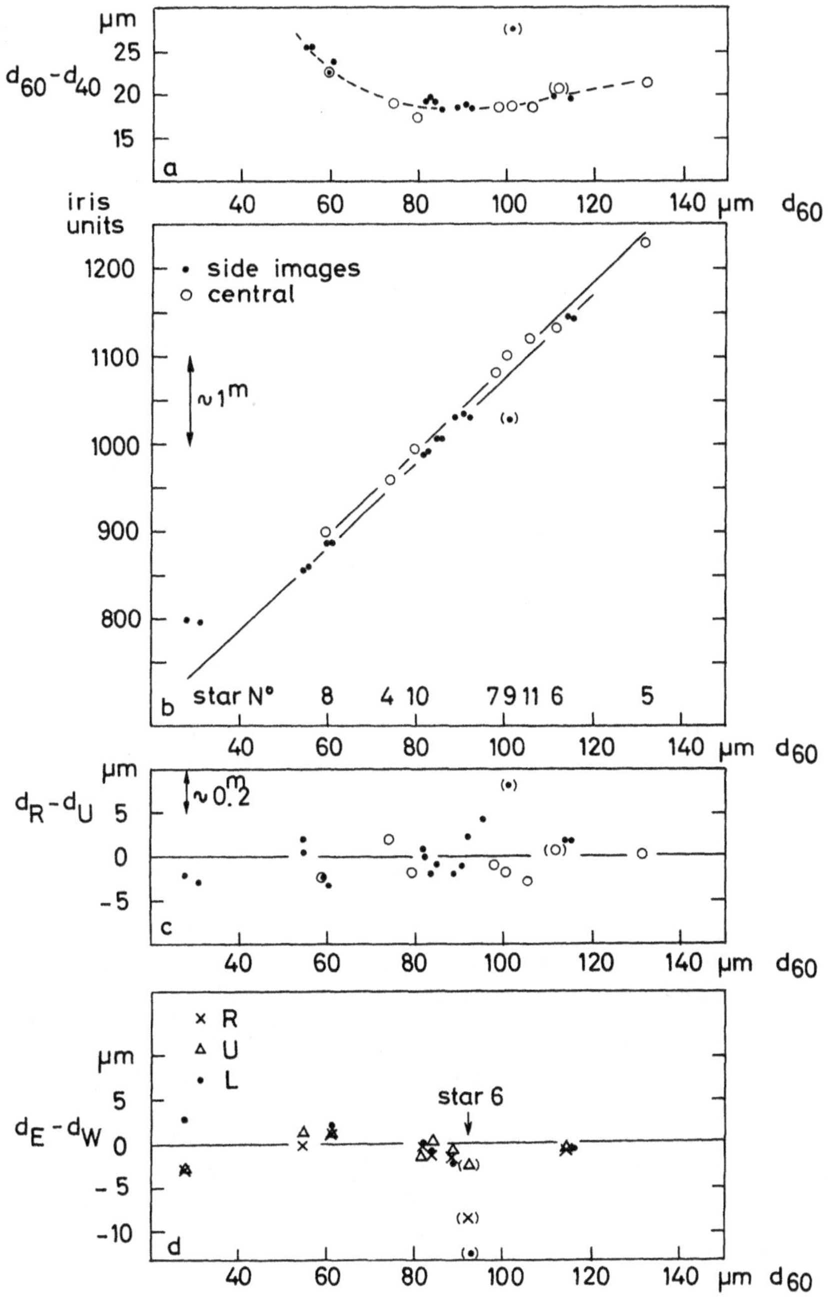

Fig. 3. Diameters d of iso-transmission curves.

other iso-transmission curves, so that checking the agreement between the parameters from several curves permits automatic recognition of the disturbed images of star No. 6.

The elongation of the first order side images was clearly recognized, both in size and direction, by means of the second harmonics of the 60% transmission curve.

The image diameters for transmission T = 60 and 40% were called d_{60} and d_{40} respectively. Their differences express the diffuseness of stellar images and are plotted in Fig. 3a as a function of d_{60}. The small scatter indicates that star images may be distinguished from slightly diffuse images as galaxies or planetary nebulae, and from the sharper images of some kinds of plate flaws.

The diameter d_{60} is an approximately linear function of the readings obtained with an iris photometer on the same images, as appears from Fig. 3b. The expected systematic difference between the round central and the elongated side images from the two kinds of measurement is seen. This difference disappeared when a numerically simulated iris diameter D was evaluated instead of the diameter of the iso-transmission curve. Also the deviation for the faint side images of star No. 8 and for the disturbed side image of No. 6 disappeared. This is shown in Fig. 4.10 of Paper I - in which the description of the horizontal axis should correctly read $\bar{D} = (D_R+D_U)/2$. The scatter in the two cases was $0\overset{m}{.}07$ and $0\overset{m}{.}05$, quite satisfactory for this first attempt, but still somewhat larger than expected.

The difference d_R-d_U of the diameters obtained in two orientations of the plate is plotted in Fig. 3c and indicates again an unexplained high scatter of $0\overset{m}{.}07$ for d_{60}.

The difference d_E-d_W between the East and West side images obtained in the three plate orientations (Fig. 3d) indicates however, a repeatability of $\pm0\overset{m}{.}03$ for a single photometric measurement and we believe that photometric repeatability adequate to match the accuracy inherent in the plate itself can be obtained in a production version of the machine.

The evaluation of four iso-transmission curves for an image takes 200 ms with an Algol program in a HP2100 computer, which may be reduced to 40 ms with Assembler programming, and even less with Microprogramming. If less accuracy is acceptable and higher speed required, a larger raster size of e.g. 20 x 20 μm^2 may be adopted.

IMAGE PROCESSING SYSTEMS DEVELOPED AT MIT TO HANDLE SILICON VIDICON IMAGES AND SPECTRA OF ASTRONOMICAL OBJECTS*

T.B. McCord, P.F. Kinnucan, and G. Fawcett, Jr.

Center for Remote Sensing, Department of Earth and Planetary Sciences, Massachusetts Institute of Technology, Cambridge, Massachusetts 02139

ABSTRACT. Several image processing systems have been developed and used at MIT to calibrate and analyze two dimensional images and spectra of astronomical objects. An interactive system allows creative processing under direct control of the astronomer and a batch system enables the astronomer to process large numbers of images while requiring little of his time. System descriptions and a list of processing routines are given.

A two-dimensional integrating vidicon imaging system has been developed at MIT (see paper 3.2 of this Proceeding) and it has been used at the telescope to produce thousands of images of planets, galaxies, and stars, and images of their spectra. To calibrate and analyze these images, we have developed two image processing systems: a batch and an interactive system. The systems are run on an IBM 360/65 computer and can be used by students and astronomers with no knowledge of programming.

The images are stored on 9-track magnetic tape and consist of arrays of 12 bit picture elements (0-4095 intensity levels). The arrays may be up to 1024 x 1000 elements. The usual size is 256 x 250. All images are initially copied from instrument tapes to library tapes for permanent storage. The images are individually labeled on the library tapes for identification. The label contains descriptive information from the observing run log. A printed catalogue of the image library is maintained for locating images.

Images selected for processing are transferred to a disk pack for fast access. The disk pack can hold 25 floating point images (256 x 250) or 50 integer format. The two formats may be freely intermixed on a single pack. An image directory is maintained on

*This research was supported by NSF Grant No. GP31516.

de Jager/Nieuwenhuijzen (eds.), Image Processing Techniques in Astronomy. 329-334.

the pack, and images are referenced by an 8 character name. Images are kept on disk during processing, which may extend over several runs, and then are copied back to tape for permanent storage.

Four kinds of processing are used to reduce images: (1) calibration, (2) noise removal, (3) analysis, and (4) display. In general, these involve such operations as addition, subtraction, and division of one image by another, digital and Fourier filtering, registering, and contrast enhancing images. A film-converter is used to obtain film hardcopy of images. Contour plots, and other graphic displays, are produced on a Calcomp plotter. Both the film-converter and the plotter are stand-alone devices driven by intermediate tapes. Images and plots may also be viewed on the interactive system's TV monitor.

The batch system was developed first. It reads control statements from card input and performs the functions indicated. The control statements, consisting of function names followed by keyword parameters, are punched in a free format. Functions may be grouped together and performed repeatedly on sets of images. Examples of functions are moving an image from tape to disk, subtracting or dividing two images, producing a printed number listing of an image, transferring an image to a film-converter tape, etc. To run a job, the user submits the control deck, inputs images on tape and disk to the computer operator and picks up the output when the job is completed.

The great advantage of the batch system is that many operations can be performed on many images automatically while the astronomer is doing other things.

The interactive system was developed to allow us to process some images more quickly and to allow the astronomer to analyze his data more creatively. The user communicates with the system via a Princeton 801 CRT terminal, which can display graphs and images of up to 32 grey levels on a TV screen. The system is controlled by commands, which are interpreted, checked for errors, and executed as they are entered from the terminal. Commands can be aborted at any point. A "help" command is available for displaying a list and description of commands at the terminal.

The user can display images and graphs side by side on the terminal screen for comparison. Up to six images can be displayed simultaneously and erased on the screen independently of each other. A cursor can be used to point to features in a frame. The system can then read the location of the feature directly from the terminal screen. For example, suppose the user wishes to know the brightness of a star. He displays the frame containing the star and points to it with the cursor. The system then determines the area of the star and integrates the intensity over the area to give the total brightness.

To protect the user from accidentally destroying an image, all operations are actually performed on a temporary copy, called the active image. Before he can begin processing an image, the user must declare the image active, using the "process" command, which creates the temporary copy. Thereafter, all operations are performed on the copy. When the user is done processing the image, he can do one of three things with the temporary copy: (1) replace the original, (2) save it as a new image on the disk pack, or (3) delete it.

The batch system and interactive system are completely compatible. Images can be entirely or partially processed on either system. Usually the batch system is used for such routine operations as transferring images between tape and disk, obtaining number listings and film hardcopy, and standard calibrations.

The interactive system is used when it is necessary to process a few images quickly or when doing special processing requiring judgments at intermediate stages. It enables the user to scan a file of raw images, select the best, process them, correct errors, visually enhance the final images, all in one session. Registration of images which are to be divided is a particular example. The same features in the numerator and denominator frames are usually offset with respect to each other. It is necessary to shift the numerator image so that it correctly overlays the denominator. To compute the amount of the shift, the user displays the two frames on the terminal screen, points to a feature in one frame, a lunar crater, for example, and points to the same feature in the other frame. The computer then computes and displays the offset between the two features.

The processing routines available to the users of these image processing systems are listed in the table along with the execution and CPU times required to carry out the process on a 256 x 250 element image. New processes are being added continually. The routines and the systems were designed and written by us. The processing routines are written in PL/I.

These are working image processing systems actually used to analyze astronomical data. The batch system operation was begun four years ago and the interaction system was first used about one year ago.

Both the interactive and batch systems are needed for processing astronomical data. Each has its important capabilities. Interactive systems alone can take too much astronomer time to process larger numbers of images in a routine manner; batch systems do not allow creative data manipulation.

A third type of system, hardware processing, also has its uses. A machine wired to do routine processes of any of the four types mentioned earlier can be very efficient, cheap and small. We have so far only used this type of system for display purposes (Brooks, 1975) but other systems are contemplated.

Interactive Image Processing Functions

COMMAND MNEMONIC	FUNCTION	EXECUTION TIME (sec)	CPU (sec)
ADDC	Adds a constant to every element of	2	.25

	an image.		
AVG	Produces an image which is the average of two or more images.		
CONTR	Draws an intensity contour map of an image.		5
DELETE	Deletes an image from the image file.	<1	.01
DISP	Scales (linearly or logarithmically) and displays an image in any one of six sectors on the terminal screen.	2	.5
DIST	Draws a graph of the distribution of intensities in an image.		
DIV	Produces an image which is the quotient of two images. Shifts the numerator image vertically and horizontally with respect to the denominator before dividing, so that identical features overlay each other.	15	2
DIVC	Divides every element of an image by a constant.	2	.25
ERASE	Erases any area of the display screen.	2	.01
FLAT	Applies a flat field correction to an image.	15	.25
FLUX	Computes star intensity.	10	.5
FFT	Computes the Fourier or inverse Fourier transform of an image.	30	15
HELP	Displays a description of the syntax, operands and function of any command.	2	.01
HPFILT	Applies a line-by-line high pass filter to an image. Used to emphasize local features in an image at the expense of overall shading variations.	15	2
LISTD	Displays the directory to the images in the image file.	5	.01
LOC	Displays the screen coordinates of point on the screen pointed to by the user, using the terminal cursor.	1	.01
LOGOFF	Terminates the interactive system program.	<1	.01
MATCH	Given the location of a feature in one image, finds the location of the same feature in a second image.	15	
MULTC	Multiplies each element of an image by a constant.	2	.25
PEND	Terminates process mode for currently active image. See PROC command.	1	.01
PROC	Starts process mode for an image. The image is copied into a temporary disk file. This copy is called the active image. All subsequent commands	2	.5

	not explicitly naming an image are performed on the active image.		
SAVE	Save the active image in the image file, either in the space occupied by the original of the active image, or in a new location with a new name.	2	.5
SCALE	Linearly scales an image from one intensity range to another.	2	.5
SMOOTH	Applies a simple noise filter to an image.	15	1
STAT	Computes the standard statistical parameter over any rectangular region of an image.	2	.5
SUB	Subtracts one image from another.	15	.25
SUBC	Subtracts a constant from every element of an image.	2	.25
TEST	Dummy command used for testing new functions.		
TRAV	Draws a plot of intensity versus position on any straight line across an image.	2-15	.25
TTOD	Transfers an image from tape.	30	1

Batch System Functions

NAME	DESCRIPTION	EXECUTION TIME (sec)	CPU (sec)
COPY	Copies an image on the disk.	20	.5
DIV	Divides one image by another. The numerator image is shifted so that it is in register with the denominator image.	60	2
DIST	Prints a histogram of the distribution of intensities in an image.	15	.5
ALTE	Deletes an image from the disk.	<1	.01
FILM	Copies an image onto the film-converter intermediate tape.	15	.5
LIST	Prints a list of the images on the disk pack.	<1	.01
MON	Copies an image onto an intermediate tape for viewing on the vidicon photometer monitor screen.	15	.5
PRINT	Prints a number listing of an image.	100s	5
RSTR	Restores an image to the disk pack from a save tape.	15	.5
SAVE	Saves an image from the disk on a save tape.	15	.5
SCALE	Scales an image from one range of intensities to another.	20s	2
SMOOTH	Filters a noisy image.	20s	2

SPEC	Calibrates and ratios images of spectra.		2
SUB	Subtracts one image from another.	20s	2
TTOD	Transfers an image from a raw data tape to a disk pack.	15	.5
TTOD1	Transfers an image from an image library tape to a disk pack.	15	.5

*In seconds. 0.85/min.

HANDLING OF 'GALAXY' PHOTOMETRY

P. A. Wayman and M. J. Stift

Dunsink Observatory, Dublin Institute for Advanced Studies, Dublin, Ireland

ABSTRACT. The properties of photographic images present certain well-known difficulties if accurate photometry is to be carried out. In order to derive light curves of variable stars to the limit of long-exposure photographs, means have been specially developed to convert 'Galaxy' measures of 607 stars, including 150 photoelectric standards, on a total of 93 plates, to UBVR magnitudes. For the range B = 11 to 17, the standard errors of the individual measured magnitudes are around 0.07 mag.

1. THE PHOTOGRAPHIC PROCESS AND 'GALAXY' MACHINE

The properties of the normal photographic emulsions that hinder accurate stellar photometry are (i) *low quantum efficiency* and (ii) the rapid onset with increasing exposure of *saturation* and hence of *non-linearity* in response. These two factors are interrelated, being dependent on current photographic technology, and choice of emulsion should ideally always be matched to the procedures of observation, measurement, and analysis. The overall requirement is to find the best way of utilising the chief virtue of the photograph, that is the cheapness and fidelity with which "information" in great quantity is obtained and stored. Automatic measuring machines such as 'Galaxy' are designed to provide efficient access to the relevant stored information. Restricting ourselves here to stellar (i.e., "zero-dimensional") images, the two properties, *position* and *integrated size*, of such images are those provided

* This work has been supported in part by a Research Grant of the National Science Council of Ireland.

de Jager/Nieuwenhuijzen (eds.), Image Processing Techniques in Astronomy. 335-340. *All Rights Reserved.*

specifically by 'Galaxy' machine.

In the literature concerning the use of 'Galaxy', little has been published on results of accurate photometry of stellar images. The quality of the magnitude measurements derived is often treated only briefly (*see e.g.*, Sherwood, 1974), and it may be suspected that, in performing photometric work with automatic measuring machines it is difficult to extract results of traditional accuracy (errors ~0.05 mag.). Iris-diaphragm photometers, operated by hand, have provided this accuracy only after adopting careful techniques of measurement and reduction. It would seem *prima facie* that an automatic machine with a well-designed sequence of operations would provide ideal operating standards, but the photographic material must be suitably selected and the subsequent reduction-processes must be performed satisfactorily. We report here some results that have been obtained using IIa-O and 103a-D emulsions on plates taken with the 81-91-305 cm. ADH Baker Schmidt telescope of Boyden Observatory, where the image-quality usually varies slightly over the 2° field, and where, with exposure-times ~45 mins., the sky-background reaches a density 0.2, or in crowded regions even more. In measuring to the limit of these plates, B~18.5, and aiming at 0.05 mag. accuracy, we are attempting an inherently difficult task.

2. PRECEDING WORK

The present work follows the investigation by C. J. Butler of cephei variable stars in two regions of the Magellanic Cloud (Butler, 1971, 1972a, 1972b, 1975) where he used least-squares solutions for the representation of magnitude in terms of iris-diaphragm measurements to derive light curves in B and V of about 160 cepheids. We estimate that Butler's systematic errors of photometric scale probably did nc exceed 0.1 mag. to the faint limit of B=18.0 or V=17.5. Butler used the plates of a third area, in the L.M.C., differently. In 1973, wit the generous co-operation of Mr. C. A. Murray of the Royal Greenwich Observatory, he obtained 'Galaxy' data from 93 plates for 451 variables or suspected variables and 156 standard stars within a 2° fiel (labelled LMC I). It is the handling of these data that is now in progress.

The Magellanic Clouds are crowded regions where the effective background density is high and strongly variable across the plate. Hence the difficulties in extracting accurate data on the brightness of faint stars are severe. We have to be aware of the inherent "inversi presented by the task of measuring black stellar images, when what we are aiming at is the brightness of a point of light, but perhaps one so faint that it is barely distinguishable above the sky background. With an iris-diaphragm photometer, the brightness level can be adjusted so that readings are taken with the diaphragm tightly surrounding a stellar image. If this is done, the influence of the variations in sky background is diminished. Much of the information in the bright images is wasted, but this wastage can be tolerated when the aim is to maintain a certain accuracy expressed in magnitude (i.e., light-*ratio*) and when it is possible to keep optimum measuring conditions applicable to the difficult faint images. The "tight" diaphragm helps to ensure a reasonable independence with respect to varying sky-background, although Butler (1972a) found a method of incorporating background density into his least-squares reduction-process.

3. CHOICE OF 'GALAXY' WEIGHTING FUNCTION

Methods of analysis and procedure that have proved useful for iris-diaphragm measurements can have parallels when considering the choice of library profiles and weighting function with 'Galaxy' machine (Walker, 1971). But complete adaptation to the requirements of individual plates is not readily determined during measuring runs. Thus there is a natural tendency to adopt one set of library profiles covering, in 1024 steps, a large magnitude range, and one may find in practice that almost all measurements of interest come between, say, M = 100 and M = 300, and that one step in the scale of M corresponds to 0.03 mag. This is just adequate, even though a more expanded measure would sometimes be helpful. It was found that there was a difficulty in setting the weighting-function limits adequately far in because the same setting alters the amount of light available to actuate the servo controls of the centering mechanism. If the weighting-function is well into the central parts of the stellar images, as is helpful in producing less dependence on background fog (*see* Figure 1), the diminished amount of light available for the servos may result in a mechanical instability during the centering process, causing the machine to "loop". More sophisticated versions of 'Galaxy' should incorporate greater flexibility in the setting of the weighting function when making the photometric match with the library profiles.

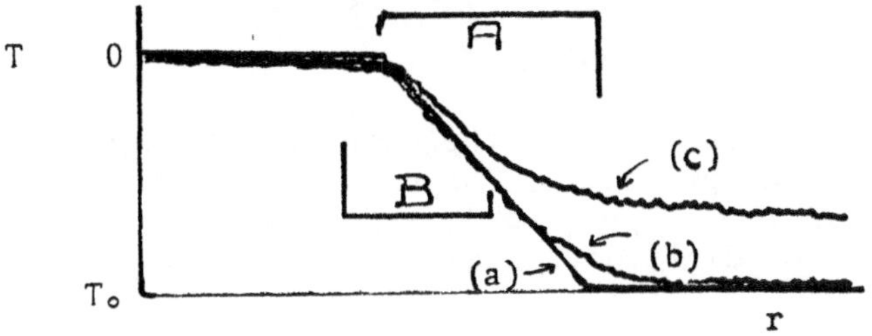

Fig. 1. Matching an image with the library profile. (a) Library profile. (b) Stellar image profile with normal background. (c) Image profile of similar star with enhanced background. A is the weighting function limit for maximum information. B is preferable if we seek to avoid the influence of background fog.

4. REDUCTION PROCESS

A trial-and-error procedure resulted in a reduction process being evolved which derived from an analysis of the background-fog correction implicit in a consideration of Figure 1. The influence of background fog on the selected library profile (or M-value) was estimated by numerical integrations, using an assumed function to represent the transmitted light between the weighting-function limits. To do this, a radial Gaussian form

$$I(r) \propto e^{-r^2/\alpha^2}$$

represented the light-distribution in the image, and Elder's equation

$$D(r) = D_{max}\left(1 - e^{-\kappa I(r)}\right)$$

represented the characteristic curve of the emulsion. Through these representations and by carrying out many numerical integrations a reasonably satisfactory simulation of the action of 'Galaxy' machine was arrived at. If the transmission T in the neighbourhood of every stellar image could be measured, a correction ΔM to Galaxy brightness measure M could in principle be calculated, in order to find a way of referring to a standardised background transmission T_o. In practice, relying on previous experience, it was only possible to measure, with the help of a calibrated grid, the value of T for a few plates, using the assumption that on other plates the transmission would be scaled similarly. This procedure had previously been tested for plates in all colours, U, B, V, and R, and had been found to hold approximately. Then a series of coefficients were determined in the least-squares solution for the photometric calibration of each plate on the basis that a calculated correction is required to the measured M-value of which the terms linear in T are

$$\Delta M = a(T_o - T) + b(T_o - T).\exp(-M_c)$$

$$\text{where} \quad M_c = (1/p).(M/100)^p$$

This form of ΔM was arrived at from the numerical integrations which represented the matching procedure. The exponent p, which was found in this way to lie in the range 2 to 3, represents the rather sudden onset of background fog difficulties as M becomes smaller for the faintest images.

In the actual reduction-process, a least-squares solution was performed for the photoelectric magnitudes of the standard stars in terms of a series of 15 unknown coefficients. A term for colour-system and 5 terms in X,Y position (X, Y, X^2, XY, Y^2) to deal with vignetting and other field errors, were incorporated; also, a power series in M only up to M^2 was found to be satisfactory. Then, with a constant term to be determined, the representation was

$$\text{Mag.} = a + bM + cM^2 + d.\text{Col} + (e_1X + e_2Y + e_3X^2 + e_4XY + e_5Y^2) + f(M,\Delta M)$$

and since the quadratic terms in M suffice it was possible to have the bckground-fog correction of the form

$$f(M,\Delta M) = f_1T + f_2T\exp(-M_c) + f_3MT + f_4MT\exp(-M_c) + f_5T^2 + f_6T^2\exp(-M_c)$$

the terms involving $\exp(-M_c)$ squared being dropped.

After a few trials, it was found that this procedure was satisfactory for the whole series of plates if p=2.5. The fitting of the curve to the individual points used a rejection criterion δ = 0.25 mag. for excluding stars from the solution, and a series of successive approximations guided by the computer operator converged rapidly to a definite solution.

5. RESULTS OBTAINED

Figure 2 gives the systematic residuals and the mean standard errors (corresponding to a single measurement) for those stars which were measured without any rejection on the 39 B-plates. The system of the reductions, so represented, does not go fainter than B = 17.0 since every star fainter than this is rejected at least once. Generally speaking, it might be estimated that over the range B = 11 to 17 the systematic errors probably do not exceed 0.05 mag. and the accidental errors of the individual measurements are well represented by σ = 0.065 mag. While these accuracies are sufficient for our purpose, there must be some anxiety as to whether the system is as well defined as, say,

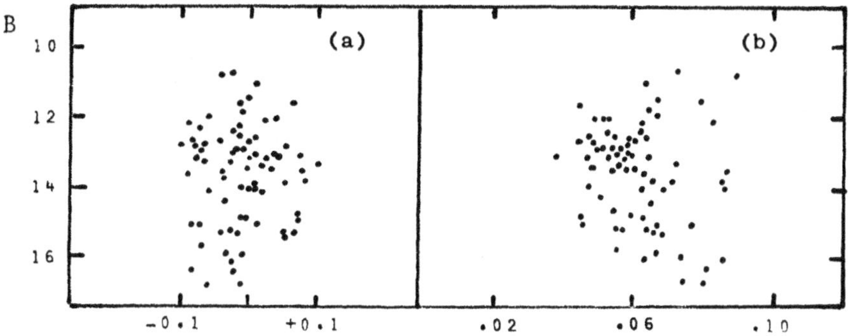

Figure 2 (a) Residuals pg - pe and (b) mean standard errors for the reductions of standard stars on 39 B-plates. Stars not measured on every plate are omitted.

with carefully-executed iris-diaphragm photometry. In our 'Galaxy' work there was a certain amount of difficulty with the identification of the measured stars (Sherwood, 1974, speaks of "only a probability of measuring any given star") and the reduction process exhibited an appreciable chance that an inexplicable error had occurred. Overall, however, we have been greatly impressed by the ability of the data to stand up well to analysis, even when reaching down to the faint stellar images which are barely discernible against an appreciable background fog.

REFERENCES

Butler, C.J., 1971. Third Colloquium in Astrophysics, Trieste. p.128
- 1972a. Irish Astr. Journ. 10, 251.
- 1972b. Dunsink Observatory Publications, 1, 133.
- 1975. Astronomy and Astrophysics (in press).

Sherwood, W.A., 1974. Publ. Royal Obs. Edinburgh, 9, 85.
Walker, G.S., 1971. Publ. Royal Obs. Edinburgh, 8, 103.

P A R T 5

APPLICATIONS TO ASTRONOMICAL PROBLEMS

SOME PRELIMINARY RESULTS OBTAINED WITH A TV CAMERA IN THE FIELD OF THE OBSERVATION OF VISUAL DOUBLE STARS.

J. Dommanget

Observatoire Royal de Belgique
Uccle, Bruxelles.

ABSTRACT. A description of a TV camera used for double star observations is given. Some very preliminary registrations of images of binaries, Saturn and the Moon, are shown.

At the Royal Observatory of Belgium, we are interested in systematic extended programmes of observation of visual double stars, more than in the observation of particular binaries such as very close or very faint pairs.

These last ones are of course of very high importance for many reasons but this may not allow us to forget the other more common double stars for which a great number of accurate measurements are urgently required in order not to lose the homogeneity of the observational material.

We must for instance recall that binaries separated by distances of 1" to 5" are generally not sufficiently observed by the visual observers as well as by others using different methods, e.g. photographic ones.

For that type of binaries a simple TV technique could be useful, a technique using not necessarily a very sophisticated automatic reduction device but only a simple commercial, not too expensive one and based on a " point by point " analysis of the images.

de Jager/Nieuwenhuijzen (eds.), Image Processing Techniques in Astronomy. 343-345.

Therefore we acquired at the Royal Observatory, a FERNSEH camera which is now mounted on our COOKE-ZEISS refractor of 45 cm aperture and 700 cm focal length. We are presently elaborating a simple automatic reduction equipment which we hope will soon be operational.

The camera is of the EBS type, has an image intensifier of 30 times and is preceded by a magnification lens of 7 times. Color filters are placed in the light beam in order to select a spectral field of the order of 3 to 400 Å band width, centered on 5600 Å, corresponding with the top of the focal curve of the refractor which is corrected for the visual region.

The EBS tube has been chosen because :

- it is approximately ten times more sensitive than a SEC tube, given that all other elements are the same ;
- it has higher resolving power than a SEC tube.

It is also less dammageable by a sudden illumination from a strong light source, such as the moon.

On the other hand and at the contrary of what is possible with a SEC tube, the EBS tube does not allow image integration over an interval of time longer than the normal exposure time of 1/25 of a second.

But we have in view the registration of star images in very short exposure times, in order to remove the broadening effect of the image motions and to assure the best similarity of the blurring effect on both images of the binaries components (See : J. Rösch, Proceedings of " Le Choix des Sites d'Observatoires Astronomiques " I.A.U. Symposium n° 19, p. 74, 1963).

A few slides are shown of the refractor and the TV camera mounted on it. The small weight (5 kg) and the compact size of the camera make it a very easy equipment to use on any refractor or telescope.

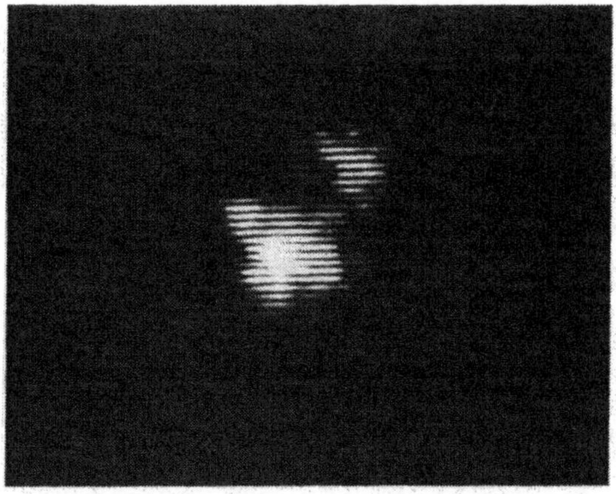

One magnetoscope registred image (1/25 of a second) of the binary : Index n° 05591 N 1819, for which : $\rho = 2\overset{''}{.}2$, $m_A = 9,1$ and $m_B = 10,1$

Some binaries having a separation of 2" and magnitudes of the order of ten, have been easily observed and recorded on a magnetoscope. As can be seen on the figure, such images cover some ten to twenty TV lines. Analysis made with an oscilloscope of the recorded images shows that separation of 1" will be measured in the best conditions, especially with an " on line " equipment for automatic reduction.

Recorded images of Saturn and of the Moon show the wide range of sensitivity of the whole equipment.

ECHELLE SPECTRA IMAGE PROCESSING FOR THE INTERNATIONAL ULTRAVIOLET EXPLORER

D. A. Klinglesmith and E. Dunford

Goddard Space Flight Center, Greenbelt, Maryland, U.S.A.
Appleton Laboratory, Slough, England

ABSTRACT. This paper will present the techniques needed to convert the International Ultraviolet Explorer (IUE) high resolution echelle spectra into useful scientific data.

The image is obtained with a digitally controlled SEC Vidicon camera system. The processes that must be carried out on the image include: noise removal, correction for geometric and optical distortion, non-uniform photometric corrections and two dimensional wavelength determination. Results from the breadboard camera system will be presented.

1. THE IUE CONCEPT

The basic concept underlying the design of the IUE is that of an ultraviolet astronomical observatory intended primarily for use as an international research facility. Astronomers should be able to come to this observatory and carry out their own observing programs without going through a tedious training course in the specialised techniques of operating a telescope in earth orbit. The IUE will contain a 45 cm telescope which will be used exclusively for spectroscopy. An echelle spectrograph has been selected to obtain the high resolution spectra desired for brighter objects. With this type of instrument a high dispersion is easily achieved, and there is the additional advantage that the format of the spectrum will consist of a series of adjacent spectral orders displaced one above another in the raster-like pattern illustrated in Figure 1. This format makes efficient use of the sensitive area of the SEC Vidicon television tubes which will be used to integrate and record the spectrum. Since the

de Jager/Nieuwenhuijzen (eds.), Image Processing Techniques in Astronomy. 347-352.

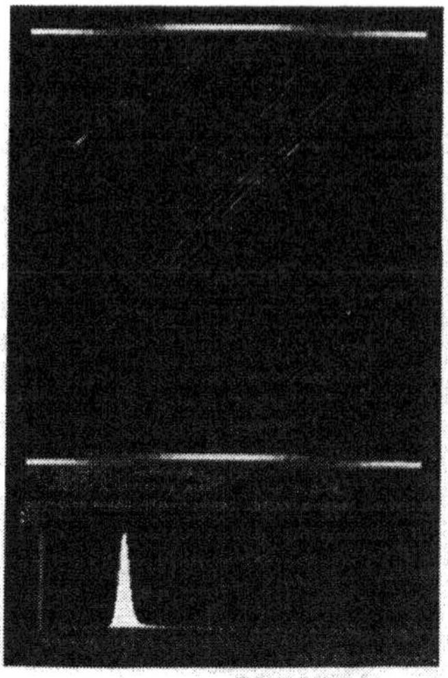

Fig. 1 IUE echelle format

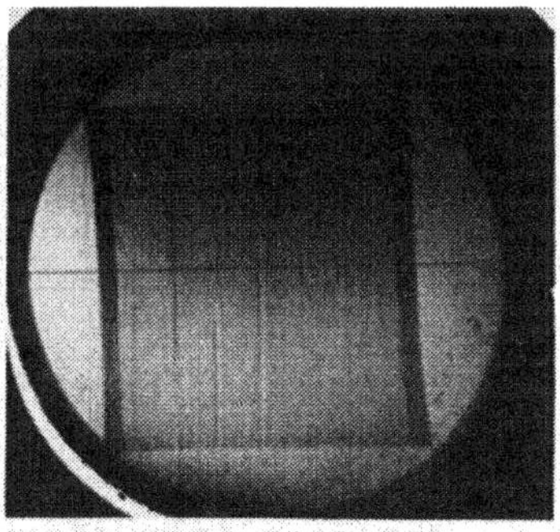

Fig. 2 Raw test pattern

echelle spectrograph design contains a high dispersion echelle grating in series with a low dispersion grating, the instrument is easily converted into a low resolution spectrograph by simply inserting a plane mirror in front of the echelle, leaving the low dispersion grating to act alone.

In order to achieve the dispersion required in the high resolution mode, it has been necessary to split the spectrum from 1150 to 3200 Å into two ranges, and two exposures are required to record the entire spectrum. Observing efficiency is not significantly affected, however, for the grating blaze angles and other design parameters can be separately optimized for the two spectral ranges, greatly improving the overall optical efficiency. Two exposures would frequently be required in any case in order to make optimal exposures of both the short and long wavelength portions of the ultraviolet spectrum.

The telescope is to be placed in a synchronous orbit such that it can be in continuous contact with the operations centre located at Goddard Space Flight Center. The satellite can also be controlled from a second centre near Madrid, Spain, and observing time will be shared by the two centres.

2. INSTRUMENTAL EFFECTS

Since the IUE system produces a digital television image of the echelle spectrum, it is our intention to follow the methods used by the Mariner Project (1). We are developing an image processing system that has converted the concepts of I/O handling

Fig. 3 Geometrically corrected image

Fig. 4 Geometrically and photometrically corrected image

and sub-tasking used by the VICAR system on an IBM 360/75 to a similar system on a XEROX Sigma 9. A significant number of the applications programs will be copied directly, thereby reducing the development effort.

The instrumental effects fall into five major areas: noise, geometric distortion, loss of resolution, wavelength scale and photometry. These will be discussed separately and then an overall flow of the data reduction system will be given as we see it at the present time.

The noise in the system comes from several sources: electronic, photon, physical, transmission and spacecraft. The electronic noise is noise introduced into the video chain starting in the SEC Vidicon itself and going through to the analogue to digital converter. The photon noise is the noise inherent in the stellar signal itself. The physical noise is of two kinds: blemishes and read gun pointing errors. The blemishes are faults in either the SEC Vidicon photocathode or target or in the UV converter photocathode or phosphor. The transmission noise results from burst errors and/or telemetry synchronisation errors between the spacecraft and the ground station. The spacecraft noise is of two types: interference and microphonics. The interference results from clock pulses or other types of periodic electrical impulses being picked up on the video line. The microphonics result from mechanical vibrations in the spacecraft or scientific instruments during readout of the television camera.

The best way to remove the noise effects is to remove the source of noise before launch. Every effort is being made by the engineers to provide a clean, quiet spacecraft and instrument. However, things do go wrong, so we must be prepared to handle any or all of the effects.

There are several techniques that can be used for the various types of noise removal. The first is a nearest neighbour type of replacement, in which a small area of the image that deviates from the average of the surrounding points by a large amount is replaced by the average of the surrounding points. This type of correction will be used on blemishes and most transmission errors. A second type of correction is smoothing by averaging adjacent points. This will be done to suppress the photon and electronic type noise effects. The third noise removal technique that can be used is Fourier filtering. This can be used to remove periodic noise patterns seen in the data. Read beam pointing errors, microphonics and periodic electrical interference are examples of this type of noise. However this type of noise removal must be done with extreme care and understanding in order to avoid any smearing or removing of the real image information.

The television camera system and the optical system both contribute to geometric distortion and point spread function smear, but the main sources of geometrical distortion are the pincushion and S-distortion arising in the gun section of the SEC Vidicon. Both types of distortion can be seen in Figure 2, which is a test pattern image of 4.6 lp/mm. The IUE camera system has reseaux deposited on the faceplate, some of which can be seen in Figure 2, and these are located using a cross correlation technique. Curve fitting is used to improve the accuracy of these apparent positions and the difference between these and the real positions is used to perform a rubber sheet stretch and return the distorted image to its true shape, as shown in Figure 3.

The other type of distortion inherent in the IUE system is point spread function smear. Besides the point spread functions of the UV converter and the SEC Vidicon, the optical aberrations in the telescope and spectrograph and the instantaneous pointing errors, which will effectively fill the entrance aperture, also contribute to spreading the image of the star at each wavelength. The amount of the spread will be a function of both position on the television camera and the wavelength of the image. These effects can be corrected to some extent by convolving the correct sharpening filter or by Fourier transform techniques. Again extreme care must be taken with this type of correction. It is currently planned not to do any PSF enhancement or correcting, since this can be a highly personal type of correction, subject to the scientific criteria and objectives of the astronomer.

The wavelength scale has to be determined from either the stellar spectrum itself or a comparison spectrum taken either before or after the stellar spectrum. Using a rectangular co-ordinate system (x, y) the method adopted will be to solve equations of the form

$$x = A + B\lambda + Cm + Dm\lambda$$

$$y = E + F\lambda + Gm + Hm\lambda$$

where λ is wavelength and m is the echelle order.

This will separate the problem into two relatively independent one dimensional problems rather than a complicated two dimensional problem. Using this type of technique on the hydrogen spectrum shown in Figure 1 we have been able to obtain a positional accuracy of better than 0.1 Å across the twenty or so orders seen.

The photometric effects of the system fall into several categories: blaze angle efficiency, wavelength dependence of the UV converter, non-uniform response of the SEC Vidicon and background and scattered light contaminations. The wavelength dependence of the UV converter and the non-uniform response of the SEC Vidicon combine to require that we do a pixel by pixel calibration, i.e. a correction table for each pixel (about 600,000 tables). Such a table has been constructed for the IUE breadboard camera and the image in Figure 3 has then been corrected. The result is shown in Figure 4. The background and scattered light corrections can be made by extrapolating from between the spectral orders and then subtracting the proper amount of energy in the orders. This type of correction will depend on the amount of background and scattered light and the ability of the system to separate the orders from each other. The blaze angle efficiency, which produces a 60% variation between the centre and edge of each order can be corrected by either of two methods: first by correcting to some standard star system that has been previously observed or by putting the orders together and requiring a smooth distribution across the overlapping part of the orders.

A flow chart of the data reduction steps as we currently see them is shown in Figure 5. Several points can be made about this flow chart. First, we are planning a near real time interactive data reduction system. The images will be reduced within 24 hours of the observation by a data reduction specialist at an interactive graphics type terminal, where he will be able to monitor the image as it goes through the various stages of data reduction. He will need to interact directly in the wavelength determinations. Second, it will be possible for the guest astronomer to view the images in a quick look fashion as the data

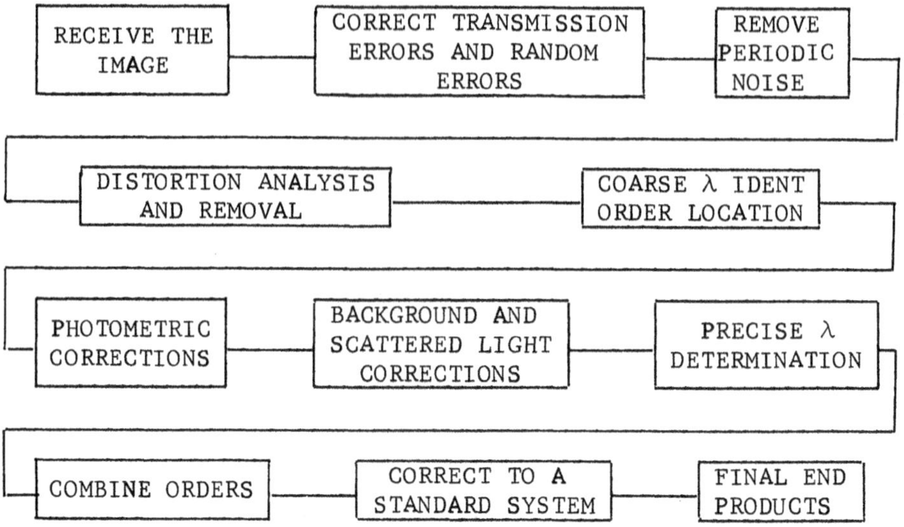

Fig. 5 Processing flow chart

comes from the spacecraft, at which time he and the night assistant will be able to determine if all or part of the data should be reduced or not. Obvious reasons for not reducing the data might be over- or under-exposure, transmission error problems, wrong star or any of the many other types of problems that we are all familiar with at a ground based observatory.

We are not trying to contend two years before launch that we understand the entire data reduction task. However we are contending that we are designing a flexible enough system that will permit fundamental changes to the actual data reduction algorithms to be made without affecting the basic image processing task of starting with a poor image and ending with a clean, correct, ready-for-scientific-use image.

The techniques described here are being developed as a joint undertaking between Goddard Space Flight Center, Maryland and the Appleton Laboratory, Slough, involving close collaboration between the two groups. The individuals most closely concerned are J. Niebur, B. Sada, E. Glover and D. Fischl in the U.S. and T. Dimbylow, E. Golton, M. Kendall and M. Lawden in the U.K.

REFERENCE

1. W. B. Green, et al., Applied Optics, 14, 105, 1975.

SEMI-AUTOMATIC DETERMINATION OF EQUIVALENT WIDTHS

A. Cassatella°, L. Gratton°° and R. Nesci°°

° Present address: ESRO-Fellow-Paris, Observatoire de Meudon, Equipe étoiles froides, France

°° Cattedra di Astrofisica, Università di Roma, Laboratorio di Astrofisico Spaziale C.N.R., Frascati, Italia

1. INTRODUCTION

This is essentially a progress report concerning an ambitious plan of determination of chemical abundances in K-giants, in which about 50 stars will be observed at rather high dispersion (3.2 A/mm). The full team at Frascati is composed by A. Cassatella, L. Gratton, G. Natali, R. Nesci, R. Viotti; some observations have been made also by Miss C. Rossi.

All the plates have been taken at the French national observatory in Haute Provence; we wish to express our gratitude to our French colleagues and especially to Professors C. Fehrenbach and Y. Andrillat for the use of the magnificent equipment and organization at St. Michel. All the plates for this investigation have been already obtained; the measurements are under way.

Due to the enormous amount of work of processing high dispersion spectra, we planned from the start to employ a semiautomatic method, which however evolved considerably and is still evolving while the work is in progress. We wish here to report only upon the method adopted for the determination of astrophysically significant quantities (equivalent widths or their substitutes) and refer to a Publication of the Laboratorio di Astrofisica Spaziale (Gratton et al 1975) for technical details concerning observations and processing.

A summarized description of the processing is, however, given in Fig. 1, which is self-explanatory. We emphasize only the use of Shannon's interpolation equation to express the intensity (normalized to the "continuum") as a function of wave-length. The result of the processing is the storage in digitized form on magnetic tape

de Jager/Nieuwenhuijzen (eds.), Image Processing Techniques in Astronomy. 353-365.

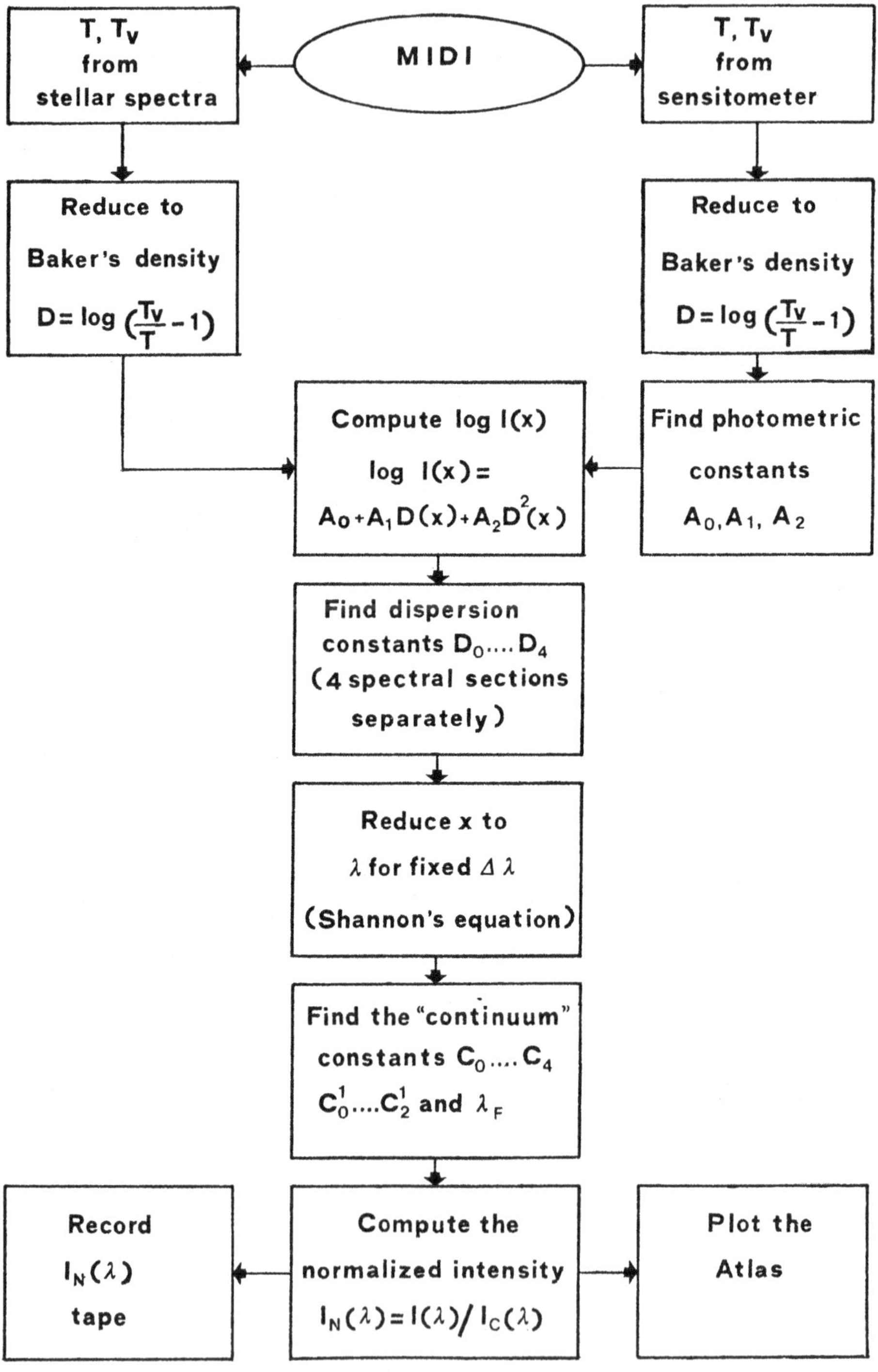

Fig.1. Flow diagram for the "Atlas"

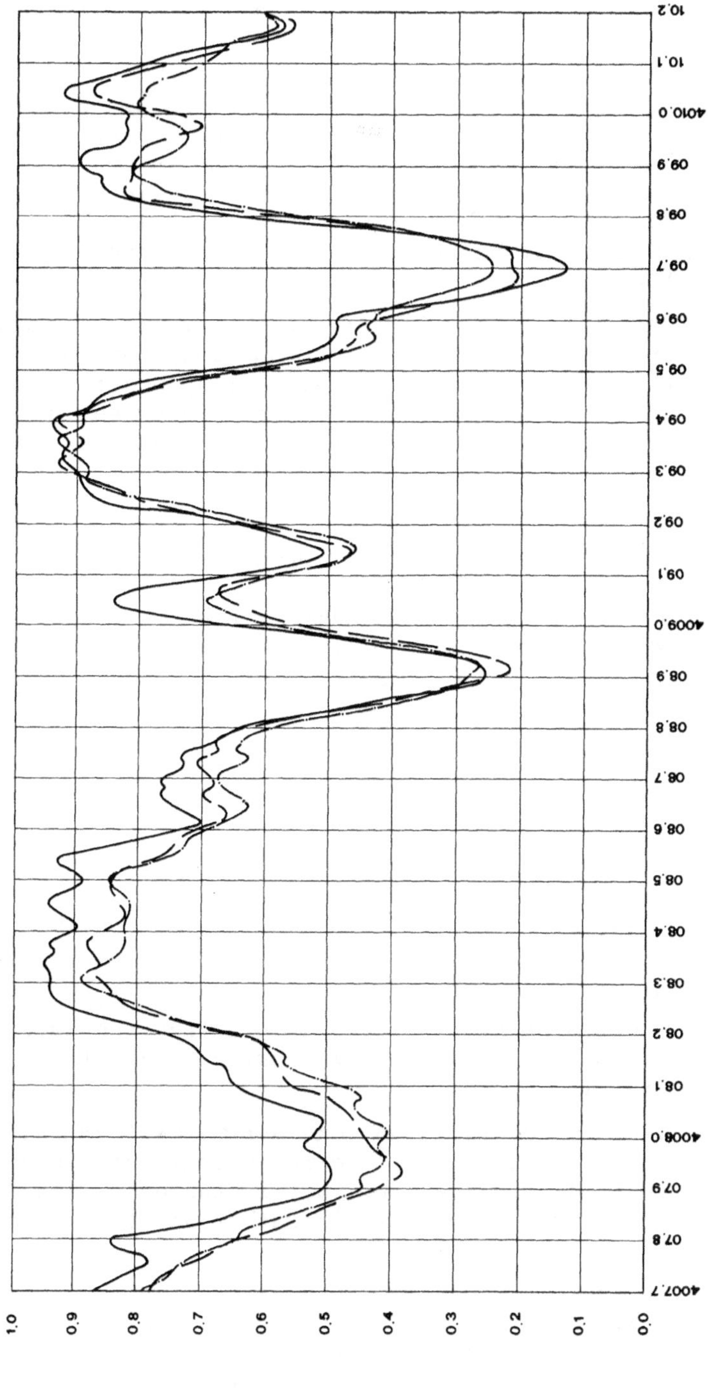

Fig.2. A portion of the spectra of η Cep (————), γ Tau(— — — — —) and δ Tau (—.—.—.—.—)

of the flux $I(\lambda)$ in units of the continuum at the same wave-length λ, as a function of λ, at points spaced 0.02 A from λ 3985 to 4810 ; in all a total of 41250 points for each star.

Apart from the usual photometric precautions to insure a good accuracy - we estimate the average mean error in our photometry to be around $\pm$ 3% at each point - we have tried to attain a rather high accuracy in the wave-length of each point in the spectrum. By comparing the results from different stars we believe that the accuracy of each wavelength is better than one half the interval between two points, viz. better than $\pm$ 0.01 A, and perhaps as good as $\pm$ 0.005 A.

Fig. 2 shows a small portion of the spectra of γ Tau, δ Tau and η Cep superimposed; it must be emphasized that the distance between two consecutive points (0.02 A) corresponds to about 6 μm on the plate. We believe that the differences between γ Tau and δ Tau approach the limit of our accuracy, but those of these two stars from η Cep are quite significant.

2. EXPANSION IN THE SUM OF GAUSSIAN COMPONENTS

In order to obtain astrophysically interesting results, the usual procedure is, now, to obtain from the spectrum the line intensities or, better, the equivalent widths. For an isolated line this is defined as

$$W = \int r(\lambda)\, d\lambda \,, \tag{1}$$

where

$$r = 1 - I(\lambda) \tag{2}$$

is the oberved depth inside the line; in practice the integration is extended to a small interval of λ inside which $I(\lambda)$ differs significantly from 1.

When the line is not isolated, one estimates from that part of the profile which is assumed to be relatively uninfluenced by nearby lines, what the profile would be, if the other lines did not disturb it. The whole procedure is rather arbitrary snd does not lend itself to machine calculations.

We tried, therefore, to find a more satisfactory procedure, in which is assumed for each line a certain typical profile - depending upon a small number of parameters - the resultant profile being due to the contribution of all the lines. The first profile which comes to mind is that of a gaussian curve because the majority of the lines is broadened by Doppler and instrumental effects, both of which give an approximate gaussian profile. The main objections to the use of gaussian profiles is the saturation of the core of most of the lines, and eventually a different method was used (see below). Nevertheless we give here a very short account of the expansion of the spectrum in a sum of gaussian functions, because this expansion may be useful in other problems.

Let us assume then that

$$r(\lambda) = \Sigma_i A_i e^{-h_i(\lambda-\lambda_i)^2} ,$$

or, for a digitized spectrum

$$r_j = \Sigma_j A_i e^{-h_i(\lambda_j-\lambda_i)^2} . \tag{3}$$

Here r_j is the depth at wavelength λ_j . λ_i is the central wavelength of line i , A_i its central depth, h_i the (reciprocal) width; the sum must be extended to all the lines. The parameters λ_i , A_i and h_i are computed by least squares, from the observed r_j . In practice a finite length of spectrum (window) corresponding to $j = 1, 2, \ldots N$ and containing M lines $(i = 1, 2, \ldots M << N)$ is discussed at a time, N being of the order of 2000 (40 A).

If a good "system" of wave-lengths λ_i is available, that is if we have previously identified <u>all</u> the lines, then the parameters λ_i may be assumed as known quantities, so that the unknowns reduce to the A_i and h_i . The procedure is to assume preliminary values of the unknowns, say $A_i^{(0)}$, $h_i^{(0)}$ and linearize equations (3) by putting

$$A_i = A_i^{(0)} + x_i , \quad h_i = h_i^{(0)} - y_i/A_i , \tag{4}$$

treating the new unknowns x_i and y_i as "small" quantities whose squares may be neglected. This of course is a classical procedure and it would not be worth mentioning, except for a few details which greatly simplify the solution.

If we write the linearized condition equations under the form

$$\begin{cases} \Sigma_i A_{ik} x_i + \Sigma_i B_{ik} y_i = P_k , \\ \Sigma_i C_{ik} x_i + \Sigma_i D_{ik} y_i = Q_k , \end{cases} \tag{5}$$

it may be found after a few elementary reductions that

$$A_{ik} = \Sigma_j e^{- h_j^{(0)}(\lambda_j-\lambda_i) - h_k^{(0)}(\lambda_j-\lambda_k)} , \tag{6}$$

and

$$\begin{cases} P_k = \Sigma_j X_j e^{- h_k^{(0)}(\lambda_j-\lambda_k)^2} \\ Q_k = \Sigma_j (\lambda_j-\lambda_k)^2 X_j e^{- h_k^{(0)}(\lambda_j-\lambda_k)^2} , \end{cases} \tag{7}$$

where the X_j are the errors of the first approximation, or

$$X_j = r_j - \Sigma_i A_i^{(0)} e^{- h_i^{(0)}(\lambda_j-\lambda_i)^2} . \tag{8}$$

The remaining coefficients may be found from the A_{ik} by noting that

$$\begin{cases} B_{ik} = - \dfrac{\partial A_{ik}}{\partial h_i^{(0)}} \\ C_{ik} = - \dfrac{\partial A_{ik}}{\partial h_k^{(0)}} \\ D_{ik} = \dfrac{\partial^2 A_{ik}}{\partial h_i^{(0)} \partial h_k^{(0)}} \end{cases} \tag{9}$$

The A_{ik} themselves are computed by noting that, from (6)

$$A_{ik} = e^{-\frac{h_i^{(0)} h_k^{(0)}}{h_i^{(0)}+h_k^{(0)}}(\lambda_k-\lambda_i)^2} \Sigma_j\, e^{-(h_i^{(0)}+h_k^{(0)})(\lambda_j-\frac{h_i^{(0)}\lambda_i+h_k^{(0)}\lambda_k}{h_i+h_k})^2}$$

$$\simeq e^{-\frac{h_i^{(0)} h_k^{(0)}}{h_i^{(0)}+h_k^{(0)}}(\lambda_k-\lambda_i)^2} \frac{1}{\Delta\lambda}\int_{-\infty}^{+\infty} e^{-(h_i^{(0)}+h_k^{(0)})\,\lambda^2}\, d\lambda \qquad (10)$$

$\Delta\lambda$ being the spacing or the interval in λ at which the spectrum was measured.
Hence

$$A_{ik} = \frac{\sqrt{\pi}}{\Delta\lambda}\,\frac{1}{\sqrt{h_i^{(0)} + h_k^{(0)}}}\; e^{-\frac{h_i^{(0)} h_k^{(0)}}{h_i^{(0)} + h_k^{(0)}}(\lambda_k-\lambda_i)^2} \qquad (11)$$

The expressions for the B_{ik} , C_{ik} and D_{ik} are then easily obtained from (9) and (11).

Even with a moderately large window the solution of the system (5), is rather cumbersome; for instance for a 50 A wide window, $N = 2500$, $M \simeq 300$ in our spectra. Fortunately there is an enormous simplification due to the fact that $A_{ik} \to 0$ very quickly when $|\lambda_k-\lambda_i|$ increases; in other words a line is disturbed only by the nearby lines.

Clearly, the dominant terms of the x_i , y_i are obtained by solving the "diagonal" systems

$$\begin{cases} A_{ii}\, x_i + B_{ii}\, y_i = P_i \,, \\ C_{ii}\, x_i + D_{ii}\, y_i = R_i \,. \end{cases} \qquad (12)$$

If we call $x_i^{(0)}$, $y_i^{(0)}$ the solutions of (12) or

$$\left\{\begin{aligned} x_i^{(0)} &= \frac{D_{ii}\,P_i - B_{ii}\,Q_i}{A_{ii}\,D_i - C_{ii}\,B_{ii}} \\ y_i^{(0)} &= \frac{A_{ii}\,Q_i - C_{ii}\,P_i}{A_{ii}\,D_i - C_{ii}\,B_{ii}}, \end{aligned}\right. \tag{13}$$

we may thus put

$$\left\{\begin{aligned} x_i &= x_i^{(0)} + \xi_i\,, \\ y_i &= y_i^{(0)} + \eta_i\,, \end{aligned}\right. \tag{14}$$

and note that the corrections ξ_i , η_i should be as a rule smaller than the first term. By replacing (14) into the original system (5) we obtain, then, another system having the same coefficients of the unknowns ξ_i , η_i , but much smaller right members because the dominant parts have been subtracted. The solution can then be repeated in the same way leading to an iterative procedure which converges very rapidly. In practice it is useless to go beyond, the 3rd iteration, but the x_i and y_i computed by means of these iterations must be used to compute new values of the A_i and h_i , which are employed to start the whole computations once again.

The initial values $A_i^{(0)}$ and $h_i^{(0)}$ are not critical and c an be easily estimated from the spectra; even better one may use for them the values previously found from another star if available.

3. SATURATED LINES

The representation of the observed spectrum by means of the sum of gaussian profiles is entirely satisfactory from a purely empirical point of view. Clearly for an isolated line the equivalent width would be given simply by

$$W_i = \sqrt{\pi h_i}\, A_i\,, \tag{15}$$

but for blended lines this is certainly not true and it is not very easy to find how the areas under the single profiles combine together to give the combined profile. We tried therefore a different kind of expansion, which worked rather satisfactorily and is now currently used for processing our spectra.

For this procedure, we start from the well known empirical formula for the profile of a line which was proposed by Minnaert and by Unsöld (1955) and is employed in the theory of the curve of growth,

$$r(\lambda) = \frac{1}{\frac{1}{R_c} + \frac{1}{\eta(\lambda)}} , \tag{16}$$

where R_c is a constant - the central depth of the strongest lines of the spectrum - and $\eta(\lambda)$ is the ratio of the line absorption coefficient to the continuum absorption

$$y(\lambda) = \frac{\kappa_\ell(\lambda)}{\kappa_c} . \tag{17}$$

We note that both κ_ℓ and κ_c are certain averages taken through the atmosphere. κ_c is essentially a constant for a given line, but varies slowly throughout the spectrum and $\kappa(\lambda)$ depends on λ through the well known Voigt function. However if we exclude that part of the profile (wings) lying outside 3 or 4 times the Doppler width D ,

$$D = \frac{v}{c} \lambda , \tag{18}$$

where v is the turbulent velocity and λ the wavelength of the line, we may neglect the damping and write simply

$$\eta = A_e^{-(\lambda-\lambda_i)^2 / D^2} \tag{19}$$

λ_i being the wavelength of the center of the line and A a constant. Since in the atmosphere of a giant K star, v is of the order of 10 km/s , at 4000 A we have

$$D = 0.13 \text{ A}$$

and therefore equation (19) can be used until about 0.4 or 0.5 A from the center of a line. This means that with very few exceptions, like the 4227 Ca I line and another dozen of very strong lines, all the lines in our spectra can be represented by equations (16), (17) and (19) even when the core is already strongly saturated.

Now, if at a certain wave-length λ, we find that several lines contribute to the absorption, it seems reasonable to sum all the corresponding κ_ℓ and write

$$\eta(\lambda) = \Sigma_i A_i e^{-(\lambda-\lambda_i)^2 / D^2} , \tag{20}$$

the sum being extended to all contributing lines.

For a digitized spectrum we thus obtain the expansion

$$r_j = \cfrac{1}{\cfrac{1}{R_c} + \cfrac{1}{\Sigma_i A_i e^{-(\lambda_j-\lambda_i)^2 / D^2}}} \tag{21}$$

which is the equation we have adopted for processing our spectra. Again, the procedure is classical and we do not need to go into details. The following points however must be noted.

(a) Although equation (21) depends upon equation (16) which is essentially empirical, in the sense that it lacks a full theoretical justification, still it is greatly to be preferred to equation (9) on theoretical grounds. Indeed it is equivalent to the use of a curve of growth, without going through the trouble of obtaining the equivalent widths for all the lines and constructing the curve of growth itself; in fact the A_i's are precisely the abscissae of the curve of growth, if one leaves out the few lines lying upon the damping branch. We are confident that moderately blended lines will still give reasonably good results and the saturation effect is taken into account just as well as in computing the curve of growth.

(b) R_c and D (or better v) are constant quantities for a given star, that is they are the same for all the lines of a given spectrum. The number of parameters to be computed is, therefore one half that of the parameters in equation (9). Clearly, due to this circumstance, the residuals between the observed r_j's and those computed with the final values of the parameters may be somewhat less good if equation (21) instead of (9) is used; but this disadvantage is only apparent; in fact the physically meaningful quantities (the A_i's) have a greater weight (due to their smaller number).

(c) We must warn against an apparent advantage of equation (21); clearly, putting

$$r_j' = \frac{1}{\frac{1}{r_j} - \frac{1}{R_c}}, \qquad (22)$$

equation (21) may be written

$$r_j' = \Sigma_i A_i e^{-(\lambda_j-\lambda_i)^2 / D^2} \qquad (23)$$

which is already linear in the unknowns A_i . But the errors in the r_j' are not simply proportional to those in the r_j and therefore in applying the least squares method one ought to give a different weight to each equation. The resulting complication of the equations is much greater than that corresponding to a linearization of the equations like that used for the case of the gaussian profile, so that on this respect the two expansions are equivalent.

Fig. 3 shows a small part of the spectrum of γ Tau reduced by the method of this section. As it was expected, the agreement between the observed and computed spectrum is somewhat less good than that obtained by means of gaussian profiles, but we are convinced that the discrepancies are well inside the observational errors. It is very remarkable that strong greatly saturated lines like 4005 Fe I are represented by the same value of D just as well as the faint lines, although the A_i's differ by a factor of 50 or more.

(d) According to our experience a very important point is to employ a list of lines as complete as possible. Omitting a line means distorting the computed profiles for nearby lines in a complicated fashion due to a kind of propagation of errors. We are solving this problem by including - as a start - all the lines in the Rowland's revised Tables (Moore et al. 1966) and adding here and there, by trials, a line where we feel there is a need for it. We intend to prepare by this way a complete list of all the lines in the spectral region λλ 3985 - 4810 of K-giants, then be used for all the remaining stars. It is quite possible that this list will contain some unexisting lines, but we consider this as a minor disadvantage.

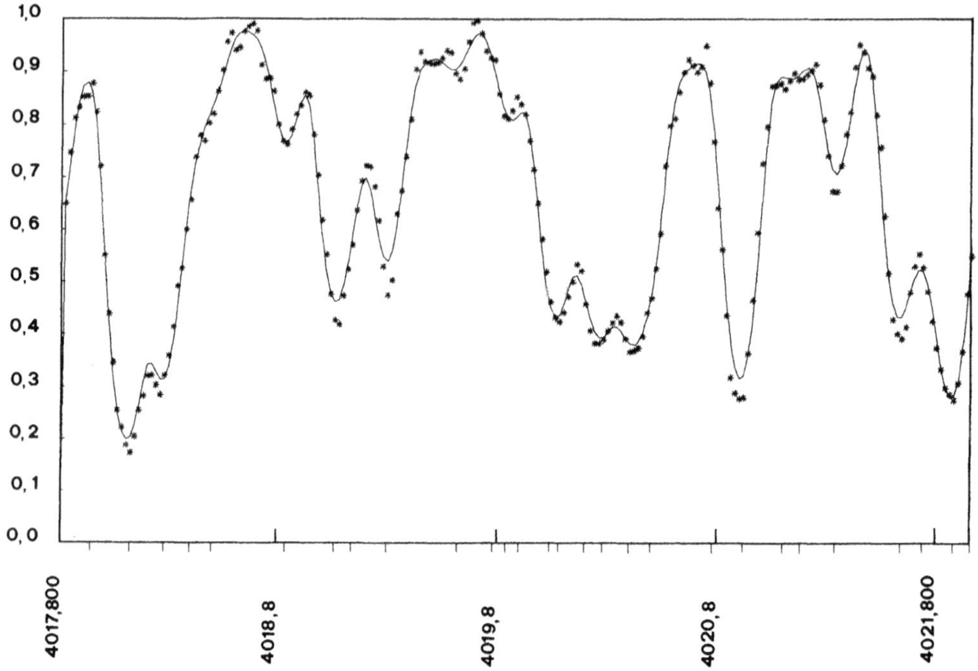

Fig.3. Fitting of a portion of the spectrum of γ Tau. The full line is the curve computed by means of equation (21); the marks correspond to the observed points; the small lines along the λ-axis indicate the wavelengths of the spectra lines used.

(e) Another point which was carefully evaluated was that of filtering the observed spectra for plate noise and deconvoluting for the instrumental broadening. This can be done numerically without complicating too much the processing program. But eventually we decided to use the raw spectra, without filtering and deconvoluting them, on the following grounds.

The method of least squares in itself takes into account accidental errors due to the noise; in fact for each line we average upon a "theoretical" profile and fast oscillations due to the noise are eliminated. But the use of theoretical profile implies that the observed profiles be corrected for instrumental broadening.

However, the use of a transfer function corresponding to some instrumental profile would be equivalent to filtering in the opposite way, that is to increasing the amplitude of the high-frequency Fourier components, which are precisely those which are more affected by noise. Also it is very doubtful whether it is correct to consider the resulting profile as a simple convolution of the "real" spectrum with the instrumental response, in the case of a non-linear detector like the photographic plate.

On the other side (Gratton et al. 1975) the instrumental broadening in our spectra was found to be of the order of $\pm$ 0.045 A; this is not negligeable, but is considerably less than the expected Doppler half-width, $0.13 / \sqrt{2} = 0.090$.

With a middle-size computer (an IBM 370/125) the processing of a 35 A window containing about 200 lines takes about 15 minutes, including the printing of the result and the plot of the observed and computed spectra (double precision is necessary). With a larger computer this time is considerably shortened; also wider windows can be used. In principle it is possible to process the whole spectrum at once.

Our program in FORTRAN may be made, of course, available to everybody who might be interested in them.

REFERENCES

1. L. Gratton, G. Natali and R. Nesci, A spectrophotometric Atlas of K giants, Laboratorio di Astrof. Spaz. Franscati, in press (1975).
2. C.E. Moore, M.G.J. Minnaert and J. Houtgast, 2nd revision of Rowland's Table, Nat. Bur. of Standards, Mon. 61, 1966.
3. A. Unsöld, Sternatmosphären, page 407, 1955.

ACCURACY AND RESULTS OF SPECTROPHOTOMETRY WITH OBJECTIVE PRISM PLATES

A. Cassatella* and R. Viotti

Laboratorio Astrofisica Spaziale, Frascati, Italy

1. INTRODUCTION

For its potential application to a large variety of astronomical investigations, objective prism is of fundamental importance among the astronomical techniques of data acquisition involving photographic plates. Reference may be made to the investigation on the interstellar extinction by Nandy (1964), on Nova Del 67 (Woszczyk et al., 1968), and to the calibration methods developed by Nandy and Smriglio (1970), Ardeberg and Virdefors (1972), and others.

New methods have been developed at Frascati for the rapid reduction of objective prism plates for the purpose of providing energy distribution of peculiar stars (Baratta et al., 1974) and of the Hyades stars of intermediate brightness (Cassatella et al., 1973). The accuracy of the derived monochromatic magnitudes is of $\pm 0^{m}.03$ to $0^{m}.1$ depending on the quality of the plates and on the number of standard stars used for the plate calibration. In the following we shall describe our method of calibration based on the use of standard stars in the field of the plate, and discuss the effect of hypersensitizing the IN plates. We show that objective prism spectrophotometry is still a very promising technique for the measure of low dispersion absolute monochromatic fluxes of a very large number of stars down to quite faint objects.

---

* Presently ESRO fellow, Observatoire de Meudon, Equipe Etoiles Froides, France.

de Jager/Nieuwenhuijzen (eds.), Image Processing Techniques in Astronomy. 367-372.

2. SPECTROPHOTOMETRIC CALIBRATION OF OBJECTIVE PRISM PLATES

A set of unfiltered IN plates of the Hyades Cluster was secured with the Schmidt telescope of the Asiago Astrophysical Observatory during 1970-71. All the spectra were recorded with our MI.DI. digitized microphotometer. For the plate calibration we used the available scanner measurements of cluster stars in the field of our plates (Oke and Conti, 1966). For each plate characteristic curves at 20 wavelength bands 50 to 100 A wide were obtained by comparing the scanner monochromatic magnitudes m_λ of 3 to 6 stars with the corresponding average Baker densities Δ_λ measured on the objective prism spectra. In general the derived curves are remarkably linear down to low densities. Fig.1 shows the calibration of a IN plate of good quality in the spectral range from 3636 to 8400 A. The abscissae are scanner m_λ plus a term $2.5 \log(d\lambda/dx)$ which takes into account the spectral thickering towards longer wavelengths due to the selective prism dispersion. The r.m.s. deviation of the data from the curves in Fig.1 is between $\pm 0^m.03$ in the red and $\pm 0^m.1$ for the less exposed spectral regions. The derived gradients $\gamma = d\Delta/dm$ have no large wavelength dependence, and are in fair agreement with the results of an independent sensitometer calibration of the plates.

Once the plates were calibrated, we derived the energy distribution of several stars down to 11^m. Details on the computer programs and some results are given in Cassatella et al.(1973) and Cassatella and Viotti (1974). For the more exposed plates on which Oke and Conti's stars are generally overexposed, the calibration was obtained by using as secondary standards the stars of intermediate brightness measured on the less exposed plates. By comparing the m_λ's of the same stars derived from different plates, we found a r.m.s. dispersion between $\pm 0^m.05$ and $0^m.1$ for $8-9^m$ stars. The computed energy distribution of an A6V star of the Hyades (BD+15°637) is presented in Fig.2.

To reach the faintest (late-type) stars we hypersensitized part of the IN plates with a pre-exposition to a flash lamp. The use of hyp. emulsions has the advantage of a gain in limiting magnitude and in exposure time. The main disavantage is the rapid increase in plate background with exposure time. Assuming that the minimum measurable level should be five times the noise, we found a gain in limiting magnitude of $1^m.5$. Similarly, we derived that the same information - i.e. the same signal to noise ratio - can be reached with hyp. plates and exposures 6 to 9 times shorter (Cassatella and Viotti, 1972).

In the case that too few or no spectrophotometric standard stars are present in the field of the plate, we used the following procedure*. Stars are selected and recorded for which a good multi-

* This method was also used to improve the photometric calibration of some of the Hyades plates.

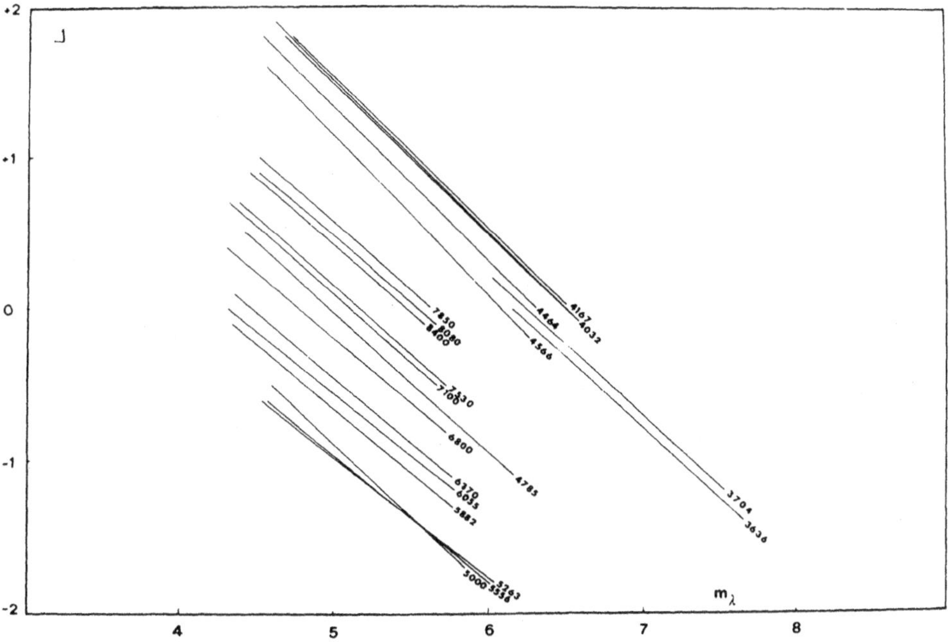

Fig. 1. Spectrophotometric calibration of a IN objective prism plate of the Hyades using the scanner measurement of six stars made by Oke and Conti (1966).

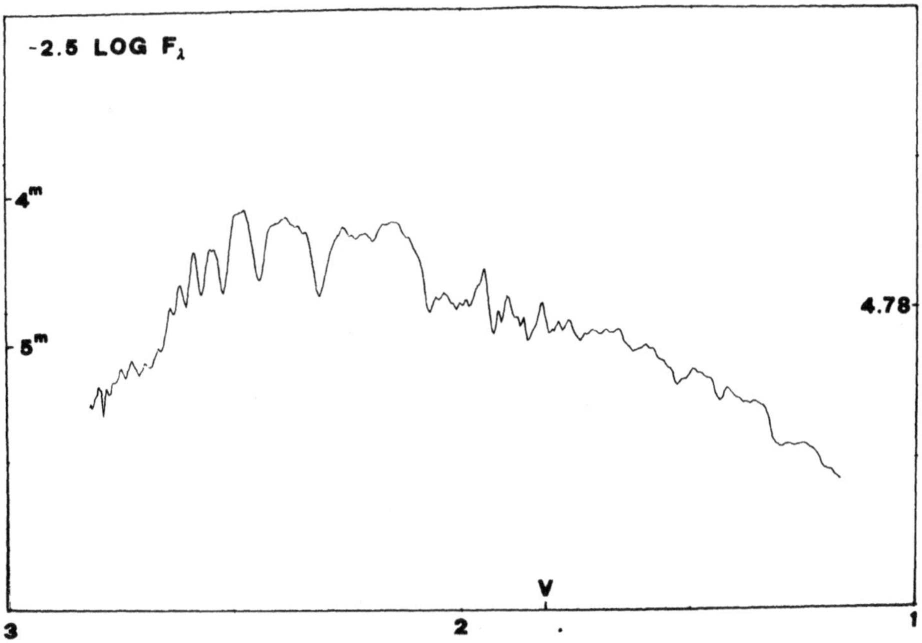

FIG. 2. The computed energy spectrum of BD+15°637.

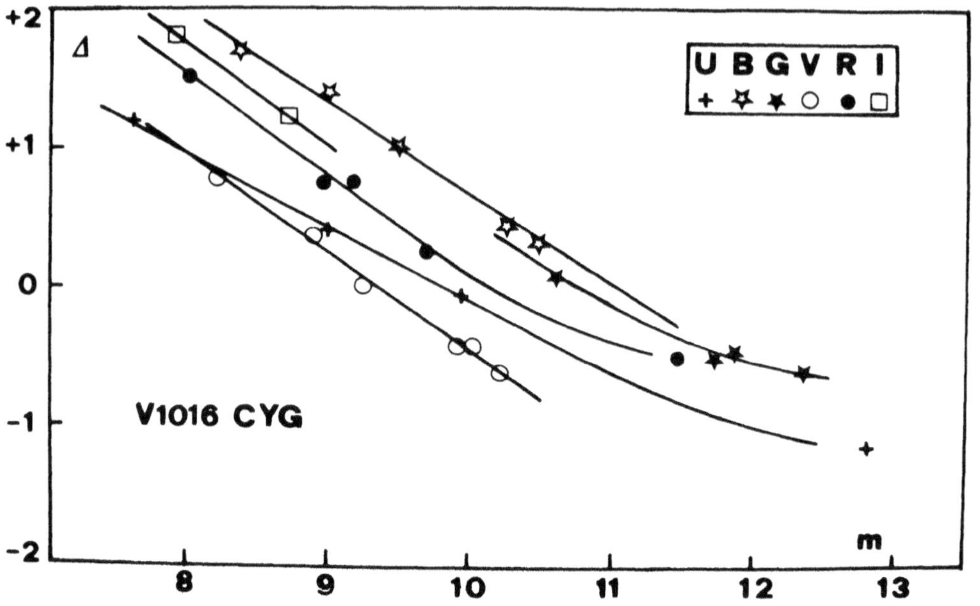

Fig. 3. Six colours calibration of a IN objective prism plate (see Baratta et al., 1974).

colour photometry is available. Then a photometric calibration is obtained by plotting the star colours against the "density colours" Δ_i defined by

$$\Delta_i = -2.5\,\gamma_i \log\left(\int 10^{-0.4\,\Delta(x)/\gamma_i}\, S_i(x)dx \Big/ \int S_i(x)dx\right)$$

where S_i is the response curve of the colour i divided by the standard plate sensitivity, x is the linear displacement as measured on the photographic plate, and γ_i is the mean plate gradient derived through an iterative procedure. As an illustration of this method Fig.3 gives the UBGVRI characteristic curves for one IN plate used for a study of the peculiar star V1016 Cyg (Baratta et al., 1974). Next we select in the plate field some early-type stars with known colours and well classified spectral types (for instance from the objective prism plate itself), and compute the monochromatic magnitudes uncorrected for the plate sensitivity and atmospheric extinction. These corrections are determined by comparison with standard m_λ for stars of the same spectral type, and accounting for the interstellar reddening. In the case of the above investigation the plate corrections derived from four early B-type stars with $E_{B-V}= 0^m.44$ to $0^m.56$ agree each others within $\pm 0^m.03 \div 0^m.05$.

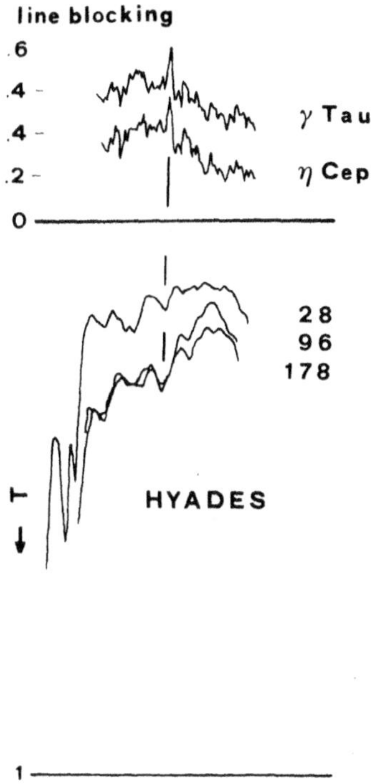

Fig. 4. Line blocking of γ Tau and η Cep (top), and transmission tracings of three Hyades stars (bottom) in the blue region.

3. DISCUSSION AND APPLICATION

We have shown that an accurate absolute spectrophotometry can be made using objective prism plates, as already confirmed by the results of other investigations. Our method is based on the presence of several stars of known m_λ in the plate field, but a reliable calibration can also be obtained from photoelectric magnitudes and spectral types. A special care should be taken of field effects (vignetting, plate distorsion, variable fog level) which were not discussed here.

Objective prism spectrophotometry could provide simultaneously the low resolution ($\lambda/\Delta\lambda \sim 10^2-10^3$) energy distribution of several stars of intermediate brightness with an accuracy high enough for most of the astrophysical problems. The information present in the spectra can be pulled out by comparison with computed synthetic spectra or with the results of high resolution spectrophotometry. An example is given in Fig.4 where the line blocking derived from two 3.2 A/mm spectrograms of the giant star γ Tau and the subgiant η Cep (see Cassatella, Gratton and Nesci, communication presented at this Conference) is compared with objective prism spectrograms of three Hyades stars. Star number 28 is γ Tau. Once the combined effect of effective temperature, gravity and metallicity on the line blocking is known from spectrum synthesis or from high resolution spectrophotometry, objective prism spectra can be used to analyze a large number of stars, in the same way as it can be done with interference filter photometry using several spectral bands.

REFERENCES

Ardeberg, A. and Virdefors, B.: 1972, Astron. and Astrophys. **20**, 177.

Baratta, G.B., Cassatella, A. and Viotti, R.: 1974, Astrophys.J. **187**, 651.

Cassatella, A., Maffei, P. and Viotti, R.: 1973, in Spectral Classification and Multicolour Photometry, ed. by Ch. Fehrenbach and B.E. Westerlund, D. Reidel Pub. Co., Dordrecht, p. 127.

Cassatella, A. and Viotti, R.: 1972, Rapporto di Ricerca, Osservatorio Astronomico di Roma, N.3.

Cassatella, A. and Viotti, R.: 1974, Rapporto Interno, Laboratorio Astrofisica Spaziale, Frascati, N.5.

Nandy, K.: 1964, Publ. Roy. Obs. Edinburgh, **3**, 142.

Nandy, K. and Smriglio, F.: 1970, Observatory, **90**, 114.

Oke, J.B. and Conti, P.S.: 1966, Astrophys.J. **143**, 135.

Woszczyk, A., Smolinski, J., Maron, N., Strobel, A. and Krempec, J.: 1968, Astrophys. and Space Science **1**, 264.

LE CENTRE DE DEPOUILLEMENT DE CLICHES ASTRONOMIQUES

M. Auvergne - A. Bijaoui - J.L. Heudier - J. Marchal
C. Ounnas - J. Postel - M. Schneider
Observatoire de Nice, Le Mont Gros
06300 - NICE - France

ABSTRACT. The "Centre de Dépouillement des Clichés Astronomiques" opened during the year I974 is equiped with a PDS microdensitometer connected to a computer PDP 11/40. We develope an interactive sofware for astronomical purpose such as classical spectrography, astronomical photometry, astrometry. This machine and its sofware are used by all french astronomers. To let them know the development of the center we publish a bulletin.

Le développement de l'informatique a ouvert de très nombreuses possibilités dans le traitement des données d'observation.
Afin que tous les astronomes français puissent bénéficier d'un microphotomètre digitalisé piloté par ordinateur et qu'un logiciel de base soit constitué, l'Institut National d'Astronomie et de Géophysique (I.N.A.G.) créa un Centre de Dépouillement des Clichés Astronomique (C.D.C.A.) et l'installa à l'Observatoire de Nice.

LE MATERIEL

Nous avons choisi pour l'équipement du centre un microphotomètre PDS, piloté par un PDP 11/40, équipé de divers périphériques.
La machine PDS IOIOA a été choisie principalement en raison :

- d'une platine à deux dimensions.
- d'une mesure des coordonnées par des règles à franges de Moiré, permettant une bonne répétitivité des mesures.
- d'une grande cadence de prises de l'information.
- de sa connection à un ordinateur de qualité possèdant un

de Jager/Nieuwenhuijzen (eds.), Image Processing Techniques in Astronomy. 373-378.

système évolué.

- de l'existence d'un logiciel permettant de traiter les clichés les plus divers.
- de la possibilité offerte par l'interface de piloter la platine simultanément en x et y.

Divers tests ont été effectués afin de contrôler ses qualités et de déterminer les meilleures conditions d'utilisation. En ce qui concerne les qualités photomètriques les mesures obtenues nous ont semblé très correctes, pour une tension du photomultiplicateur telle qu'il n'y ait pas de fatigue de cathode. La cadence de prise de mesure doit être en accord avec la bande passante de l'amplificateur. Selon les gradients de densité, on pourra augmenter ou diminuer cette cadence. Le passage d'une mire de 40 dt/mm nous a permis de mettre en relief cet effet en fonction de la vitesse de la platine.

Du point de vue mécanique, les tests effectués sur des mesures d'étoiles, donnent des valeurs inférieures au micron.
Le microphotomètre PDS nous semble donc une très bonne machine adaptée au besoin des astronomes.

Pour piloter cet instrument, nous avons choisi un PDP II/40 de 56 Koctets pour les raisons suivantes :

- il existe un interface adapté à cet ordinateur pour le PDS
- le temps de base de l'unité centrale est très faible (0,9 μs)
- les mots de I6 bits permettent des enregistrements suffisamment longs et une mémoire importante.
- il existe un système suffisant, en particulier un compilateur Fortran.

Du point de vue purement pilotage le PDP II/40 ne s'imposait pas mais nous désirions pouvoir traiter les données directement après enregistrement.

Pour cela nous avons adjoint une unité de disque de 2,5 millions d'octets, une unité de bande magnétique 7 pistes et une console graphique Tektronix 40I0. Celle-ci joue un rôle fondamental dans le traitement grâce à la rapidité des entrées sorties et à la possibilité de tracer courbes, isophotes,... Nous l'avons équipée d'un reprographe électronique afin de conserver, le cas échéant, un document après utilisation.

L'expérience ayant montré qu'une très faible partie du temps d'unité centrale était utilisée nous allons installer un système permettant le travail temps partagé avec enregistrement parallèlement au traitement de données obtenues auparavant.

LE LOGICIEL

En ce qui concerne l'enregistrement, le programme PDS nous permet la plupart des travaux de dépouillement. Nous avons donc développé des programmes de traitement des données.

L'aspect fondamental de notre équipement est la possibilité de construire un système conversationnel de traitement : le microphotomètre acquiert l'information, l'ordinateur effectue les diverses opérations demandées, mais l'astronome garde une grande liberté dans le choix des paramètres à introduire effectue des mesures grâce à la console graphique, demande l'exécution de la série de programmes qu'il désire et intéragit en permanence avec le programme en cours d'exécution.

L'écriture sous forme conversationnelle des programmes est légèrement différente de celle d'un programme écrit pour cartes perforées. L'introduction d'options permettant de reprendre le programme à n'importe quel stade de son exécution avec modification du minimum de paramètres est un des aspects principaux.

Le logiciel que nous développons est axé sur trois directions : astromètrie, spectrographie classique, traitement de clichés astronomiques à deux dimensions.

Du point de vue astromètrique, une première série de programmes a été écrite permettant la détermination des coordonnées (x,y) de l'étoile sur le cliché puis l'obtention des (α,δ) à partir des étoiles de référence. Pour la détermination des (x,y), on enregistre une matrice de densité, puis on calcule le centre de gravité en ayant bien soin,d'éliminer le fond de ciel. A partir de ce point, on construit une seconde matrice l'ayant pour centre et inscrite dans la première. On renouvelle le processus jusqu'à ce que deux valeurs succéssives du barycentre ne diffèrent pas de plus de I/IO de micron sur chacune des coordonnées. Cette opération se fait automatiquement dans le programme XY. A partir du fichier de positions et d'un fichier d'étoiles de référence, REDUCI, sous forme conversationnelle, réduit le cliché.

En ce qui concerne la spectrographie classique, nous développons une série de programmes permettant, dans une première phase, d'obtenir le I (λ). Actuellement, il existe un programme conversationnel de traitement en vitesse radiale (RAVEL) des enregistrements. Un programme de détermination des courbes de calibrations D = f(logI) pour différentes longueurs d'onde en cours de mise au point.

Pour les clichés divers bi-dimensionnels, un certain nombre de programmes généraux ont été mis au point permettant par appels successifs, la réalisation d'un nombre important de traitements de clichés. Tous ces programmes ont été écrits sous forme conversationnelle, telle que l'utilisateur dispose de nombreuses possibilités. Ces premiers programmes permettent :

- de tracer sur l'écran de la console une ligne ou portion de ligne d'enregistrement, de déterminer l'abscisse et la densité de points choisis par un réticule électronique (CESAR).
- de tracer des isophotes selon un code ternaire cyclique à partir d'une zone de l'enregistrement, et de déterminer les coordonnées de points repérés (ISIS).
- de transformer les densités de tout un champ en intensité I, logI, ou magnitude m au dessus du fond de ciel (CALIB).
- de filtrer les mesures avec un opérateur choisi par l'astronome (VENUS).
- d'effectuer des opérations arithmétiques entre des fichiers (ARIFIL).

D'autres programmes généraux sont en cours de mise au point.

DIRECTIONS DE TRAVAIL

Pour résoudre de nombreux problèmes de dépouillement de clichés, nous avons entrepris de développer des programmes d'enregistrement très différents du programme PDS. Parmi les buts nous avons l'analyse suivant une courbe quelconque pour spectres échelles, la prise des mesures conditionnelles à une valeur de densité, le déplacement de la platine à une position x,y calculée puis prise des mesures dans une zone déterminée. Les deux derniers programmes sont nécessaires pour la détection et la mesure des étoiles d'un cliché stellaire.

De même nous avons l'intention de développer des programmes de réduction de clichés plus complexes que celui existant, en intégrant par exemple des fichiers contenant des catalogues stellaires pour repérer les étoiles à mesurer.

Dans le domaine de la spectrographie, nous pensons prolonger notre étude par l'obtention des I (λ), tracé du continu, profil des raies, largeurs équivalentes..., les programmes étant toujours écrits sous forme conversationnelle. Grâce à des fichiers classiques (table de Moore, atlas d'Utrecht,...) on identifiera les raies importantes selon le type de l'étoile...

Dans le domaine bi-dimentionnel, après écriture des programmes généraux, nous développerons principalement la photomètrie stellaire électronographique et le traitement des clichés provenant de spectrographes échelles.

LE CENTRE DE DEPOUILLEMENT

Le centre a été ouvert le I6 Septembre I974 aux astronomes français. C'est principalement l'aspect conversationnel du dépouillement qui soulève le plus grand intérêt : un certain conservatisme de la part des observateurs allait à l'encontre des traitements automatiques. Le traitement conversationnel correspond à la transition entre le dépouillement traditionnel à la main et celui entièrement automatique. Il évite l'écriture de programmes complètement figés qui otent toute souplesse au traitement des observations.

Au 24 Mars I975 c'est près d'une quarantaine d'astronomes qui ont pu bénéficier de l'équipement du C.D.C.A.

Afin d'informer les astronomes français de l'évolution de l'opération, nous éditons régulièrement un bulletin. Le N° 3 de ce bulletin est paru en Février I975. Ce bulletin comprend principalement :

- un état de l'opération, donnant l'évolution du matériel et du logiciel
- une étude des programmes mis au point, illustrée d'un exemple d'utilisation.
- un récapitulatif de l'utilisation de la machine, en indiquant les sujets traités.

CONCLUSION

Après très peu de temps d'expérimentation le centre a été saturé par les utilisateurs. Nombreux sont ceux ne désirant plus réutiliser des microphotomètres traditionnels même digitalisés.

Bien qu'actuellement l'aspect centre de calcul apparaît comme le plus important, il nous semble que ce n'est que transitoire et que très vite nous pourrons aborder de nouvelles étapes importantes dans le traitement de données après avoir construit un système classique satisfaisant notre communauté.

En effet très nombreux sont les problèmes de traitement de données qui n'ont pu être abordés par manque de ce moyen d'analyse. En particulier :

- utilisation des méthodes d'analyse factorielle pour déterminer la structure d'un objet.
- les méthodes de classification automatique.

Par le traitement de nombreux clichés et la création de fichiers de plusieurs paramètres d'intérêt astrophysique (W_λ , paramètres de structure de raies ou de continus,...), on peut élargir les domaines d'investigation. Dans ces conditions l'extraction de l'information sera de plus en plus totale.

Le développement des tubes TV en astronomie et particulièrement celui des méthodes de comptage. permettent d'imaginer la mise en action d'un tel système de traitement des données directement à la sortie des grands télescopes. Le C.D.C.A. peut être considéré comme une étape pour réaliser cette opération.

INTERCOMPARISON OF GROUND BASED AND SATELLITE PICTURES OF THE SUN

Robert D. Chapman, Gabriel L. Epstein, Robert W. Hobbs, Werner M. Neupert and Roger J. Thomas

Laboratory for Solar Physics and Astrophysics, NASA-Goddard Space Flight Center, Greenbelt, Maryland 20771

ABSTRACT. Using NASA facilities in space (OSO-7) and on the ground (Goddard Multi-Channel Spectrophotometer at Sacramento Peak, New Mexico) we have mapped an active region, and by combining these ultraviolet, X-ray and visible data, have constructed a physical picture of this structured region from the photosphere to the corona, corresponding to temperature regimes over the range 4500 K to 4 000 000 K.

The morphology of the active region was then studied by comparing grey-shaded images in which fine details stand out more clearly than in the contour plots. One result of the study is that gross similarities persist from the low photosphere up to high in the transition region while some changes occur in the corona.

We have demonstrated the capability for studying the complete morphology of an active region, and this capability can now be extended to investigate evolutionary histories of such structures.

1. Introduction

Various observations of solar active regions show that they manifest themselves over great ranges of space and temperature. To fully characterize active regions, therefore, it is necessary to study them simultaneously throughout those ranges in which they have significant existence. To this end, we carried out a program of nearly simultaneous coordinated observations of solar

de Jager/Nieuwenhuijzen (eds.), Image Processing Techniques in Astronomy. 379-384.

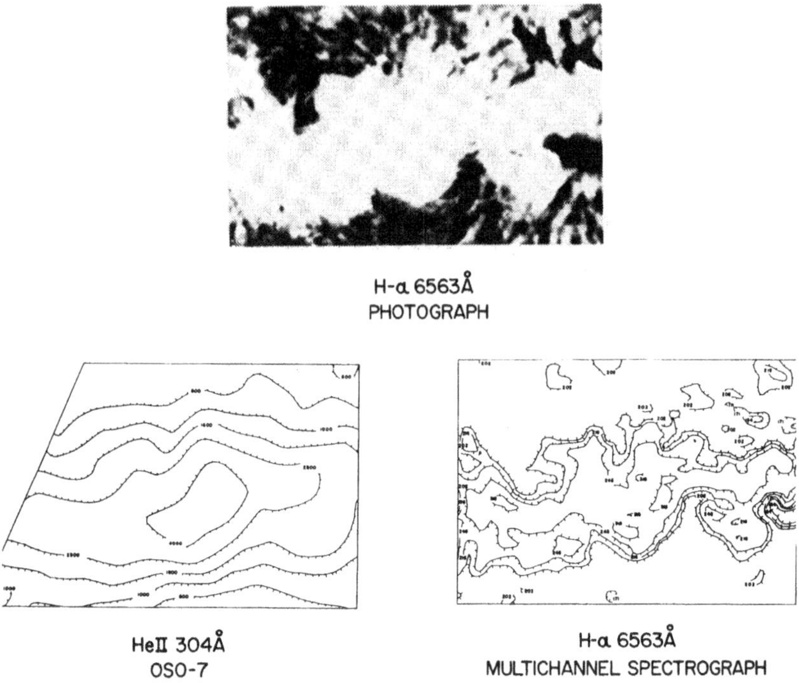

Figure 1. Co-alignment of ground-based and OSO-7 images.

active regions at over 35 wavelengths ranging from red-visible through the extreme ultraviolet to x-rays. The observations span the solar atmosphere from the photosphere, through the chromosphere and transition region into the corona.

The observations in the visible-near infrared portion of the spectrum were made with the Goddard Space Flight Center (GSFC) multi-line spectrophotometer located at Sacramento Peak Observatory, New Mexico. This instrument has been described by Hobbs et al. (1972). The ground-based observations consist of raster scans of a 3½ x 5 arc minute area of the solar disk made at up to 28 wavelengths simultaneously using a 5 x 10 arc second aperture. Each set of rasters was completed in roughly 4 minutes.

The x-ray and EUV observations were made with the GSFC spectroheliograph flown on-board OSO-7. Details of this instrument have been published by Neupert et al. (1974) and by Neupert and Underwood (1974). The x-ray and EUV observations consist of raster scans of a 5 x 5 arc minute area of the solar disk made using a 10 x 20 arc second aperture. Several OSO-7 rasters made over a 15 minute time period were averaged together to produce one raster in each of the observed wavelengths.

The orientation and overlap of the regions rastered with the two instruments were determined as follows. An OSO-7 He II (304Å) contour map was visually overlaid on a photographic H-α spectroheliogram of the general area and then, independently, an H-α contour map made with the multi-line spectrophotometer data was visually overlaid on the same H-α spectroheliogram. The result was checked and confirmed by calculation of the position of the two raster centers using pointing data from the two instruments -- the positions agreed to within 10" in the EW direction and 6" in the NS direction and the rotational orientation was confirmed to within $1/2^{o}$. Figure 1 shows the results of the fitting.

2. Discussion

Figure 2 shows the active region as seen in various lines, grouped according to the region of the solar atmosphere in which they are formed. The magnetograph was taken at the Sacramento Peak Observatory about two hours after the other observations. However, a comparison of it with our data for the Ca II H and K lines shows that, at least to our spatial resolution, there was no significant change in the magnetic structure of the region. The data for the various lines are presented in grey-scale rendition on a linear scale for most of the lower lying lines and on a logarithmic scale for the higher excitation lines. The grey-scale photos for the Si I (4103Å) line and the He D3 (5876Å) line have had their intensity scale distorted in a manner that enhances the contrast in the range from the "quiet sun" intensity up to the maximum.

On a gross scale the active region retains its identity as an East-West band of varying width all the way up to those heights at which Fe XVI is formed. Although there is considerable variation in the detailed distribution of intensity, the crude dimensions of the active region do not change significantly; any change is certainly less than 50%.

The large leading sunspot is of positive magnetic polarity and is about 50" wide by 70" long at the photospheric level. In the chromospheric Hydrogen Balmer lines and the Ca II lines, plage material can be seen to have closed in over the sunspot making its apparent size only about 20" x 60". At the height of the transition region (He II 304Å) plage material almost completely overlies the spot giving only the slightest indication of its existence. In the Mg VIII and Mg IX lines there is, once again, an indication of a feature at this location, but it appears as a knot of emission rather than as a dark spot. In the coronal lines of Fe XV and Fe XVI the location of the sunspot is again indicated by the absence of emission.

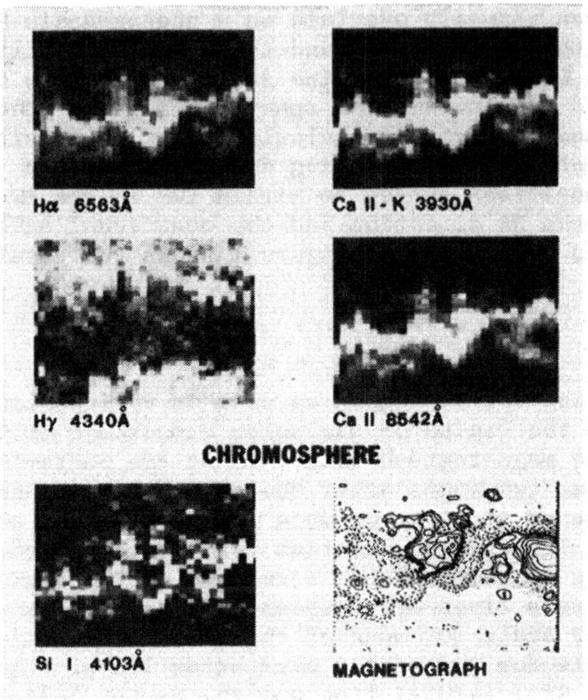

Figure 2. McMath Region 11930 in various wavelengths June 21, 1972 15:25 to 16:00 U.T. Solar North is up and east to the left (a) above photospheric (Si I) and chromospheric images together with a Sacramento Peak Observatory magnetograph. (b) next page. Transition region and coronal images.

In the center of the raster area is a heart shaped region, the westmost "chamber" of which is magnetically negative and the eastmost "chamber" magnetically positive. Within the positive area is a group of sunspots that can be seen ringed with plage material in the Si I image. The brightest part of this material delineates the magnetically negative region. Although there is bright plate material overlying the sunspots, as seen in H-α and the Ca II lines, far brighter material overlies the magnetically negative area. In He 304Å the brightest area retains the shape seen in the chromospheric lines but is broader and is centered over the neutral line separating our two regions. The greatest intensity in the Mg lines overlies the magnetically positive region, roughly mimicking the magnetic distribution. In both the Mg VIII and the Mg IX lines the very brightest spot is directly over the sunspots. In the iron lines the situation has changed and the intensity distribution looks much more like the chromospheric lines. The "dark" lane to the north of the

Fe XV 284Å
Fe XVI 335Å
Mg VIII 315Å
Mg IX 368Å
CORONA
He I - D3 5876Å
He II 304Å
TRANSITION

sunspots is again deep and the brightest regions are directly above the magnetically negative area.

In the southeast corner of the raster there are some embedded magnetically positive features. There are no easily discernible corresponding features in Si I 4103Å but in the chromosphere there is emission (or absorption in the case of H-α) distributed in the same way as are the features. Interestingly, H-α does not show this area -- at least not its lower portion. It can be seen in the He II 304Å image, although it is not very prominent. It is quite prominent in the grey-scale photos for Mg VIII and Mg IX -- and indeed, the brightest points of these two rasters lie in this area. As with the areas previously discussed, we see that emission directly above the Mg bright regions is almost gone in the coronal height range represented by the Fe XV and XVI lines.

In conclusion, we can say that although details of structure within the plage change markedly with height, overall structure, namely the outer boundaries of the active region, remain remarkedly stable and uniform over an extremely wide range of heights and physical conditions. Features showing "weak" chromospheric emission (e.g., sunspots) show strong transition region emission (Mg VIII and IX) and weak or no coronal emission (Fe XV and XVI). Features showing strong chromospheric emission show relatively weak emission in the transition region and very strong coronal emission.

References

Hobbs, R. W., Harris, G. D., and Epstein, G. L.: 1972, Publ. Astron. Soc. Pac., **84**, 74.

Neupert, W. M., Thomas, R. J., and Chapman, R. D.: 1974, Solar Phys., **34**, 349.

Neupert, W. M. and Underwood, J. H.: 1974, Solar Phys., **34**, 241.

FLYING-SPOT ANALYSIS OF SOLAR IMAGES

L. Azzarelli*, C. Carlesi*, R. Falciani°, M. Giordano°, R. Panicucci*, M. Rigutti°, G. Roberti°.

*Institute for Elaboration of Informations - C.N.R. - Pisa (Italy)
°Capodimonte Astronomical Observatory - Naples (Italy).

1.- INTRODUCTION

This work has been performed to test the new and interesting results obtained previously with a photographic isodensitometric method about the photometric evolution of solar flares (Falciani et al., 1972; Falciani and Rigutti, 1972 a,b) and to study the degree of utility and reliability, and the general performances of high speed, computer controlled devices in the photometric analysis of extended sources. We studied some series of good Hα solar filtergrams, obtained during 1969 (May 15-16-17-25 and Oct. 25-27), at the Athens National Observatory (we warmly thank Dr. C.J. Macris for having kindly put at our disposal such a material), with time resolution of about 30 sec, with uniform exposure and high photometric accuracy (≈ 5000 filtergrams).

2.- INSTRUMENT DESCRIPTION

To study the huge amount of the above material we used the S.A.D.A.F. computer controlled flying-spot digitized machine of the Istituto di Elaborazione dell'Informazione (Pisa). This instrument has been extensively described elsewhere (Azzarelli and Panicucci, 1972; Carlesi and Montanari, 1973; Carlesi, 1975). It is a conventional type of flying-spot photometer (see the general plan in

de Jager/Nieuwenhuijzen (eds.), Image Processing Techniques in Astronomy. 385-390.

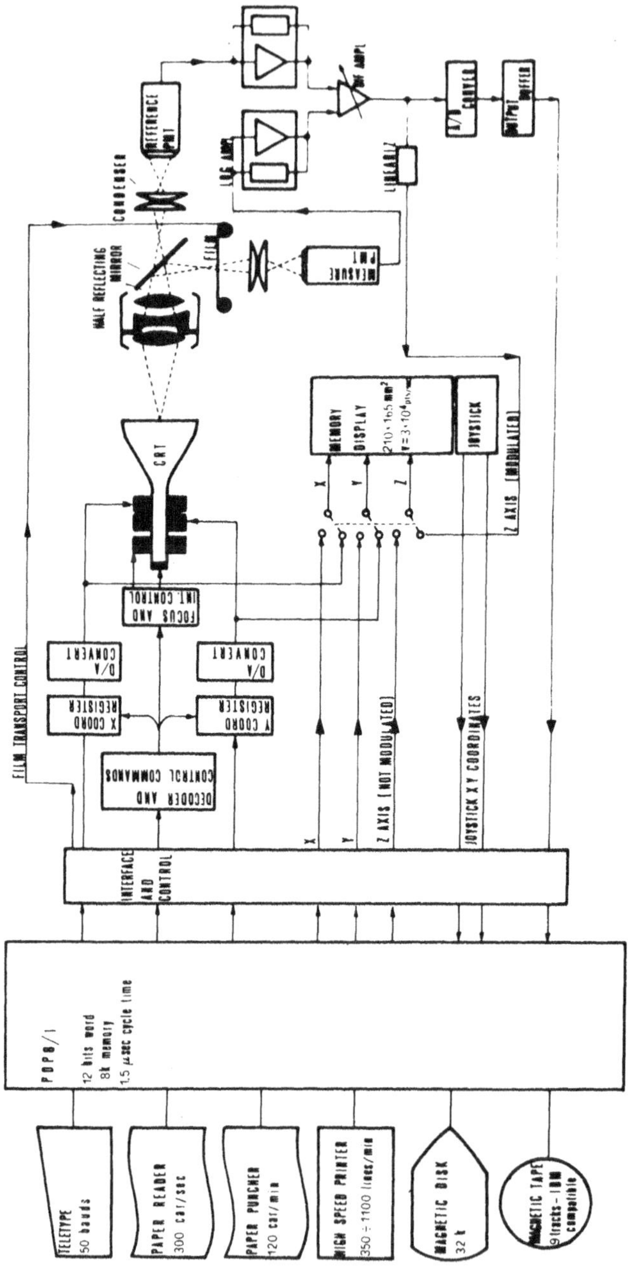

Fig.1.Block diagram of S.A.D.A.F. flying-spot.

Fig.1), the main features of which are given below:

photograms dimensions	24 x 36 mm^2
scanned area	24 x 24 mm^2
maximum lenght of automatically scanned roll film	45 m.
positioning precision of automatic film transport	$\pm$ 0.2 mm
maximum number of random access points in the scanned area	(1024 x 1024)
densitometric resolution	64 gray - levels (6 bits digitized value) in the density range 0.05 $\div$ 2.2
time for the acquisition of the data from one point (random access)	$\simeq$ 40 μ sec.
scanning velocity	30 photograms per hour at maximum resolution of 10^6 points with sequence scanning.

It is possible to monitor on-line on a CRT memory display all the scanned regions and through a joystick we can select some particularly interesting areas to further detailed reductions. A flexible, interactive program enables to run and controll all the working procedures of the equipment (interactive scanning parameters determinations, densitometric scale and digitization minors selections, output peripherals, magnetic tape memorization options, automatic film transport, jump of not interesting photograms etc.). We also developed a program to reduce the obtained informations with an off-line computer. The raw data are converted, through the calibration curves of the instrument and the photographic emulsion, into intensity values, measured in unit of the mean indisturbed chromosphere, surrounding our events. The histogram of the counts per digitized gray level is shown in Fig.2. From the mean dimension of the spot we get by interpolation the projected area of the phenomenon at a given intensity levels j and then deduce the energy E_j emitted from a selected intensity level upward. E_j's are measured in terms of the energy emitted by a unit surface of the indisturbed chromosphere. We can select various output possibilities (print, write, plot, punch).

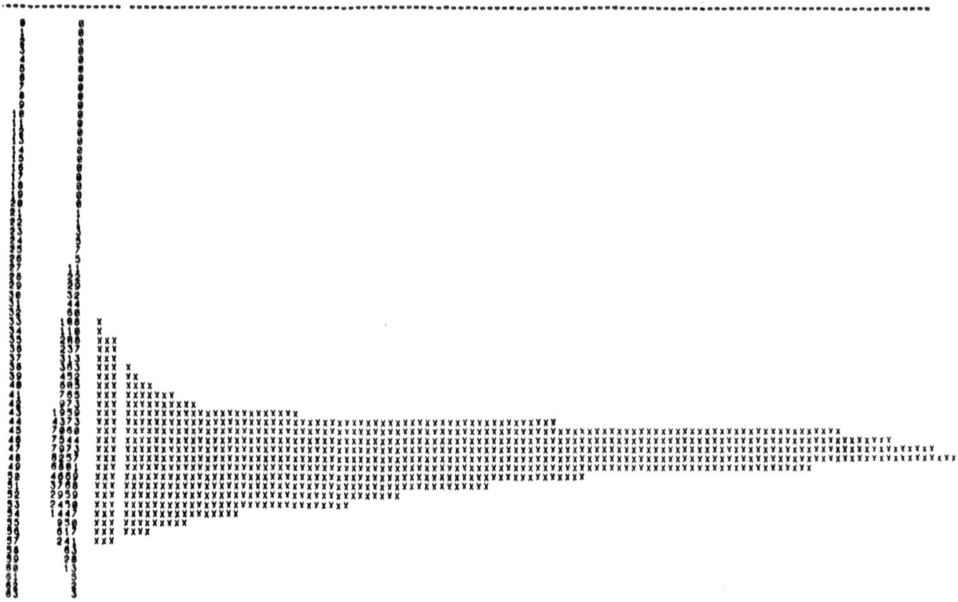

Fig.2.Histogram of the counts vs. gray levels.

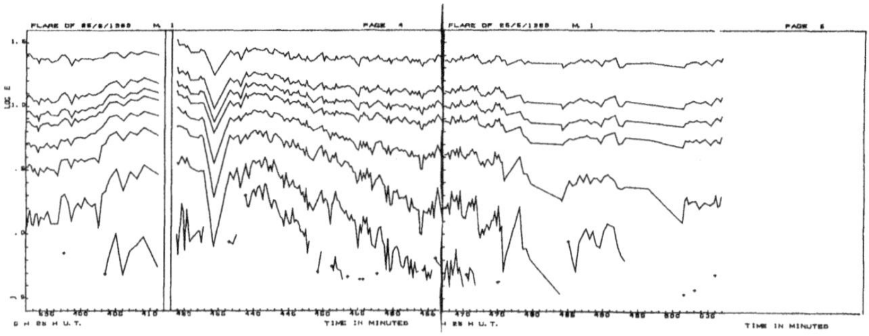

Fig.3.Evolutive curves of May 25, 1969 solar flare.

3.- EQUIPMENT PERFORMANCES AND TESTS

The step filter for the densitometric comparison and test of the machine has been calibrated in the same density scale D of the calibration curve of the analyzed photographic emulsion.
Checkes for the photometric stability of the flying-spot gave results of $\sim 2‰$ for $D < 0.6$, $\sim 8‰$ for $0.6 \leq D < 1.3$ and $\sim 1 \div 2\%$ for $D \geq 1.3$ (the latter is due only to the difficulties to disentangle and read exactly neighbouring hystograms at high D values).
We calibrated with extreme care the mean value of the scanning spot area and tested the reproducibility of all the photometric analysis (hard + soft) by scanning the same photogram many times and with various working conditions during several days. We obtained a rms error in the area determinations ranging from 2.3% to 7% for areas of 20 mm^2 and 0.3 mm^2 respectively.

To check the differences between this analysis method and the photographic isodensitometric one, we measured many filtergrams of our previous work (Falciani and Rigutti, 1972 a) and obtained a rms agreement of the order of $2 \div 10\%$ and a small systematic deviation of $1 \div 2\%$ in the mean values for $1.1 < j < 1.6$.
Finally we like to stress how powerful and flexible in comparison of similar ones is the present method of analysis, which supply all the wanted data (areas, isophotes, points determinations, etc.), keeps in a numerical matrix the information of the original photograms, viz. in a form particularly suitable to any further elaborations, and presents very simply any feed-back procedure between the first approximation results and the next steps of the work.

4.- SOME PRELIMINARY RESULTS

With the analysis of a series of solar Hα filtergrams we obtained the evolutive curves of Ej vs. time. Fig.3 show one example. From a very preliminary examination of the obtained evolutive curves we can confirm our previous results: before the flash phase of the flare a contraction of the whole active area, preceded by a sort of instability intensity fluctuations of the sorrounding plage, takes place. After the flash phase (characterized by the birth of small bright points inside the active region), an exponential decay of the emitted intensity vs. time, with higher time constants for higher j values is observed. However we can emphasize that intensity fluctuations of $\approx 26\ \%$ are always present in the studied active regions, in "quiet" plages too. The autocorrelations of these fluctuating evolutive curves show clearly a periodicity of about 2 min (1.5 ÷ 2.5). It should be necessary more uniform material to establish with higher precision this feature.

The isophotes maps confirm that the fluctuating points inside the active region are the same and there is a sort of diffusion of the perturbation from these fluctuating points to the outer parts of the plage. These very rough results are in agreement with the conclusions obtained through direct scanning of the solar images by Argo et al. (1973).

REFERENCES

Argo, P.E., Wulfing D.A., and Hildebrand, V.E.: 1973, NELC. Tech.Rep.1890, Naval Electr. Lab. Center, San Diego.

Azzarelli, L., and Panicucci,R.: 1972, Techn.Note C72-2, Institute for Elaboration of Informations-C.N.R. Pisa.

Carlesi, C., and Montanari, U.: 1973, Techn.Note C73-11, Institute for Elaboration of Informations - CNR Pisa.

Carlesi, C.: 1975, Int.Note C75-2, Institute for Elaboration of Informations - CNR, Pisa.

Falciani, R., Macris, C.J., and Rigutti, M.: 1972, Solar Phys. 26, 108.

Falciani, R., and Rigutti, M.: 1972a, Solar Phys. 26, 114.

Falciani, R., and Rigutti, M.: 1972b, Solar Phys. 28, 539.

REMOVAL OF VIGNETTING FROM APOLLO LOW LIGHT LEVEL PHOTOGRAPHS

Gregory C. Alvord

State University of New York at Albany, U.S.A.

Daniel A. Klinglesmith, Lawrence Dunkelman

Goddard Space Flight Center, U.S.A.

Robert M. Mercer

Dudley Observatory, U.S.A.

Abstract

The comparison of the vignetting function for Apollo 15, 16 and 17 35mm Nikon cameras is presented. A technique for removing this effect is described, and demonstrated on an image of the L_4 point of the Earth Moon system.

1. statement of the problem

In the collection of certain scientific data, it is convenient to use readily available film, lenses and cameras. When the phenomena is of such a light level as to require that the lens be wide open, an effect known as vignetting is observed. The effect changes the relative photometry at different points on the film. This effect must be accounted for before acceptable photometric analysis can be performed.

During Apollos 15, 16 and 17 under S211 and S158 pictures were taken while in lunar orbit with a 35mm Nikon camera with a f1.2 lens. The camera was mounted on a clamp in front of the window and exposed for from 15 to 240 sec., depending upon the

de Jager/Nieuwenhuijzen (eds.), Image Processing Techniques in Astronomy. 391-396.

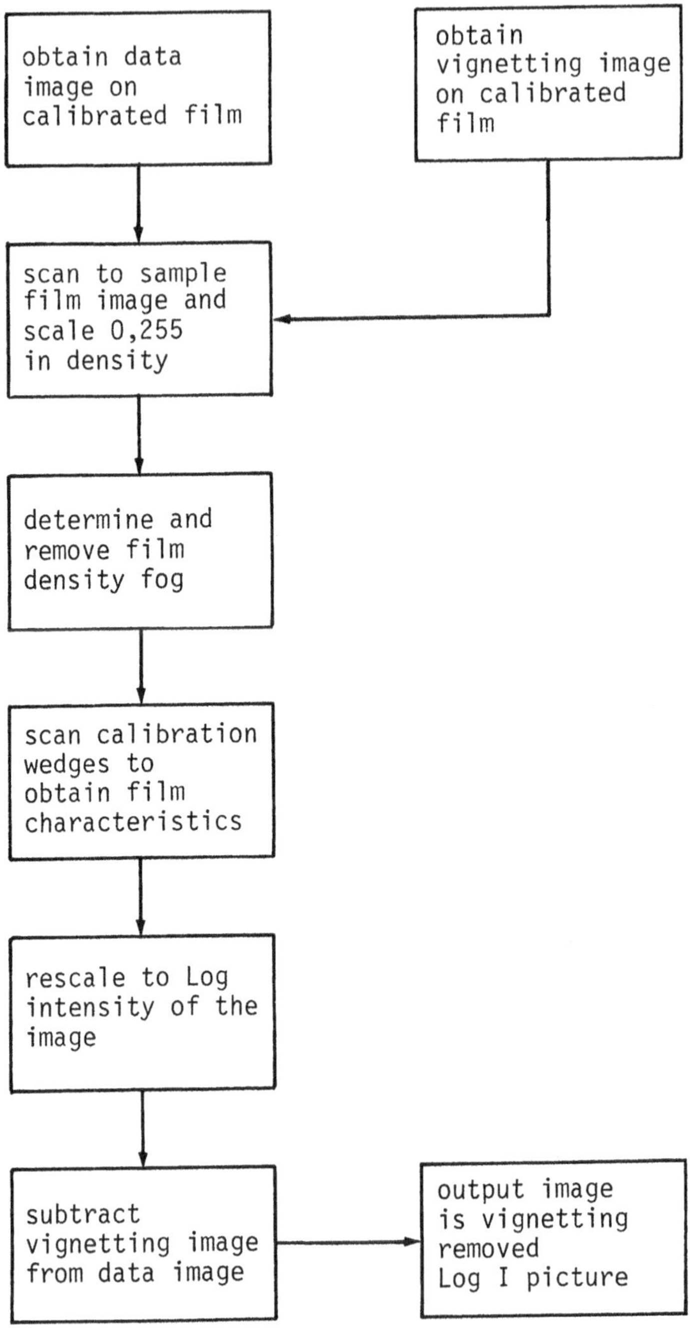

Figure 1. Flow chart for vignetting removal.

phenomena. The camera was loaded with Kodak 2485 film with an estimated equivalent ASA rating of 8000.

We sought an algorithm for removing vignetting which could be brought to fruition on the VICAR system. It is our intent to present that algorithm. We also shall present the properties of vignetting in general, and the properties which specifically relate to our test data.

2. technique for vignetting removal

Within the VICAR system we are restricted to 256 grey levels (eight bits of accuracy). We choose to make the best use of this limited accuracy by scaling the logrithmic intensity instead of the intensity. This provided us with the benefits that the simulation of position dependent attenuation could be accomplished by subtraction instead of integer devision. We can still maintain the absolute intensity definition by recording the intensity levels assigned to the scaled values. We can also do interframe comparisons by maintaining Log I constant for each unit grey level. Figure 1 shows the processing flow for the removal of vignetting.

3. properties of the vignetting function

Vignetting acts as if it were a neutral density filter uniformly varying with position. The net result is to rescale the intensity image as a function of position before the intensity image is recorded as density on film. To remove the effect of vignetting we must replace the uniform scaling which was in the intensity image before the image entered the lens.

We measured the vignetting function by photographing a uniform flat field. Figure 2 is a digitized display of the Log intensity images. Figure 2 has been contoured by setting every eighth grey level to 255. Note the shape of the contours. The shape is not the smooth circle that theory predicts. The contours are every .17 in Log Intensity. Note the wide band caused by very grainy film used to capture the vignetting function. Figure 2a is a single frame from Apollo 15 whereas 2b, 2c and 2d are averages of several frames.

It was our hope to see the same shape with the three lenses used. However, Figure 2b, c and d show the different shapes of the Apollo 15, 16 and 17 lenses respectively. These figures are contoured averaged vignetting functions. Averages were made between several exposures from the same flight. This averaging reduced the grain noise as evidenced by the thinner contour bands. It was concluded that only the average vignetting for a single flight could be used to remove vignetting.

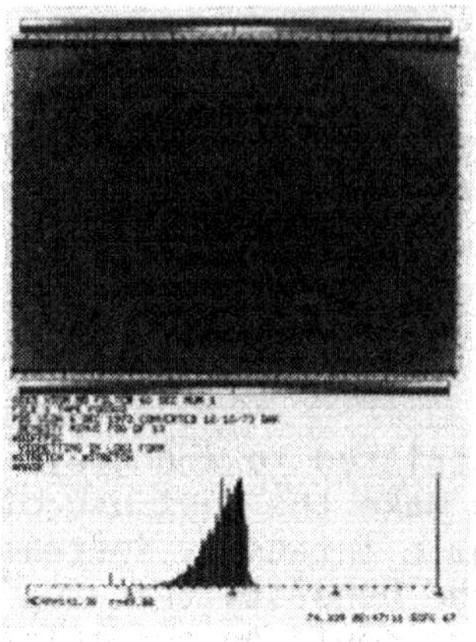

(a)

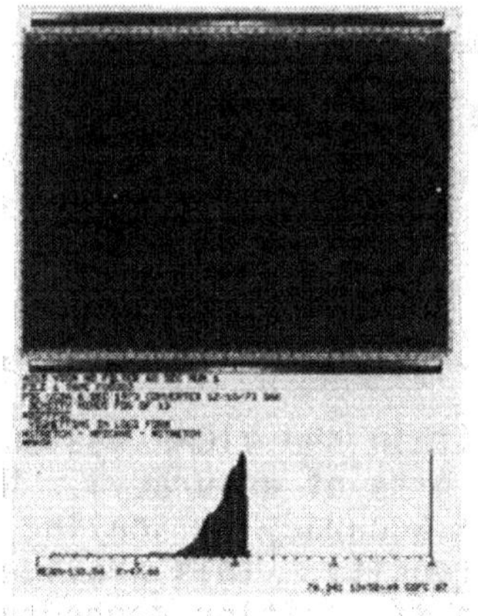

(b)

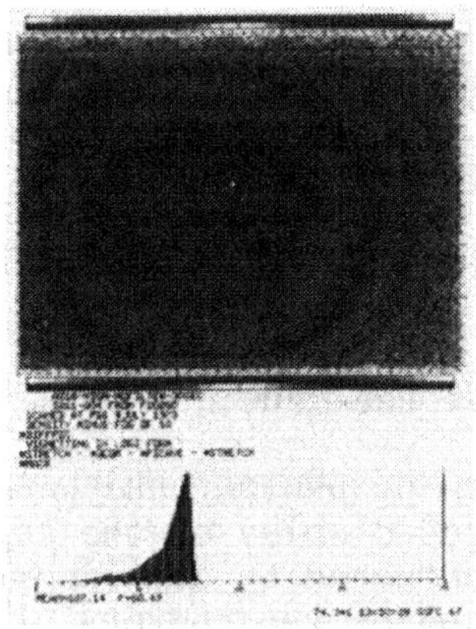

(c)

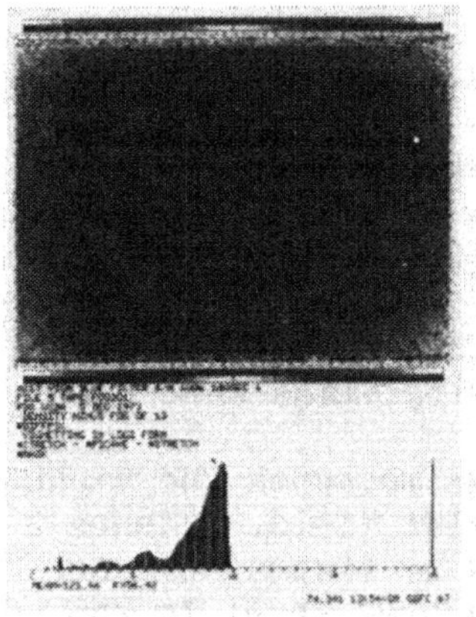

(d)

Figure 2. Contoured Vignetting Images: 2a, single frame, note the noise level in the contour lines. 2b, an average of 6 frames fram Apollo 15. 2c, an average of 2 frames from Apollo 16. 2d, an average of 3 frames from Apollo 17.

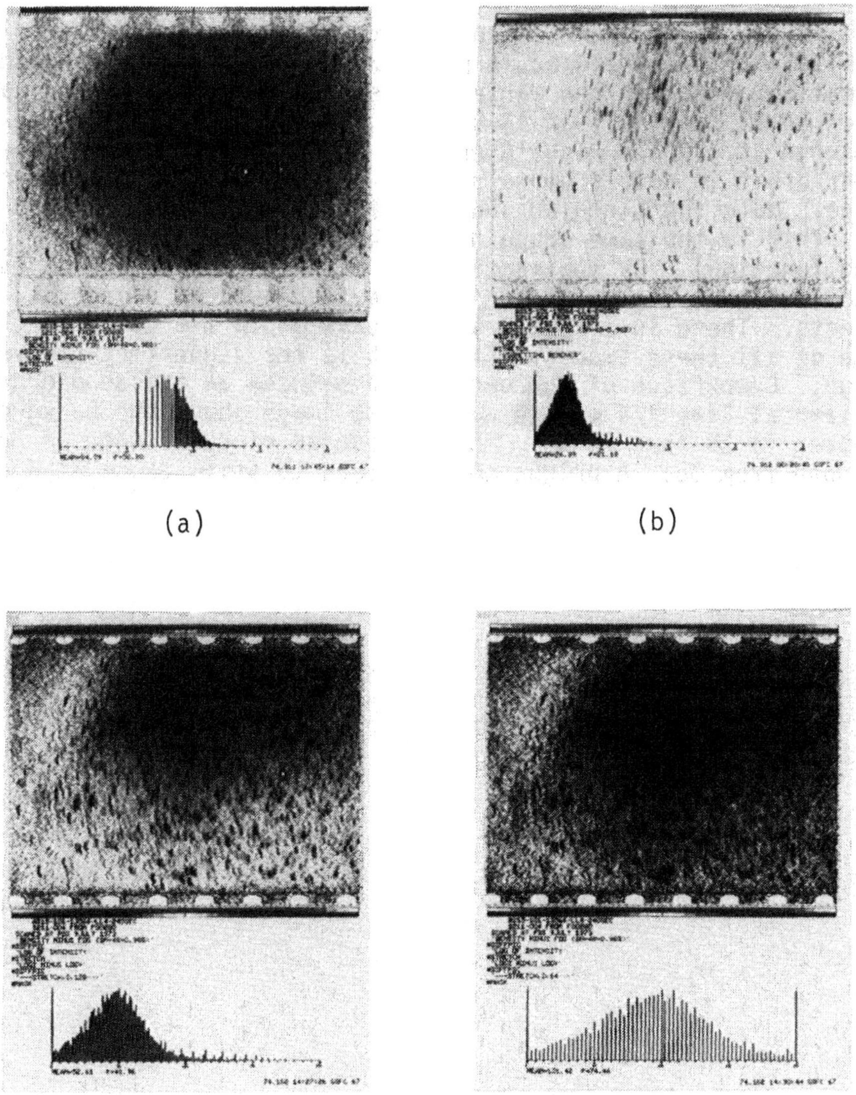

(a) (b)

(c) (d)

Figure 3. The effects of Vignetting removal. 2a, an uncorrected image of Lunar Libration point L_4. The digital numbers are linearly related to the logarithm of the sky brightness. 3b, vignetting removed image. 3c and 3d contrast stretched images of 3b. The lack of image in the upper left hand corner is due to an obstruction in the spacecraft window.

4. examples of results

Figure 3a through 3d are results of processing frame AS15-101-13566-LL4-240sec, a photograph of the fourth langrange libration point for the earth moon system. The photograph was taken on 31 July 1971 at 13:37 GMT. The 240 second exposure was centered at $23^h\ 15^m$ Right Asension -3° declination taken from lunar orbit in double umbra shadow. Figure 3a is a Log intensity image. Note the circular shape to the "birght" (dark) region in 3a. This is the same shape one sees in figure 2 (the A15 vignetting function). By subtracting the data in Figure 2b from the data in Figure 3a, we effectively devided out the vignetting effects. There is a spacecraft oscuration in the upper left side of all these images. Figure 3b is the vignetting removed image. Comparison of the background between 3a and 3b along a horizontal line 1/4 of the way up the image shows the background flatter in 3b than in 3a. The pronounced circular edge of 3a is gone from 3b. A bright spot appears to still exist at center in the upper portion at 3b. This region is asymetric and off axis so thus may well be the L4 dust cloud.

SEARCH OF SIGNIFICANT FEATURES IN A DIRECT NON PARAMETRIC PATTERN RECOGNITION METHOD. APPLICATION TO THE CLASSIFICATION OF A MULTIWIRE SPARK CHAMBER PICTURES.

R. Buccheri[°+], P. Coffaro[°+], G. Colomba[*], V. Di Gesù[°], S. Salemi[°]

° Istituto di Fisica dell'Università di Palermo
+ Laboratorio di Fisica Cosmica e Tecnologie Relative C.N.R.
* Cassa di Risparmio, Palermo

ABSTRACT. We give preliminary results of the application of a direct non parametric pattern recognition method to the classification of the pictures of a multiwire spark chamber. The method, developped in an earlier work. for an optical spark chamber, looks promising. The picture sample used has, respect to the previous one, the following characteristics: a) the event pictures have a more complicated structure; b) the amount of background sparks in an event is greater; c) there exists a kind of noise which is almost always present in some structured way (double sparkling, bursts, ...). New features have been used to characterize the event pictures; the results show that the method could be also used as a super filter to reduce the cost of further analysis.

1. INTRODUCTION

The analysis of the data obtained from spark chamber experiments used to detect gamma-ray events by the electron pair resulting from the materialization process, is generally performed in two fundamental steps. In the first a selection is made in order to discriminate between good gamma rays and background events. In the second step the good gamma pictures are used to infer numerical values for the evaluation of the physical parameters. The necessity of the automatization of this second step has been realized rather quickly due to the really complicated number of operations to be performed. Conversely in the first step the phylosophy of the "manual editing" has been adopted for a long

de Jager/Nieuwenhuijzen (eds.), Image Processing Techniques in Astronomy. 397-402.

time because the cost of the automatization of the selection procedure could possibly not justify the loss in efficiency respect to the human scanning;this can be true if the amount of data is not so large as it is in the case of balloon borne experiments.
In gamma-ray astronomy the problem of the automatic selection of pictures has increased in importance when satellite experiments using very long observation times, began to supply experimenters with very large amount of data. As an example, the COS-B satellite will transmit, in two years lifetime, something like one or two millions of pictures in which the percent of contamination by spurious events is foreseen very high. In this case the "visual selection" by human scanners is certainly extremely costly and long;automatic procedures could compensate with their high speed of analysis the eventual loss in efficiency of selection.

2. METHODS OF SELECTION

The problem of the automatic selection of pictures obtained from spark chamber experiments can be approached in two ways :
i) with "parametric methods" by using the knowledge of the physical process involved in order to obtain sintactical rules useful to describe the event pictures. These rules are then applied to the unknown events to perform the selection;
ii) with "non parametric methods" by using a set of pre-classified events from which the characterizing features are extracted.
The classification of the unknown events is then performed by means of a decision function which operates on the values of the features extracted.
The "non parametric methods" have the following main characteristics :
- modularity of the computer programs in the sense that they can be adapted with small changes to a wide class of experiments or to the same experiment if variations are introduced;
- possibility of shortening the computation times by using a simple decision function (linear for example) without sensible losses in efficiency.
The possibility of reaching satisfactory results in the selection efficiency is related to the finding of mathematical parameters which optimize the separation between the two classes of events (good events, here denoted as γ and noise events, here denoted as $\bar{\gamma}$). On the other hand for parametric methods the performance is conditioned by the complexity of the process;in fact too complicated physical processes could require heavy schematizations leading to a loss in efficiency.

3. APPLICATION OF A PARAMETRIC METHOD TO A SAMPLE OF DATA OF THE COS-B WIRE SPARK CHAMBER.

3.1. Parameters used

In the following we give some preliminary results on the application of a non parametric method of automatic selection to the data obtained during a calibration run of the COS-B spark chamber For information about COS-B spark chamber see Bignami et al. (1974). Two classes of parameters have been used in this application. The first already applied to the data obtained from an optical spark chamber by Buccheri et al. (1975), takes into account combination of sparks in a section of the event picture. This kind of parameter can be considered "robust" because its definition is independent on the particular structure of the image and we can expect rather equivalent selection efficiency when applied to different classes of data. The second class of parameters used are "ad hoc", defined in function of the double track structure of the gamma ray event picture : for these we can expect good selection efficiency only in very restricted cases.

3.2. The method

The method emploied here is fully explained in Buccheri et al. (1975) ; we will briefly recall its fundamental phases.

- The first phase consists in a preprocessing of the event pictures in order to find the most significant zone. In our case we have defined as most significant zone the fraction of the whole image consisting in a 6x12 matrix in which is contained the highest number of pairs of sparks (see figs. 1 and 2). The condition to find in the 6x12 matrix a minimum of four sparks and at least two pairs of sparks is used as a first raw filter to eliminate clearly spurious events.

- In the second phase a number of preprocessed events pre-classified by a human scanner, are used to extract numerical values for the parameters chosen. In the case of the parameter "combination of sparks" the 6x12 matrix is subdivided in 12 segments six bits long and the combination of sparks inside each segment, coded as a number varying between zero and 2^6-1, is recorded. From the analysis of a number N1 of γ events and N2 of $\bar{\gamma}$, we have derived a series of 12 training hystograms (one for each of the 12 segments) in which the abscissa gives the code of the combination of sparks and the ordinate the difference

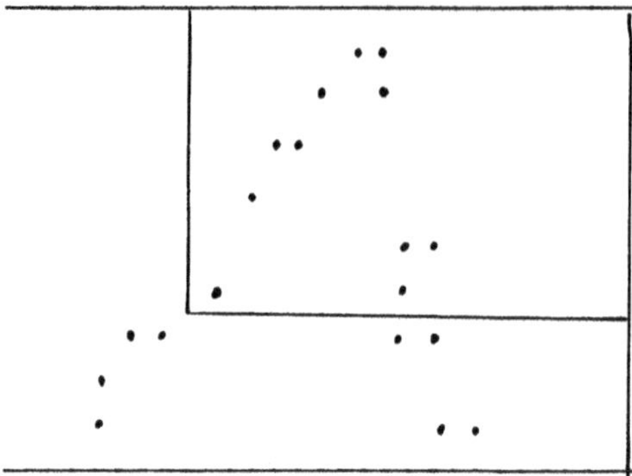

Fig. 1. Tipical γ-event of the sample used

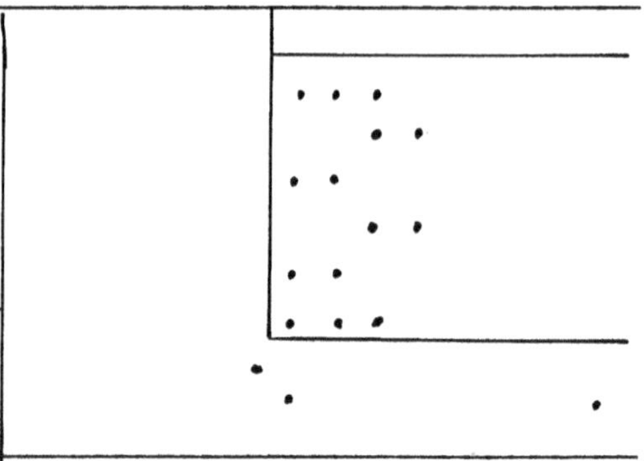

Fig. 2. Tipical $\bar{\gamma}$-event of the sample used

($a_{\gamma} - a_{\bar{\gamma}}$) between the frequency of occurrence of a given code for the γ-events and for $\bar{\gamma}$-events. Fig. 3 gives an example of training hystogram. In the case of the parameter "Number of sparks" the whole image is analised. Here the training hystogram has in abscissa the number of sparks in the whole image and in ordinate the difference ($b_{\gamma} - b_{\bar{\gamma}}$) between the frequency of the γ-events having a given number of sparks and the frequency of the $\bar{\gamma}$-events having the same number of sparks. In the case of the parameter "Chi-square" the whole image is subdivided in 8 equal parts and the number of sparks N_i in each of the parts is recorded; the chi-square value is computed

$$\chi^2 = \sum_{i,1}^{8} \frac{(N_i - \bar{N})^2}{\bar{N}}$$

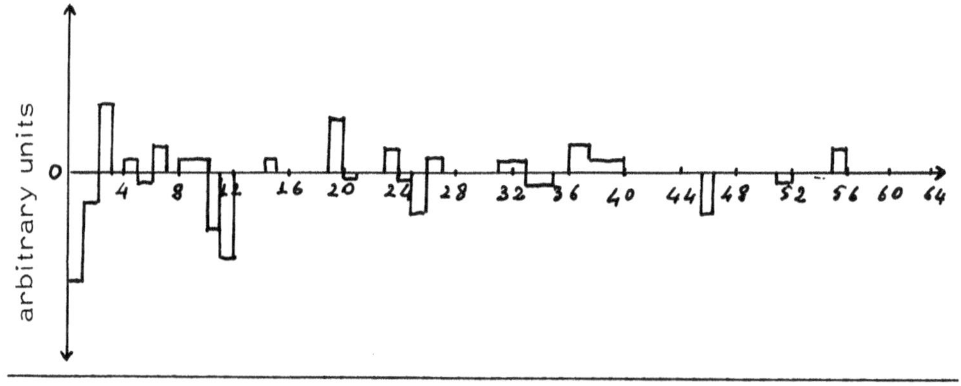

Fig. 3. Training hystogram for the parameter "Combination of sparks" relative to the segment n. 1. It has been derived analysing 243 γ-events and 243 $\bar{\gamma}$-events.

where $\bar{N}$ is the average value of the N_i's. In this case the training hystogram has in abscissa intervals of values for χ^2 and in ordinate the difference $(c_\gamma - c_{\bar{\gamma}})$ between the frequency of the γ-events for which χ^2 is contained in a fixed interval and the same frequency for the $\bar{\gamma}$-events.

- In the third phase the selection is performed by using a linear decision function and a fixed threshold value ϕ according to the following criterium

$$(d_\gamma - d_{\bar{\gamma}}) \begin{cases} \geq \phi \longrightarrow \gamma \\ < \phi \longrightarrow \bar{\gamma} \end{cases}$$

where $(d_\gamma - d_{\bar{\gamma}})$ is the recorded difference between the frequency of occurrence of the measured value of the chosen parameter in the unknown event for γ-events and the same value for $\bar{\gamma}$.

4. DATA ANALYZED AND RESULTS OBTAINED

The sample of data used for the application of the method illustrated above is made of 891 events (434 γ + 457 $\bar{\gamma}$) obtained from an accelerator calibration run of the COS-B wire spark chamber and kindly supplied by the "Caravane Collaboration". Two examples of event images are given in figs. 1 and 2. The preprocessing phase (not yet optimized) applied to the entire sample has given as result the exclusion of 269 events of which 55 γ and 214 $\bar{\gamma}$. The training hystograms have been derived for the three types of parameters illustrated in 3.2. by using the 243 $\bar{\gamma}$-events retained after preprocessing plus 243 out of the 379 γ-eventsreteined after the preprocessing; they have then been used for the classification of the complete sample of the 622 events retained after preprocessing. The results are expressed in the figs. 4, 5, 6. Fig. 4 refers to the use of the first

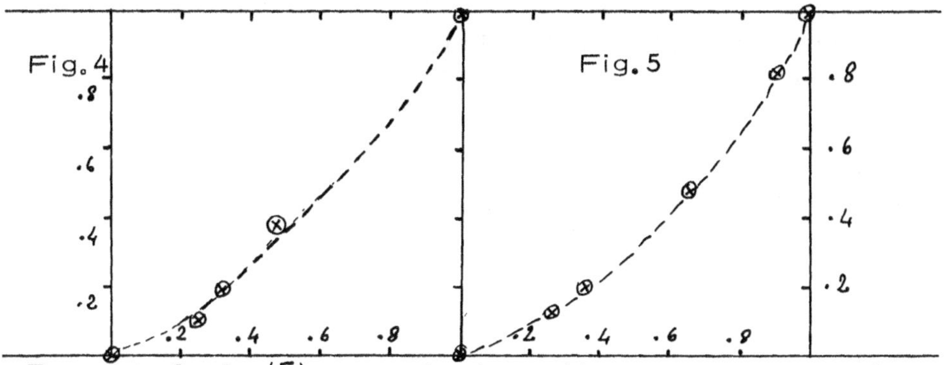

Percent of noise($\bar{\gamma}$) recognized as γ versus percent of true gammas selected as such. Fig. 4 refers to the parameter "Combination of sparks", fig. 5 to "Number of sparks".

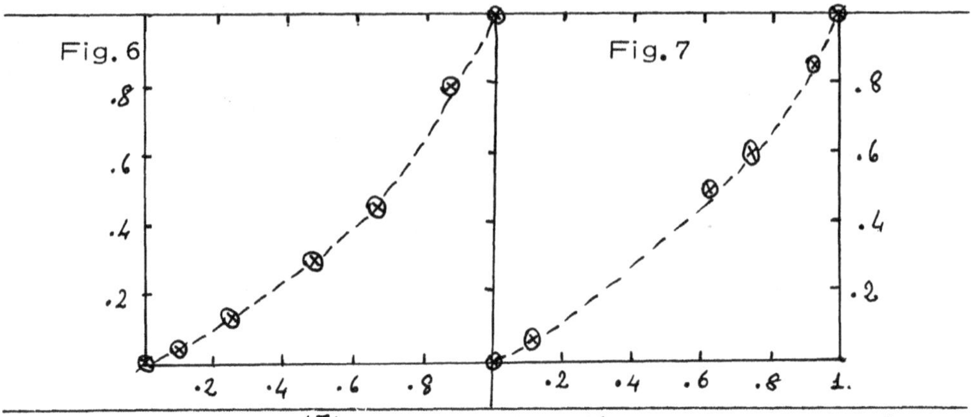

Percent of noise ($\bar{\gamma}$) recognized as γ versus percent of true gammas selected as such. Fig. 6 refers to the parameter "Chi-square", fig. 7 to "Combination of sparks" (all the segments).

one of the segments of the 6x12 matrix chosen in the preprocessing phase, fig. 5 refers to the parameter "Number of sparks" while fig. 6 refers to the parameter "Chi-square". In all curves the abscissa gives the percent of γ recognized as such while the ordinate gives the percent of $\bar{\gamma}$ classified as γ. Fig. 7 shows the selection efficiency by using the sum of the 12 segments.

REFERENCES

1. G. F. Bignami, G. Boella, J. J. Burger, P. Keirle, H. A. Mayer-Hasselwander, J. Paul, E. Pfeffermann, L. Scarsi, B. N. Swanenburg, B. G. Taylor, W. Voges, R. D. Wills. Proceedings of the 9th ESLAB Symposium on "The context and status of gamma-ray astronomy", 10-12 June, Frascati, Italy.
2. R. Buccheri, V. Di Gesù, S. Salemi. Nuclear Inst. & Meth., 123 (1975) p. 563.

LOCK-IN IMAGE SUBTRACTION: DETECTABILITY OF CIRCUMSTELLAR PLANETS WITH THE LARGE SPACE TELESCOPE*

D. Bonneau, M. Josse, and A. Labeyrie

Observatoire de Meudon, France

Faint circumstellar features in the close vicinity of stars or qso's are of considerable astrophysical interest since they can provide important informations on stellar parameters and evolutions : stellar companions give mass information ; coronae many of which are likely to be much brighter than the sun's (ref : Betelgeuse) give information on magnetic fields and fast varying phenomena at stellar surfaces ; planetary nebulae may provide information on the birth of planets ; disk structures similar to Saturn's rings provide checks on theories which predict their frequent occurence ; planets themselves of course are worth appreciable detection efforts since they may ultimatly provide hints on the existence of extra-terrestrial civilisations. In this article, we discuss on coronographic and lock-in detection techniques applicable in the near future to observations with the Large Space Telescope. Those techniques enhance the visibility of circumstellar features to such a degree that even planets should become marginally detectable in some favorable cases.

The feet of the spread function.

It is generally expected that Large Space Telescopes will be diffaction - limitted or nearly so, at visible wavelengths. This means that the central part of the spread function will somewhat resemble the Airy disk with

---

* This is a more detailed version of material previously submitted by A. Labeyrie as a proposal to NASA (1973)

de Jager/Nieuwenhuijzen (eds.), Image Processing Techniques in Astronomy. 403-409.

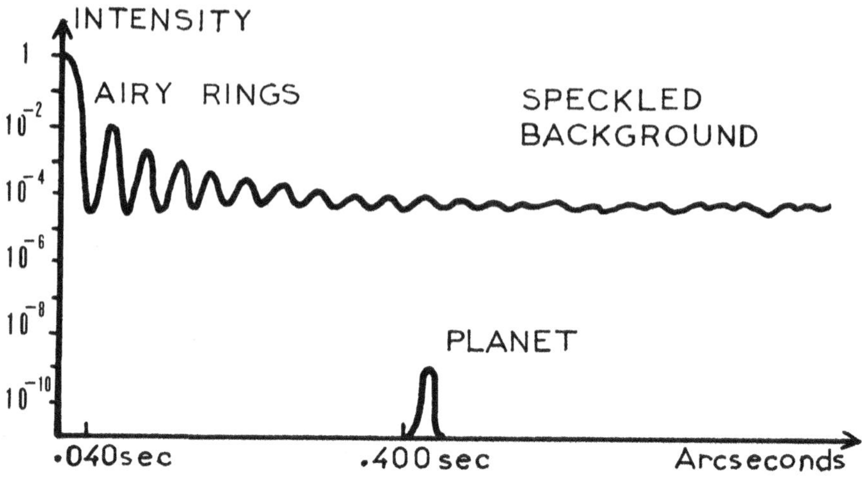

Fig. 1 : Typical spread function to be expected with the Large Space Telescope. A circumstellar planet is about 10^4 times fainter than the scattered light level.

surrounding rings. As always, the theoretical pattern of rings is increasingly affected by small scale mirror imperfections at increasing distances from the center. Eventually the ring pattern disappears among a speckled background of scattered light, having a level on the order of 4 X 10^{-5} relative to the central peak intensity with excellent surface quality. The transition between organized rings resulting from diffraction at the aperture boundaries, and random speckles resulting from orange-peel bumpiness in the $\lambda/100$ amplitude range, generally occurs in the 10-20 th ring region when dealing with a very good mirror. Scratches, dirt, pinholes and other localized defects contribute to background far from the central peak but have a negligible contribution within on arc-second. This mainly concerns the region between rings n°5 and 30 where background originates predominantly from edges effec and from middle scale bumpiness. With a 3-meter LST, 5000 light, the 8 th ring has a radius of 0.3 arc-second on the sky plane.

Lyot filtering and apodization.

Apodization techniques have been increasingly used since Lyot (1930) and demonstrated their value by observing the solar corona which is 10^6 times fainter than the photosphere. These techniques decrease the brightness of the Airy rings, at the expense of a slight resolution loss,

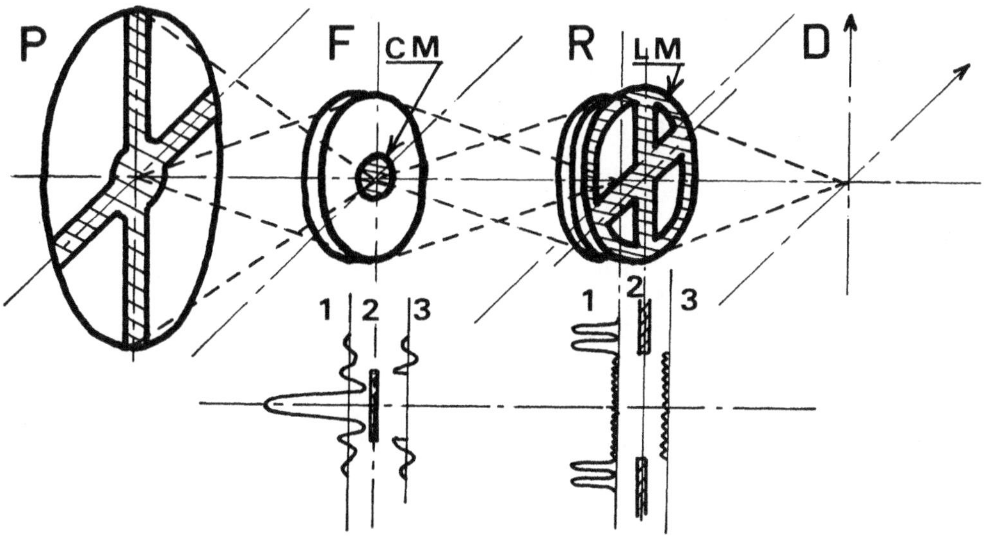

Fig. 2 : Optical arrangement of a laboratory simulation.
P : Primary mirror ; F : Field Lens ; R : Relay lens ;
D : Detector plane ; CM ; Central mask or eclipsor ;
LM ; Lyot mask or apodizer.
1 : before the mask
2 : mask position
3 : After the mask

by smoothing the sharp aperture boundaries. The spatial filtering technique evolved by Lyot achieves this result by means of two masks located respectively in the focal plane and in the relayed aperture plane. As shown in fig.2 the focal plane mask is an opaque disk or eclipsor, usually carried by a field lens or transparent plate. In our case it serves to eliminate the central peak and first few rings of the image when looking at a bright star to search faint surrounding features. The aperture mask or apodizer serves to eliminate those parts of the aperture which produce the Airy rings and diffraction spikes in the image of the bright observed star ; it may consist of a graded spatial filter or more simply a diaphragm. Other apodization techniques involve a single mask located in the aperture plane, and serving similarly to smooth down the sharp boundaries of the aperture. Handling problems with a 3-meter apodizer obviously make it difficult to use in the case of a multi-purpose space telescope, besides no appreciable again is expected with respect to Lyot-type devices. The Lyot technique is described in more detail in appendix 1. Generally speaking these techniques decrease by one or two orders of magnitude the brightness of the diffraction rings and spikes in the above defined region of the spread function. The improvement is obviously not sufficient for direct detection of circum-stellar planets, but it is a valuable first step to be performed before applying the following "blink" analysis.

Lock-in detection.

In the presence of residual scattered light surrounding the primary star image, celestial features much fainter than the local scattered light level itself can be detected by means of techniques related to the principles of the blink-microscope or lock-in detection devices wide used in infra-red research and other experimental arrange ments. Indeed, the mottled or rather speckled pattern of the scattered light is fixed in position angle with respe to the telescope, whereas any celestial feature among it rotates when the telescope is rotated in position-angle. Thus, images recorded at different position-angles can eliminate the scattered contribution from the primary sta Repeated cycles of exposure and rotation further improve the detection by eliminating possible variations of the telescope spread function, that might be caused by thermal defocusing or other alignment drifts. A typical observing procedure would thus consist in rolling the telescope back and forth between two position angles ever minute or so, guiding being maintained on the bright primary star to keep it centered on the eclipsor mask. Assuming a photon-counting television sensor and comuter integration, considerable sensivity can be achieved by respectively adding and substracting images of odd and ev orders. This has certain analogies with the use of a wobbling secondary mirror, which became to be adopted as a basic tool of infra-red astronomy.

Signal/Noise ratio estimate.

A mv = 2 star give about 10^9 photo-electrons on the catho of a photoelectric sensor, using a 3 - meter space teles-cope, 4000 Å spectral bandwidth centered at 5000 Å and a 20ms exposure, most photons are centered on the central Airy peak. The scattered light level due to residual mirror bumpiness is on the order of 4.10^4 photon-events per second per pixel. A typical planet at 10^{-9} relative luminosity thus provides 1 additionnal event in the corresponding pixel. Extracting this faint contribution from the 4×10^4 times brighter scattered background requires integrating a total of $(4 \times 10^4)^2 \simeq 10^9$ photons-events per pixel in this region. This requires slightly less than one hour of observing time. The estimate assumes a photon-limited sensor, and it should be remarked that the best photon-counting image tubes accomodate 10^3counts/pixel/frame in the digital mode. Analog operation may be preferable in this example involving a fairly bright primary star. However, the estimate remains valid assuming 6-bit digitization of the analog TV signal. Therefore, it appears that detecting circumstellar planets will not be hopeless once space telescopes in the 3-meter range of size will be available. Observations carried out according to this principle are generally applicable to binary stars detection. In addition to faint companions, they will present circumstellar matter in other forms such as disks, nebulosities, coronae, jets, etc... Applied to galaxies or QSO's, lock-in observations will also help detecting galactic disks in the vicinity of comparatively bright sources.

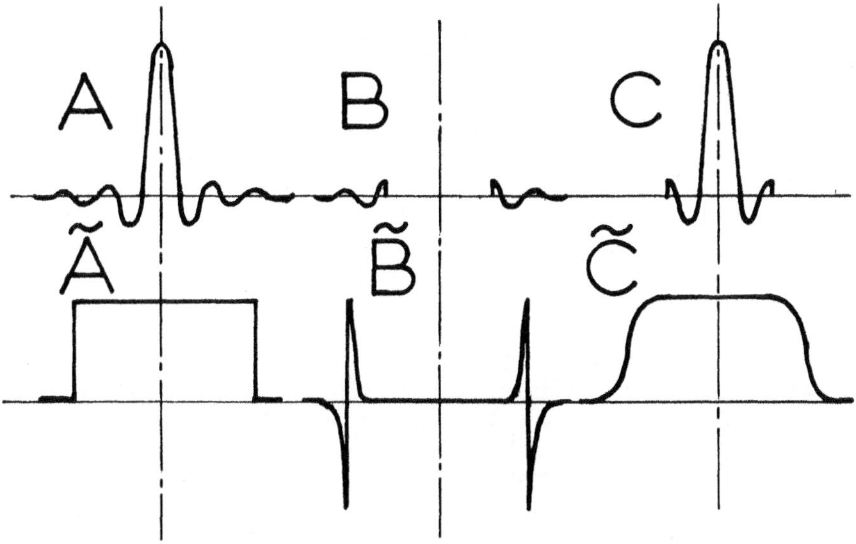

Fig. 3 : Distribution of complex amplitudes
A : Before the eclipsor; B : after the eclipsor;
C : amplitude eliminated by the eclipsor.
$\tilde{A},\tilde{B},\tilde{C}$ are the respective Fourier Transforms of A,B,C.
B = A - C the F.T. being a linear transform then :
$\tilde{B} = \tilde{A} - \tilde{C}$

APPENDIX 1

Spatial filtering analysis of the Lyot coronograph. As represented in fig. 3, the distribution of complex amplitudes in the star image occulted by the eclipsor mask may be considered as a difference of two terms. In the Fourier space, these terms correspond respectively to the system's aperture and to a convolved version of it, as imaged through a diaphragm the size of the eclipsor mask. The subtraction of these two terms gives the amplitude distribution in a relayed image of the aperture. As represented on fig. 2, this distribution is zero everywhere except along the aperture edges, a bright fringeappears on each side of the geometrical image. These edge perturbations are responsible for the Airy rings appearing around the eclipsor in a relayed image of the focal plane. The two bright edge lines are suppressed by a diaphram slightly smaller than the geometric image of the aperture. This gives a final image where Airy rings have disappeared. Field sources are not occulted by the first mask and the spread function remains a standard Airy disk with rings. Circumstellar objects, then, are unaffected by the masks, except for a slight reduction of the apparent aperture size. In the case of apertures featuring secondary obscuration and spider structure, the above arrangement also suppresses the diffraction spikes created by the spider around the primary star image. Considering utility of an eclipsor mask to avoid tube saturation or burn-out, and the fact that scattered light is cut down by one or two orders of magnitude, the coronographic masks appear as very useful accessories for operating a LST.
Following laboratory test with small spherical mirrors, one of us (A. Labeyrie) has used a 150 cm mirror figured by J. Texereau for a Schmidt telescope. Lyot filtering had been achieved in the image of a laser-illuminated pinhole, residual scattered light was found to originate mainly from air turbulence in the relatively stable laboratory. For a good estimate of the scattered light level with a state-of-the-art mirror, a vaccum tank is therefore necessary.

APPENDIX 2

The spread function at short ultra-violet wavelengths. The spatial filtering accessories described above as part of a NASA/LST proposal are intended essentially for operation at visible and near I.R. wavelengths.

In the U.V. region, scattered light from mirror bumpiness increases very fast (in principle λ^{-4}). Also, aberrations due to the low-frequency bumpiness components begin to affect the Airy peak and rings significantly, unless improbable pain is taken to insure "diffraction-limitted" quality at U.A. wavelengths. In these conditions, the Lyot mask loose much of their appeal, but a different form of image analysis becomes worth considering as discussed next.

Little experimental data is avalable on the spread function of large telescopes at short U.V. wavelengths (1000 Å). With $\lambda/10$ RMS mirror bumpiness at 5000 Å, random phase shifts approaching 180° are to be expected on the wave at 1200 Å, the Airy disk is thus likely to be completely destroyed. The angular spread can vary between wide limits depending on the power spectrum of mirror defects. In any case, however, the monochromatic spread function necessarely features speckles similar to those generated by the atmosphere in groune-based telescopes, but here they are fixed in time. Applying the principles of speckle interferometry, although in a simplified form in this case, the diffraction-limited resolution can be retrieved when dealing with objects of simple structure. In such conditions, the resolution attainable with a 3-meter telescope at 1000 Å is 0.008", five times better than at 5000 Å. Among other unexpected features, this is enough to show halos of Lyman α emission or absorption around the disk of close stars. The apparatus required for performing this experiment amounts to a focal estender and filters. Il can be combined with the Lyot filtering device and requires only one moving element.

References

Lyot : Comptes rendus, 191, 1930, p 834

A. Labeyrie : Astronomy and Astrophysics, 6, 85-87(1970)

DIGITAL IMAGE CENTERING, I.

W. F. van Altena* and L. H. Auer

Yale University Observatory, New Haven, Ct., USA

ABSTRACT. A series of parallax plates have been measured on a PDS microdensitometer to assess the possibility of using the PDS for precision relative astrometry and to investigate centering algorithms that might be used to analyze digital images obtained with the Large Space Telescope. The basic repeatability of the PDS is found to be $\pm 0.6\,\mu m$ with the potential for reaching $\pm 0.2\,\mu m$. A very efficient centering algorithm has been developed which fits the marginal density distributions of the image with a gaussian profile and a sloping background. The accuracy is comparable with the best results of the Yerkes Observatory photoelectric image bisector.

1. INTRODUCTION

Despite its comparative simplicity, the problem of accurately finding the center of an image is of vital importance for astrometry. A variety of analogue centering systems have been developed for astrometric measuring engines, but these cannot be applied to the digitized images we expect to receive from sensors on the Large Space Telescope (LST). If the astrometric potential of the LST is to be realized, it is important that we have available accurate, efficient algorithms for finding the centers of digitized images. Even for processing photographic images, digital dissection offers potential increases in speed and accuracy and, thus, deserves investigation.

---

*A portion of van Altena's research was conducted at the Yerkes Obs.

de Jager/Nieuwenhuijzen (eds.), Image Processing Techniques in Astronomy. 411-418.

In this paper we present results on the digital dissection of images measured on the PDS microdensitometer. We address two basic questions: 1) Is the PDS microdensitometer capable of doing precision differential astrometric measurements and 2) Can the numerical processing be done in reasonable amounts of computing time?

2. MEASUREMENTS

Through the courtesy of Dr. James D. Wray and Dr. G. F. Benedict the NASA-Skylab PDS microdensitometer located at the University of Texas at Austin was made available for measurement of a Yerkes Observatory 40-inch refractor parallax series previously measured with the Yerkes photoelectric image bisector. This series, which consisted of 25 images on each of 26 plates, was selected since it is one of the highest accuracy series measured so far in the new Yerkes Observatory parallax program and has a wide variation of image quality on the various plates. If a standard PDS can equal the accuracy previously obtained, then it is probable that one prepared for astrometry could be substantially better.

The plates were positioned on the PDS so that an image was always measured at the same part of the scales, within 100 μm. This procedure minimized the importance of way and scale errors on the final positions. Due to certain mechanical considerations, the pre-slit on the PDS was set for a 10 μm square thus defining the measurement axis of the system, while the post-slit acted solely as a light collector with an aperture of 100 μm square. The magnitude equation introduced by the PDS was minimized by scanning at a rather low speed, the measurement of each image taking approximately 30 seconds. The measurement time can be reduced by optimizing the scan speed and reducing the number of raster scan lines per image.

Since it is always possible to degrade the data resolution by subsequent averaging of adjacent data points, a very fine raster pattern was adopted. Each "measurement" consists of an array of 120 lines of 5 μm boxes spaced by 5 μm in the y-direction yielding an array with dimensions of 120 x 120 covering 600 x 600 μm on the photographic plate.

Initially, the choice of using the density mode instead of the transmission mode of the PDS was rather arbitrary, however, it has become apparent that the density mode is preferable. As described below, the density profiles are nearly gaussian and can be handled in an extremely stable manner by curve fitting. An experiment using transmissions as input to the curve fitting program has shown this mode to be both less accurate and less stable than the density mode. The use of the curve fitting algorithm with transmissions simulates the approach used for centering by measuring machines using scanning slits. Høg and Wiskott (1974) have previously shown the use of scanning slits to be less precise than digitally dissecting the image.

3. IMAGE ANALYSIS

Image positions were derived from the density measurements by two different techniques. The first, and most straight forward approach is to determine the moments of the density distribution, while the second approach utilizes a curve fitting technique.

3.1 The Moment Analysis

The plate background was removed by averaging the density in each corner of the array around an image and fitting a sloping background to these four points. A vector of average x-positions were computed by means of Eq. (1).

$$\bar{x}_y = \sum_x x . w(x) . D'(x,y) / \sum_x w(x) . D'(x,y) \tag{1}$$

where D'(x,y), is the density at point x, y after the background has been subtracted, and w(x) is a weighting function taken as w = 1 from the image center at 0.5 standard deviations (assuming the image density profile to be gaussian in shape) and then decreasing linearly to zero at 2.5σ. This elementary weighting function is intended only to minimize the noise in the wings of the image. The $\bar{x}_y$ are then averaged to form the mean for the whole image using Eq. (2).

$$\bar{x} = \sum_y \bar{x}_y . w(y) . N(y) / \sum_y w(y) . N(y) \tag{2}$$

where $N(y) = \sum_x w(x) . D'(x,y)$. (3)

In the first pass, all weights are set equal to 1.0 while in iterations 2-6 they are allowed to take their computed value. $\bar{y}$ is similarly computed.

The computer time associated with producing the x,y coordinates, 2nd-4th order moments, and a magnitude measure from $m = \sum_{x,y} D'(x,y)$ was 1.1 seconds per star on an IBM 370-168 computer. It appears that cell sizes larger than 5 μm may be useable without loss of accuracy, since no change in image position greater than 0.2 μm was noted for cell sizes of 5, 10, 15, and 20 μm for a sample of images. If the data were taken on 20 μm centers instead of 5 μm centers then the computing time should drop by about a factor of sixteen. One difficulty encountered with the moment analysis is that the algorithm was moderately unstable when faced with very weak images or images far from the center of the array. This may in part be due to the procedure used to remove the background, or to the use of a spatial weighting function, however neither of these possible explanations has been explored.

3.2 The Curve Fitting Analysis

The curve fitting approach involves strip summing the density measures to form marginal distributions in x and y: $\rho_x(x) = \sum_y D(x,y)$ and $\rho_y(y)=\sum_x D(x,y)$. The marginal distributions in x and y are then independently least squares fit to the sum of a linear and a gaussian function.

$$\rho_x(x) \sim a_x + b_x\, x + [N_x/(2\pi)^{1/2}\sigma_x].\ \exp[-0.5\ (x - x_c)^2/\sigma_x{}^2] \quad (4)$$

The form of Eq. 4 follows directly from the model that D(x,y) is the sum of a linear background and gaussian image; it holds even if the image axes do not lie along the x and y coordinates.

Initial values of the parameters are found by moment analysis of ρ_x and ρ_y. Corrections are found by iterative least squares solution of the linearized form of Eq. 4. Typically three iterations are sufficient. The fit is found in 0.4 sec on an IBM 370/158, which means the curve fitting approach is more than a factor of 10 faster than the moment analysis (Note the computations were run on different computers).

The curve fitting proved to be very stable on nearly all types of images irrespective of their position in the array. Fig. 1 shows a typical fit obtained by this method. The points are the observed marginal in units of average specular density and the solid line the least squares fit. The total range of the abscissa is 600 μm. The excellent fits obtained with gaussian profiles was unexpected. As with the moment analysis, degrading the resolution to 20 μm does not affect the values of x_c and y_c but does reduce the computing time.

4. POSITIONAL REPEATABILITY OF THE PDS MICRODENSITOMETER

The repeatability of the PDS has been estimated in several ways. Unfortunately, it is not possible to give a definitive result since the desirability of checks was not apparent until the measurements were completed, and, thus, only a few repeat measurements were made.

a) The first two plates measured, which were later judged to be the worst and fourth worst of the 26 plates, were measured twice in the following manner. Once an image had been raster scanned the raster scan was immediately repeated starting one line lower than before. We therefore have two raster scans of each image separated by about 30 seconds in time and offset in y by 5 μm. Therefore at a given y-coordinate where the PDS was scanning

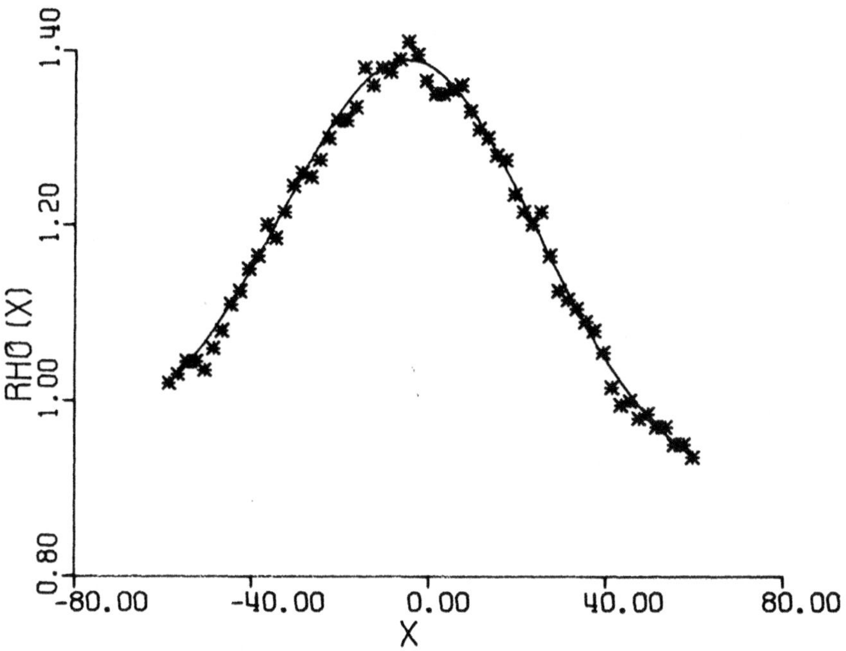

Fig. 1. Fit to an observed marginal distribution.

towards increasing x in the first raster the second raster contains data obtained scanning towards decreasing x. The average of these two lines of data would eliminate any magnitude equation introduced by the PDS. Based on 54 pair of raster scans on these two poor quality plates the error of a single measurement on the PDS on a time scale of 30 seconds is ±0.3 μm (s.e.). Approximately equal accuracy was obtained from the moment analysis and the curve fitting technique and both coordinates yielded the same repeatability.

b) The parallax star was measured both before and after the 24 reference stars, therefore we have a measurement of the repeatability on a 15-20 minute time scale. Based on repeat measures of the parallax star on 19 plates we find the error of a single measurement to be about ± 0.6 μm (s.e.) with the x-coordinates (the direction of rapid scanning) being apparently 25% more precise than the y-coordinates in spite of the fact that the images were generally elongated in the x-direction. Both the moment analysis and the curve fitting technique yielded approximately the same results.

c) The repeat measures discussed in b) include error both in the stage position and in the centering algorithm. Since we are measuring the <u>same</u> image, we can determine the <u>true</u> shift independent of any assumptions about the shape of the image. This is done by maximizing the cross correlation of the two measurements. The only assumption made is that both scans are measuring the identical image, although the two may be offset from each other. Comparing the assumption-free relative image shifts with those given by the curve fitting program, we find that the average of 12 plates in x and 16 plates in y gives $\pm 0.2\,\mu m$ for the error in both coordinates. This then in the repeatability of the centering algorithm. The $0.6\,\mu m$ figure found in b) must be almost entirely due to staging errors.

The actual situation may be worse since in computing the $0.6\,\mu m$ figure, we have deleted several extremely discrepant cases. Errors of up to $100\,\mu m$ occur. The errors appear to fall into two classes either $> 5\,\mu m$ or $< 1\,\mu m$, indicating that the largest shifts should be attributed to either human or machine errors.

5. THE ASTROMETRIC ACCURACY

The ultimate test of a measuring system is the extent to which it improves the accuracy of the astronomical results. For astrometry a valid test is to compare the relative parallaxes of the reference stars in a given field. The smaller the dispersion of relative parallaxes the better the measuring system since the cosmic dispersion of parallaxes is only about ±0".002 (s.e.) at the 14th magnitude. On the other hand, one must be careful in interpreting the results of a comparison unless many reference stars are used since small numbers often produce unreliable results. In this case we have used 22 reference stars on 24 plates which should yield a moderately reliable result.

The computed x,y positions from both the moment analysis and the curve fitting approach were run through the Yale parallax programs which are based on the central overlap method of Gatewood and Eichhorn (1973).

The results of this analysis are summarized in Table I where the relevant statistical parameters are listed for the Yerkes photoelectric image bisector (PIB), the PDS moment analysis, and the curve fitting analysis. For each solution we have listed the average absolute value of the reference star relative parallaxes, the average error of the parallaxes, the average unit weight error of the combined parallax solution, the average unit weight error of the plate constant solutions in x and y, and the number of stars deleted in the course of the analysis. The errors of the parallax and the unit weight errors of the parallax solutions have been scaled to allow for the correct number of degrees of freedom. The average unit weight error, $\langle\sigma_1(\pi_c)\rangle$, of the combined parallax solution for 332 reference stars published in the first two papers of Yerkes parallaxes (van Altena, 1971, van Altena and Vilkki, 1973) corrected for the appropriate number of degrees of freedom is ± 0".030 (s.e.). The Yerkes PIB measures of this field are the second best out of the 22 published fields.

Table I

Parameter	Yerkes PIB	PDS Moments	PDS Curve Fitting
$<\|\pi_c\|>$	0."0035	0."0041	0."0045
$<\sigma(\pi_c)>$	± 0."0051	± 0."0044	± 0."0040
$<\sigma_1(\pi_c)>$	± 0."0199	± 0."0173	± 0."0192
$<\sigma_1(pcx)>$	± 1.9 μm	± 1.6 μm	± 1.8 μm
$<\sigma_1(pcy)>$	± 1.5 μm	± 1.3 μm	± 1.3 μm
no. deleted	22	70	30

An inspection of Table I shows that most values are comparable within statistical uncertainties. It appears, therefore, that even without optimal environmental conditions during measurement, digital dissection of images by the PDS produces results at least as good as the best results produced by the Yerkes PIB. It might appear that moment analysis is slightly more accurate than curve fitting, but this conclusion is not justified. Badly centered images can not be handled by the moment analysis program but can be by the curve fitting program. Thus, forty images with measurements of questionable quality have been included in the reduction of the curve fit data. If these images are deleted, the two methods are comparable in accuracy. The factor of ten greater speed for the curve fitting method then makes it the preferable scheme.

6. CONCLUSIONS

This analysis has shown that astrometric quality relative image positions can be obtained with the PDS microdensitometer. Analysis of the dissected image by curve fitting seems preferable and can be accomplished with very small amounts of computing even on highly oversampled images. The accuracy achieved is comparable to the best results obtained with the Yerkes Observatory photoelectric image bisector. The repeatability of the University of Texas PDS is ±0.6 μm (s.e.), and the repeatability of the dissection algorithms is better than ±0.20 μm. Resolution of 20 μm is adequate so it should be possible to increase the measurement speed to 5-10 seconds per image.

Høg and Wiskott have also experimented with determining image positions from a single scan across an image with a diode line and obtained somewhat lower accuracy that we have achieved. The significance of this comparison is difficult to evaluate however, since they measured only a few stars and their Schmidt telescope images are on the average more than five times smaller than ours. It appears that the diode line might be preferable to the rapid scanning microdensitometers since the measurement time is much shorter and they are readily adaptable to conventional measuring machines which are constructed with precision carriages.

7. ACKNOWLEDGEMENTS

We would like to thank Drs. James D. Wray and G. F. Benedict for making the NASA-Skylab PDS microdensitometer available for this project and for their assistance during the measurements and Dr. R. H. Miller for many helpful discussions during the initial stages of this investigation at the University of Chicago. This research has been supported in part by NASA contract NAS5-23288 and NSF grant GP 13771 and GP 42008 at the University of Chicago and NASA contract NAS5-20841 at the Yale University.

REFERENCES

van Altena, W.F. (1971). Astron. J. 76, 932.
van Altena, W.F., and Vilkki, E.U. (1973). Astron. J. 78, 201.
Gatewood, G., and Eichhorn, H.K. (1973). Astron. J. 78, 769.
Høg, E., and Wiskott, D. (1974). European South. Obs. Tech. Report, No.5.

ASTROPHYSICS AND SPACE SCIENCE LIBRARY

1. C. de Jager (ed.), *The Solar Spectrum. Proceedings of the Symposium held at the University of Utrecht, 26–31 August, 1963.* 1965, XIV + 417 pp.
2. J. Ortner and H. Maseland (eds.), *Introduction to Solar Terrestrial Relations, Proceedings of the Summer School in Space Physics held in Alpbach, Austria, July 15–August 10, 1963 and Organized by the European Preparatory Commission for Space Research.* 1965, IX + 506 pp.
3. C. C. Chang and S. S. Huang (eds.), *Proceedings of the Plasma Space Science Symposium, held at the Catholic University of America, Washington, D.C., June 11–14, 1963.* 1965, IX + 377 pp.
4. Zdeněk Kopal, *An Introduction to the Study of the Moon.* 1966, XII + 464 pp.
5. B. M. McCormac (ed.), *Radiation Trapped in the Earth's Magnetic Field. Proceedings of the Advanced Study Institute, held at the Chr. Michelsen Institute, Bergen, Norway, August 16–September 3, 1965.* 1966, XII + 901 pp.
6. A. B. Underhill, *The Early Type Stars.* 1966, XII + 282 pp.
7. Jean Kovalevsky, *Introduction to Celestial Mechanics.* 1967, VIII + 427 pp.
8. Zdeněk Kopal and Constantine L. Goudas (eds.), *Measure of the Moon. Proceedings of the 2nd International Conference on Selenodesy and Lunar Topography, held in the University of Manchester, England, May 30–June 4, 1966.* 1967, XVIII + 479 pp.
9. J. G. Emming (ed.), *Electromagnetic Radiation in Space. Proceedings of the 3rd ESRO Summer School in Space Physics, held in Alpbach, Austria, from 19 July to 13 August, 1965.* 1968, VIII + 307 pp.
10. R. L. Carovillano, John, F. McClay, and Henry R. Radoski (eds.), *Physics of the Magnetosphere, Based upon the Proceedings of the Conference held at Boston College, June 19–28, 1967.* 1968, X + 686 pp.
11. Syun-Ichi Akasofu, *Polar and Magnetospheric Substorms.* 1968, XVIII + 280 pp.
12. Peter M. Millman (ed.), *Meteorite Research. Proceedings of a Symposium on Meteorite Research, held in Vienna, Austria, 7–13 August, 1968.* 1969, XV + 941 pp.
13. Margherita Hack (ed.), *Mass Loss from Stars. Proceedings of the 2nd Trieste Colloquium on Astrophysics, 12–17 September, 1968.* 1969, XII + 345 pp.
14. N. D'Angelo (ed.), *Low-Frequency Waves and Irregularities in the Ionosphere. Proceedings of the 2nd ESRIN-ESLAB Symposium, held in Frascati, Italy, 23–27 September, 1968.* 1969, VII + 218 pp.
15. G. A. Partel (ed.), *Space Engineering. Proceedings of the 2nd International Conference on Space Engineering, held at the Fondazione Giorgio Cini, Isola di San Giorgio, Venice, Italy, May 7–10, 1969.* 1970, XI + 728 pp.
16. S. Fred Singer (ed.), *Manned Laboratories in Space. Second International Orbital Laboratory Symposium.* 1969, XIII + 133 pp.
17. B. M. McCormac (ed.), *Particles and Fields in the Magnetosphere. Symposium Organized by the Summer Advanced Study Institute, held at the University of California, Santa Barbara, Calif., August 4–15, 1969.* 1970, XI + 450 pp.
18. Jean-Claude Pecker, *Experimental Astronomy.* 1970, X + 105 pp.
19. V. Manno and D. E. Page (eds.), *Intercorrelated Satellite Observations related to Solar Events. Proceedings of the 3rd ESLAB/ESRIN Symposium held in Noordwijk, The Netherlands, September 16–19, 1969.* 1970, XVI + 627 pp.
20. L. Mansinha, D. E. Smylie, and A. E. Beck, *Earthquake Displacement Fields and the Rotation of the Earth. A NATO Advanced Study Institute Conference Organized by the Department of Geophysics, University of Western Ontario, London, Canada, June 22–28, 1969.* 1970, XI + 308 pp.
21. Jean-Claude Pecker, *Space Observatories.* 1970, XI + 120 pp.
22. L. N. Mavridis (ed.), *Structure and Evolution of the Galaxy. Proceedings of the NATO Advanced Study Institute, held in Athens, September 8–19, 1969.* 1971, VII + 312 pp.
23. A. Muller (ed.), *The Magellanic Clouds. A European Southern Observatory Presentation: Principal Prospects, Current Observational and Theoretical Approaches, and Prospects for Future Research. Based*

on the Symposium on the Magellanic Clouds, held in Santiago de Chile, March 1969, on the Occasion of the Dedication of the European Southern Observatory. 1971, XII + 189 pp.

24. B. M. McCormac (ed.), *The Radiating Atmosphere. Proceedings of a Symposium Organized by the Summer Advanced Study Institute, held at Queen's University, Kingston, Ontario, August 3–14, 1970.* 1971, XI + 455 pp.
25. G. Fiocco (ed.), *Mesospheric Models and Related Experiments. Proceedings of the 4th ESRIN-ESLAB Symposium, held at Frascati, Italy, July 6–10, 1970.* 1971, VIII + 298 pp.
26. I. Atanasijević, *Selected Exercises in Galactic Astronomy.* 1971, XII + 144 pp.
27. C. J. Macris (ed.), *Physics of the Solar Corona. Proceedings of the NATO Advanced Study Institute on Physics of the Solar Corona, held at Cavouri-Vouliagmeni, Athens, Greece, 6–17 September 1970.* 1971, XII + 345 pp.
28. F. Delobeau, *The Environment of the Earth.* 1971, IX + 113 pp.
29. E. R. Dyer (general ed.), *Solar-Terrestrial Physics/1970. Proceedings of the International Symposium on Solar-Terrestrial Physics, held in Leningrad, U.S.S.R., 12–19 May 1970.* 1972, VIII + 938 pp.
30. V. Manno and J. Ring (eds.), *Infrared Detection Techniques for Space Research, Proceedings of the 5th ESLAB-ESRIN Symposium, held in Noordwijk, The Netherlands, June 8–11, 1971.* 1972, XII + 344 pp.
31. M. Lecar (ed.), *Gravitational N-Body Problem. Proceedings of IAU Colloquium No. 10, held in Cambridge, England, August 12–15, 1970.* 1972, XI + 441 pp.
32. B. M. McCormac (ed.), *Earth's Magnetospheric Processes. Proceedings of a Symposium Organized by the Summer Advanced Study Institute and Ninth ESRO Summer School, held in Cortina, Italy, August 30–September 10, 1971.* 1972, VIII + 417 pp.
33. Antonin Rükl, *Maps of Lunar Hemispheres.* 1972, V + 24 pp.
34. V. Kourganoff, *Introduction to the Physics of Stellar Interiors.* 1973, XI + 115 pp.
35. B. M. McCormac (ed.), *Physics and Chemistry of Upper Atmospheres. Proceedings of a Symposium Organized by the Summer Advanced Study Institute, held at the University of Orléans, France, July 31–August 11, 1972.* 1973, VIII + 389 pp.
36. J. D. Fernie (ed.), *Variable Stars in Globular Clusters and in Related Systems. Proceedings of the IAU Colloquium No. 21, held at the University of Toronto, Toronto, Canada, August 29–31, 1972.* 1973, IX + 234 pp.
37. R. J. L. Grard (ed.), *Photon and Particle Interaction with Surfaces in Space. Proceedings of the 6th ESLAB Symposium, held at Noordwijk, The Netherlands, 26–29 September, 1972.* 1973, XV + 577 pp.
38. Werner Israel (ed.), *Relativity, Astrophysics and Cosmology. Proceedings of the Summer School, held 14–26 August, 1972, at the BANFF Centre, BANFF, Alberta, Canada.* 1973, IX + 323 pp.
39. B. D. Tapley and V. Szebehely (eds.), *Recent Advances in Dynamical Astronomy, Proceedings of the NATO Advanced Study Institute in Dynamical Astronomy, held in Cortina d'Ampezzo, Italy, August 9–12, 1972.* 1973, XIII + 468 pp.
40. A. G. W. Cameron (ed.), *Cosmochemistry. Proceedings of the Symposium on Cosmochemistry, held at the Smithsonian Astrophysical Observatory, Cambridge, Mass., August 14–16, 1972.* 1973, X + 173 pp.
41. M. Golay, *Introduction to Astronomical Photometry.* 1974, IX + 364 pp.
42. D. E. Page (ed.), *Correlated Interplanetary and Magnetospheric Observations. Proceedings of the 7th ESLAB Symposium, held at Saulgau, W. Germany, 22–25 May, 1973.* 1974, XIV + 662 pp.
43. Riccardo Giacconi and Herbert Gursky (eds.), *X-Ray Astronomy.* 1974, X + 450 pp.
44. B. M. McCormac (ed.), *Magnetospheric Physics. Proceedings of the Advanced Summer Institute, held in Sheffield, U.K., August 1973.* 1974, VII + 399 pp.
45. C. B. Cosmovici (ed.), *Supernovae and Supernova Remnants. Proceedings of the International Conference on Supernovae, held in Lecce, Italy, May 7–11, 1973.* 1974, XVII + 387 pp.
46. A. P. Mitra, *Ionospheric Effects of Solar Flares.* 1974, XI + 294 pp.
49. Z. Švestka and P. Simon (eds.), *Catalog of Solar Particle Events 1955–1969. Prepared under the Auspices of Working Group 2 of the Inter-Union Commission on Solar-Terrestrial Physics.* 1975, IX + 428 pp.
50. Zdeněk Kopal and Robert W. Carder, *Mapping of the Moon.* 1974, VIII + 237 pp.
51. B. M. McCormac (ed.), *Atmospheres of Earth and the Planets. Proceedings of the Summer Advanced Study Institute, held at the University of Liège, Belgium, July 29–August 9, 1974.* 1975, VII + 454 pp.
52. V. Formisano (ed.), *The Magnetospheres of the Earth and Jupiter. Proceedings of the Neil Brice Memorial Symposium, held in Frascati, May 28–June 1, 1974.* 1975, XI + 485 pp.

GPSR Compliance

The European Union's (EU) General Product Safety Regulation (GPSR) is a set of rules that requires consumer products to be safe and our obligations to ensure this.

If you have any concerns about our products, you can contact us on ProductSafety@springernature.com

In case Publisher is established outside the EU, the EU authorized representative is:

Springer Nature Customer Service Center GmbH
Europaplatz 3
69115 Heidelberg, Germany

Batch number: 10761599

Printed by Printforce, the Netherlands